INDUSTRIAL SEPARATORS FOR GAS CLEANING

CHEMICAL ENGINEERING MONOGRAPHS

Edited by Professor S. W. CHURCHILL, Department of Chemical Engineering, University of Pennsylvania, Philadelphia, Pa. 19104, U.S.A.

Vol. 1 Polymer Engineering (Williams)

Vol. 2 Filtration Post-Treatment Processes (Wakeman)

Vol. 3 Multicomponent Diffusion (Cussler)

Vol. 4 Transport in Porous Catalysts (Jackson)

Vol. 5 Calculation of Properties Using Corresponding-State Methods (Štěrbáček et al.)

Vol. 6 Industrial Separators for Gas Cleaning (Štorch et al.)

Vol. 7 Twin Screw Extrusion (Janssen)

Vol. 8 Fault Detection and Diagnosis in Chemical and Petrochemical Processes (Himmelblau)

INDUSTRIAL SEPARATORS FOR GAS CLEANING

OTAKAR ŠTORCH et al.

Research Institute for Air Engineering,
Prague, Czechoslovakia

Contributors:
Jiří Albrecht, Jiří Hejma,
Jiří Kurfürst, Karel Pojar, Jaroslav Urban

ELSEVIER SCIENTIFIC PUBLISHING COMPANY
Amsterdam — Oxford — New York — 1979

Published in co-edition with
SNTL Publishers of Technical Literature, Prague

Distribution of this book is being handled by the following publishers

for the U.S.A and Canada
Elsevier North-Holland, Inc.
52 Vanderbilt Avenue
New York, N. Y. 10017

for East European Countries, China, Northern Korea, Cuba, Vietnam and Mongolia
SNTL Publishers of Technical Literature, Prague

for all remaining areas
Elsevier Scientific Publishing Company
335 Jan van Galenstraat
P.O. Box 211, 1000 AE Amsterdam, The Netherlands

Library of Congress Cataloging in Publication Data

Štorch, Otakar.

Industrial separators for gas cleaning.

(Chemical engineering monographs; v. 6)

Translation of Čištění průmyslových plynů a exhalací odlučovači.

Bibliography: p.

Includes index.

1. Separators (Machines) 2. Gases—Cleaning—Equipment and supplies. 3. Dust—Removal—Equipment and supplies.

TP 159. S4S7513 628.5'3'2 78-10916

ISBN 0-444-99808-X

ISBN 0-444-41295-6 (series)

Translation by Julius Freundlich

Printed in Czechoslovakia

Contents

Foreword

In these times of constant progress in technology, of rapidly expanding industries and ever new processing techniques in so many of them, industrial gas cleaning is rapidly gaining in scope and importance. Gas cleaning facilities are nowadays indispensable ancillaries of every plant where, for economic or environmental reasons or both, solid or liquid particles have to be separated out from a gas stream. Moreover, the equipment used for this purpose up to now is increasingly falling short of the ever more stringent requirements imposed by public health regulations, so that much of it is growing ripe for replacement by new, more efficient and more dependable installations. In most countries, the introduction of new and replacement of existing gas cleaning facilities is being hastened by legislation which reflects our growing awareness of the enormity of the damage caused by industrial air pollution.

All too often, however, any rational selection of gas cleaning equipment is hampered by the sheer diversity of types now on the market. The prospective buyer is confronted by a wide range of units based on entirely different physical principles, and with a huge variety of others which, although identical in their basic principles, differ substantially in detail design.

This book therefore sets out to review the existing types of gas cleaning equipment; to outline their underlying principles and principal design features; to list their chief technical data, their relative advantages and drawbacks; and to facilitate the choice by making a few recommendations about the kinds of equipment which are most suitable for some of the major industries. Most of these recommendations are based on long-term experience, but some have of necessity had to be founded on theory rather than on accumulated know-how.

Apart from this survey of the presently available gas cleaning equipment, the book also deals in brief outline with the fundamentals of the dust handling techniques involved in the operation of this equipment. This aspect is all too often underrated; yet it is a fact, borne out by dearly paid experience, that no one can design, install or operate gas cleaning facilities properly unless he is fully familiar with the problems specific to the field of dust handling.

Symbols and terminology

a	(microns), (m)	Particle diameter
a	($m\ s^{-2}$)	Acceleration
a_{mean}	(m)	Mean particle diameter
a_1	(microns)	Reduced equivalent diameter of a fictitious spherical particle
b_v	($m\ s^{-2}$)	Acceleration by centrifugal field
f	($s\ m^{-1}$)	Specific area of collecting surface
g	($m\ s^{-2}$)	Acceleration by gravity
h	(hours)	Annual utilization of separator
i	($A\ m^{-1}$)	Specific current intensity
i	($kJ\ kg^{-1}$), ($kcal\ kg^{-1}$)	Enthalpy per kg of matter
k	($kg\ m^{-3}$)	Concentration of solid or liquid particles in gas
k_0	($g\ m^{-3}$), ($kg\ m^{-3}$)	The part of k_p trapped in the separator, $k_0 = k_p - k_v$
k_p	($g\ m^{-3}$), ($kg\ m^{-3}$)	Particle concentration in gas at separator inlet
k_v	($g\ m^{-3}$), ($kg\ m^{-3}$)	Particle concentration in gas at separator outlet
l	(m)	Thickness of dust layer at time of resistance measurement
m_m	(kg)	Mass of solid particle
—	(m_n^3)	1 m^3 at +20 °C and 1 atmosphere
m_v	(kg)	Mass of macroparticle of gas
n	(kg^{-1})	Number of particles in 1 kg of dust
p	(Pa)	Gas pressure
p_0	(Pa)	Reference gas pressure
p_p	(Pa)	Partial pressure of saturated water vapour
q_p	($C\ kg^{-1}$)	Electrical charge of 1 kg of dust
r	($J\ kg^{-1}\ {}^{\circ}K^{-1}$)	Specific gas constant
r	(m)	Half diameter of cyclone outlet
r	(m)	Half diameter of wire electrode
r	(m)	Radius of curvature
r_m	(m)	Radius of solid particle trajectory
r_v	(m)	Radius of gas flow line

t	(°C)	Temperature
u'	–	Amortization factor of separating equipment
u''	–	Amortization factor of enclosed space in buildings
u'''	–	Amortization factor of floor space occupied by separators
v_i	$(m\ s^{-1})$	Average ion velocity in electric field
v_m	$(m\ s^{-1})$, $(cm\ s^{-1})$	Particle velocity
v_k	$(m\ s^{-1})$, $(cm\ s^{-1})$	Terminal particle velocity
v_p	$(m\ s^{-1})$, $(cm\ s^{-1})$	Free falling velocity of particle
v_r	$(m\ s^{-1})$	Centripetal velocity component
v_u	$(m\ s^{-1})$	Circumferential velocity component
v_v	$(m\ s^{-1})$, $(cm\ s^{-1})$	Gas flow velocity
v_v	$(m\ s^{-1})$	Absolute velocity of gas macroparticle
v_z	$(m\ s^{-1})$	Axial velocity component
v_D	$(m\ s^{-1})$	Velocity related to cross section of cyclone of diameter D
v_F	$(cm\ s^{-1})$	Flow speed of filtered gas
w	$(cm\ s^{-1})$	Mean (effective) separating brisk velocity in electrostatic precipitator
x	$(kg\ kg^{-1})$	Specific moisture content
A	$(m^3\ hour^{-1})$	Make-up water needed for separator
A	$(m^3\ kg^{-1})$	Constant for permittivity calculations, dependent on gas or particle material
C	$(N\ m^{-1})$	Surface tension
C_a	$(crowns\ m^{-3})$	Water costs
C_e	$(crowns\ kWh^{-1})$	Electricity costs
C_n	(crowns)	Annual costs of spares and replacement parts
C_{oz}	(crowns)	Cost price of separating equipment
C_p	$(crowns\ m^{-3})$	Cost of enclosed space in buildings
C_u	$(crowns\ m^{-2})$	Cost of floor space
C_{12}	$(N\ m^{-1})$	Surface tension at interface of solid and liquid
C_{13}	$(N\ m^{-1})$	Surface tension at interface of solid and gas
C_{23}	$(N\ m^{-1})$	Surface tension at interface of liquid and gas
D	$(m^2\ s^{-1})$, $(cm^2\ s^{-1})$	Diffusion coefficient
E	(kW)	Power input for fan drive
E'	(kW)	Total power input for separator and all its ancillaries
F	$(V\ m^{-1})$	Electric field intensity
F_0	$(V\ m^{-1})$	Initial critical intensity of electric field
G	$(kg\ s^{-1})$, $(kg\ hour^{-1})$	Mass flow through separator

G_c	(N)	Gravitational force of particle
G_m	(kg)	Weight of solid particles
G_o	(kg)	The part of G_p trapped in the separator, $G_0 = G_p - G_v$
G_p	(kg)	Weight of solid or liquid particles entering the separator
G_v	(kg)	Weight of solid or liquid particles leaving the separator untrapped
H_d	(hours)	Annual shutdown time for replacing components
H_u	(hours)	Annual shutdown time for maintenance
I	(A)	Electric current intensity
K_m	(%)	Cost coefficient for erection and installation work
K_p	(%)	Cost coefficient for design work
M_d	(crowns hour^{-1})	Wage rate of equipment operator, including overheads allowance
M_u	(crowns hour^{-1})	Wage rate of maintenance worker, including overheads allowance
MO	(microns), (m s^{-1})	Separation limit
N		Number of molecules per mol (*Loschmidt number*)
N_1	(crowns per 1000 m^3)*	Cost of project and design work
N_2	(crowns per 1000 m^3)*	Cost price of separating equipment
N_3	(crowns per 1000 m^3)*	Cost of erection/installation work
N_4	(crowns per 1000 m^3)*	Costs of buildings to house separating equipment
N_4'	(crowns per 1000 m^3)*	Cost of site for separating equipment
N_5	(crowns per 1000 m^3)*	Electricity costs for fan drive
N_6	(crowns per 1000 m^3)*	Total electricity costs for separator operation
N_7	(crowns per 1000 m^3)*	Water costs
N_8	(crowns per 1000 m^3)*	Costs of spares and replacements
N_9	(crowns per 1000 m^3)*	Wage bill for operation and routine maintenance
N_{10}	(crowns per 1000 m^3)*	Wage bill for installation of replacements
O_c	(%)	Overall collecting efficiency
O_d	(%)	Partial collecting efficiency
O_f	(%)	Fractional collecting efficiency
P	(N)	Force, electrical forces

*Note: Costs N_1 to N_{10} are referred to 1000 m^3 of cleaned gas.

P_m	(m^2)	Particle surface area
P_n	(N), (kp)	Force perpendicular to the gas flow
P_t	(N), (kp)	Force opposing the motion of a macroparticle of gas
P'_m	($m^2\ kg^{-1}$)	Specific surface area of particles
PMO	(microns), ($m\ s^{-1}$)	Approximate separation limit
$\bar{P}$	(Pa)	Vector of pressure gradient
$\bar{P}_n$	(Pa)	Pressure gradient component perpendicular to gas flow lines
$\bar{P}_t$	(Pa)	Pressure gradient component tangential to gas flow lines
Q	($m^3\ s^{-1}$), ($m^3\ h^{-1}$)	Volumetric flow rate
Q_c	(C)	Electric charge of particles
Q_{c1}	(C)	Electric charge of one particle
Q_h	($m^3\ s^{-1}$), ($m^3\ h^{-1}$)	Throughflow capacity of separator
Q_n	(C)	Saturated particle charge
Q_τ	(C)	Parcicle charge in time τ
R	($J\ mol^{-1}\ {}^\circ K^{-1}$)	Molar gas constant
R	(N)	Aerodynamic drag
R	(ohms)	Resistance of dust layer
R	(m)	Half diameter of centrifugal chamber in a cyclone
R	(m)	Half diameter of tubular electrode
R	(m)	Gap between wire and plate electrodes
S	(m^2)	Surface area of dust layer at time of resistance measurement
T	(K)	Thermodynamic temperature of gas
T_0	(K)	Reference absolute temperature of gas
U	(V)	Feed voltage
U_0	(V)	Initial critical voltage
U_p	(V)	Arc-over voltage
V	(m^3)	Enclosed space needed for separator operation and maintenance
V'	(m^2)	Floor space needed for separator operation and maintenance
V_m	(m^3)	Volume of solid particles
V_s	(m^3)	Apparent bulk volume
Z	(%)	Proportion of retained particles
$Z(a)$	–	Retained fractions curve related to a
$Z(v_p)$	–	Retained fractions curve related to v_p
$-\frac{dZ}{da}$	–	Particle frequency related to a

$-\frac{dZ}{dv_p}$	–	Particle frequency related to v_p
$-\frac{dZ(a)}{da}$	–	Particle size distribution curve related to a
$-\frac{dZ(v_p)}{dv_p}$	–	Particle size distribution curve related to v_p
α	(degrees)	Angle of impact
β	–	Damping coefficient
β	–	Relative charging coefficient (ratio of real to saturated particle charge)
β_{mean}	–	Mean relative charging coefficient
δ	–	Relative weight of gas in given reference state
ε	($F\ m^{-1}$)	Permittivity
ε_0	($F\ m^{-1}$)	Permittivity of vacuum ($8.855 . 10^{-12}$)
ε_r	–	Relative permittivity
η	(Pa s)	Dynamic viscosity of gas
ϑ	(degrees)	Angle between liquid/solid interface and tangent to liquid surface at point of contact between solid, liquid and gas (contact angle)
$\varkappa$	–	Charging coefficient
λ	–	Excess air ratio in combustion process
ν	($m^2\ s^{-1}$)	Kinematic viscosity of gas
ξ	–	Drag coefficient of particle
ξ_D	–	Aerodynamic resistance coefficient of separator
ϱ	($kg\ m^{-3}$)	Specific gravity
ϱ_i	($C\ m^{-3}$)	Space charge of gas ions
ϱ_{im}	($C\ m^{-3}$)	Maximum space charge of gas ions
ϱ_k	($kg\ m^{-3}$)	Specific gravity of fluid
ϱ_m	($kg\ m^{-3}$)	Specific gravity of a solid particle
ϱ_p	($C\ m^{-3}$)	Space charge of dust
ϱ_{pn}	($C\ m^{-3}$)	Saturated space charge of dust
ϱ_s	($kg\ m^{-3}$)	Bulk density
ϱ_{st}	($kg\ m^{-3}$)	Tapped density
ϱ_v	(ohmmetres)	Specific resistance of dust layer
τ	(s)	Time; time elapsed since start of charging
ω	(s^{-1})	Angular velocity
Δh	(Pa)	Pressure drop across the separator at $\varrho = 1\ kg\ m^{-3}$
Δp	(Pa)	Pressure drop across the separator
Δp_F	(Pa)	Pressure gradient across filter cloth

Explanations of some of the terms

Particle diameter a (microns or m)

This measure of solid or liquid particle sizes is either the diameter of a spherical particle which will just pass through a screen with a mesh size of a, or else the diameter of a fictitious spherical particle with the same free falling velocity and same specific gravity as the real particle under consideration (i.e the equivalent particle diameter).

Free falling velocity v_p ($m\,s^{-1}$)

This is the speed at which the particle will descend, under the effect of gravity alone, in a stationary volume of carrier gas. For Reynolds' numbers up to and including unity, it can with sufficient accuracy be established from *Stokes's expression*

$$v_p = \frac{1}{18} a^2 \frac{\varrho_2 - \varrho_1}{\eta} g \tag{1}$$

Terminal velocity v_k ($m\,s^{-1}$)

This is the ultimate speed which the particle will attain, relative to the carrier gas, under the effects of the equilibrium of forces acting upon it in the separator.

Proportion of retained particles Z (%)

This is the numerical proportion, or proportion by weight, of particles which exceed a specified value of the parameter which characterizes the particle size; in other words, which exceed a given mesh size, equivalent diameter, free falling velocity, or actual particle size, in terms of a or v_p.

Retained fractions curve $Z(a)$ *or* $Z(v_p)$

This curve (Fig. 1) shows how the magnitude of Z varies with the value of the characteristic parameter a or v_p.

Particle incidence $-\dfrac{dZ}{da}$ *or* $-\dfrac{dZ}{dv_p}$

This quantity, related either to the number of particles or to their weight, is the negative value of the figure obtained by differentiation of the retained particle proportion Z by the size-governing parameter (a or v_p).

Particle size distribution curve $-\frac{dZ(a)}{da}$ *or* $-\frac{dZ(v_p)}{dv_p}$

This curve (Fig. 1) indicates how the frequency of incidence of any given particle size varies with the value of the size-governing parameter (such as a or v_p).

Overall collecting efficiency O_c (%)

This is the ratio (multiplied by 100) between the weight of solid or liquid particles trapped in the separator (G_0) and the weight of these particles (G_p) entering that separator in the carrier gas within a certain span of time, at a certain flow rate and physical state of the carrier gas, and given a certain concentration and a certain set of physical properties of the particles,

$$O_c = \frac{G_o}{G_p} 100 \tag{2}$$

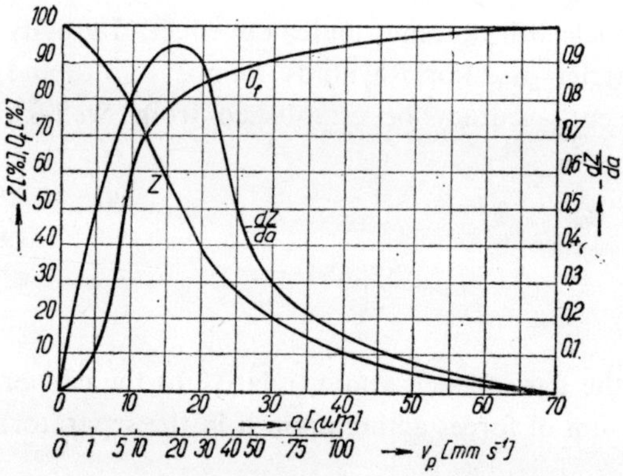

Fig. 1 How the proportion of retained particles Z, the particle frequency $-\frac{dZ}{da}$, and the fractional collecting efficiency O_f vary with the equivalent particle size a.

Since the weight of solid or liquid particles entering the separator equals the sum of the weight of particles trapped (G_0) and the weight of particles leaving the separator untrapped (G_v), i.e. $G_p = G_0 + G_v$, we can also define O_c as

$$O_c = \frac{G_o}{G_o + G_v} 100 = \frac{G_p - G_v}{G_p} 100 \tag{3}$$

Fractional collecting efficiency O_f (%)

This is the ratio (multiplied by 100) between the weight of solid or liquid particles with a certain magnitude of the size-governing parameter which are trapped in the separator, and the weight of particles with the same magnitude of that parameter which enter the separator in the carrier gas, again at a certain flow rate and physical state of that gas and a certain concentration and set of physical properties of the particles. For particles of a certain size a (or free falling velocity v_p), the fractional collecting efficiency is determined as

$$O_f = O_c \frac{\frac{dZ_o}{da}}{\frac{dZ_p}{da}} \tag{4}$$

In this expression, the retained fraction curves for the trapped and for the incoming dust are differentiated at the points corresponding to the given values of a or v_p.

Fractional collecting efficiency curves $O_f(a)$ or $O_f(v_p)$

These curves (Fig. 1) show how the fractional collecting efficiency alters with the magnitude of the size-governing parameter a or v_p.

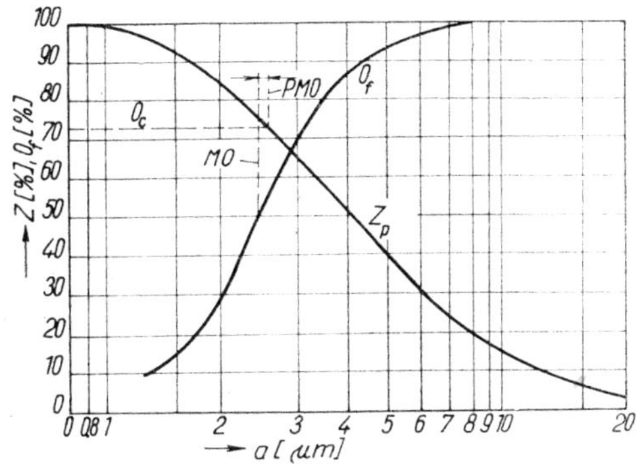

Fig. 2 Graphic representation of the separation limit MO and the approximate separation limit PMO.

Partial collecting efficiency O_d (%)

This is the trapped percentage of all solid or liquid particles whose size-governing parameter (a or v_p) lies within a specified interval, in relation to the amount of particles of this specified size range which enter the separator at a certain flow rate and physical state of the carrier gas.

Separation limit MO (microns), (m s^{-1})

This limit is defined by the magnitude of the size-governing parameter of the particles which, at a certain flow rate and physical state of the carrier gas and a certain set of physical properties of the particles, are trapped with a fractional collecting efficiency of $O_f = 50\,\%$ (Fig. 2).

Approximate separation limit PMO (microns), (m s^{-1})

For particles described by a given retained fractions curve, this limit (in terms of a or v_p) is defined by the intersection of that curve with the O_c abscissa denoting the overall collecting efficiency, as evident in Fig. 2.

Flow rate through the separator Q ($m^3\,s^{-1}$), ($m^3\,h^{-1}$)
G (kg s^{-1}), (kg h^{-1})

This is the volume Q or weight G of gas passing through the separator per unit of time, at a certain state of the gas (in terms of its pressure, temperature and humidity) which is usually ascertained at the separator inlet.

Throughflow capacity of the separator Q_h ($m^3\,s^{-1}$), ($m^3\,h^{-1}$)

This is the volume of gas with a specific gravity of $\varrho = 1$ kg m^{-3} that can pass through the separator per unit of time at a pressure drop of $\Delta h = 1$ Pa

$$Q_h = \frac{Q}{\sqrt{\Delta h}} \tag{5}$$

Separator resistance coefficient ξ_D

This dimensionless coefficient is the ratio betwen the pressure drop Δp and the dynamic pressure corresponding to the flow velocity v_v in a selected cross-sectional plane (e.g. in the cylindrical part of a cyclone) at a given physical state of the gas,

$$\xi_D = \frac{2\,\Delta p}{v_D^2 \varrho} \tag{6}$$

Pressure drop across the separator Δp (Pa)
Δh (Pa) at $\varrho = 1$ kg m^{-3}

This is the difference between the total (static and dynamic) pressures, or pressure heads, at the inlet and at the outlet cross-sections of a separator. It is generally quoted for the gas flow rate applying to the state of the gas at the separator inlet. Apart from the flow rate for which it is valid, this figure must always be qualified as applying either to clean gas or to gas containing a stated concentration of solid or liquid admixtures. While the available figure will usually be the actual pressure drop Δp, a better criterion for comparative assessments of various separators is the pressure drop Δh when the specific gravity of the gas is $\varrho = 1$ kg m^{-3}. The relationship between these two figures is

$$\Delta p = \Delta h \varrho_{\text{real gas}} \tag{7}$$

1. PROPERTIES OF SOLID AND LIQUID PARTICLES

Ing. J. Kurfürst, CSc., Ing. K. Pojar, Ing. O. Štorch, CSc., Ing. J. Urban

1.1 Origins of solid particles

Solid particles are created either by natural processes or by human activities, and can thus be classified into *natural* and *man-made dusts.*

Natural dusts can be broken down into several categories:

a) Cosmic dust, containing hydrogen, potassium and sodium, and dust of meteoric origin;

b) Inorganic dusts of terrestrial origin, i.e. geological dusts arising by volcanic activity, by natural conflagrations, sandstorms, by the erosion of rocks, and by the comminuting and distributing effects of the wind;

c) Organic or biological dusts, such as pollen, plankton, spores of fungi, plant seeds, or viruses;

d) Condensation nuclei, or heavy ions, like the salt-laden mists over and near the sea, or the finest of the particles raised by sandstorms.

Man-made dusts fall into two principal classes:

a) Dusts which arise as by-products; the obvious examples are the dusts generated by abrasion, in the processing or machining of various materials or during their wear; the dust thrown up by road vehicles; combustion and sublimation products, mainly the soot and fly ashes created by the combustion of fuels; the dust content of industrial waste gases; etc.

b) Dusts which are produced deliberately, and which tend to escape into the outside atmosphere during the production and handling mainly of granular bulk materials, such as cement, gypsum, lime, artificial fertilizers, coals, the fillers used in the rubber and plastics industries (e.g. carbon black), flour, etc.

The atmosphere of our planet has apparently always contained some dust, but its dust content has become a problem only fairly recently. Owing to the increasing pace of industrial production, and especially of power and heat generation, as well as of transport, building and civil engineering activities, the concentration of man-made dusts in most residential and industrial areas now greatly exceeds that of natural dusts, and often presents an acute health hazard. Since natural dusts generally form only an insignificant proportion of the total dust concentration, and since as a rule no systematic technical measures are taken to limit their occurrence, this book will be devoted exclusively to man-made dusts. That, of course, covers all dusts except those of cosmic, geological or biological origin; in fact, it may apply even to dusts of these three categories, if they have been further affected by or involved in human activities.

In Czechoslovakia, the principal industrial sources of dust can be assembled in the following order of decreasing importance:

a) Combustion processes, mainly serving for power and steam generation, e.g. in power and/or district heating stations, garbage incinerating plants, industrial boilers, and local heating facilities.

b) The metallurgical industry, meaning the production of both ferrous and non-ferrous metals and alloys, including foundries.

c) The production of building and similar materials, including the associated quarrying operations; this applies to the cement, limestone, and magnesite industries; to the production of other building, roadmaking and insulating materials and refractories; and to the winning and processing of raw materials for the ceramics, glass and porcelain industries.

d) The chemical industries, i.e. inorganic, organic and petroleum chemistry; outstanding examples are the production of sulphuric acid, calcium carbide, soda, carbon black, artificial fertilizers, pigments, organic dyes, etc.

e) The mining or quarrying and subsequent processing of ores and coal at stone coal, brown coal and ore mines, dressing and beneficiation facilities, coking and briquetting plants, gasworks, ore pelletizing and sintering installations, etc.

f) Transport and the other branches of industry, particularly the metalworking and engineering, textile, paper, food, pharmaceutical, and woodworking industries.

The processing operations or technological activities which actually give rise to the dust can be similarly classified as follows:

a) Machining operations: drilling, planing, cutting, turning, milling, grinding, polishing, etc.

b) Comminuting operations: blasting, crushing, hammering, spraying, pulverizing, grinding, etc.

c) Thermal processing: combustion, baking and firing operations, roasting, melting, coking and carbonization processes, evaporation, drying, distillation and gasification processes, etc.

d) Handling operations: transport, loading and unloading, screening, classifying, mixing and vibrating procedures.

e) Agglomeration techniques, like briquetting, pelletizing or sintering.

f) Corrosion, which, though not a process operation, also represents a major industrial source of dust.

The amount of dust actually generated depends on a number of factors, the chief of which are:

a) The physical properties and physical chemistry of the dust;
b) The grain size distribution of the dust particles and their surface properties;
c) The flow route of the material from which the dust arises (e.g. the degree of recirculation, the height from which the material drops, changes in the direction or velocity of the flow, etc.);
d) The frequency and intensity of contacts or impacts between the individual particles;
e) The coefficient of friction between the dust and the surfaces over which that dust moves.

TABLE 1

Average dust concentration levels*

Location	Concentration ($mg\ m^{-3}$)	Location	Concentration ($mg\ m^{-3}$)
Countryside	0.02	Flour mills	45 to 50
Cities (concentration drops		Shipyards	55
exponentially as altitude		Rag sorting at paper mills	30 to 80
increases)	0.1 to 0.4	Clinker screening	100
City streets	1 to 3	Porcelain factories	120
Industrial cities:		Fine ore feeder magazines	100 to 200
18-month average	0.2	Mine roads	100 to 300
peak concentrations	5.0	Fine ore stockyards	150 to 500
during long-term		Electric furnace steelworks	200 to 300
temperature inversions	1 to 2	Breweries and malt plants	240
Spas:		Hemp hackling	420
mean concentration	0.097	Metal grinding shops	25 to 450
in winter heating season	0.07 to 0.134	Foundry fettling shops	75 to 550
Shops	5	Bag filling stations handling	
Department stores	8	powders	685
Schools	18	Cement factories:	
Metallurgical plants	over 15	average concentrations	100 to 400
Sawmills	15 to 20	clinker conveyors at rotary	
Engineering factories	20 to 25	furnaces	225 to 2000
Foundries	20 to 60	bag filling and loading	
Paper mills	25	stations	100 to 950

*The concentrations quoted for dust-generating workplaces apply when no exhausting equipment is installed.

A rough picture of the amounts of dust with which we are commonly confronted can be formed from Tables 1 to 3. These Tables list the normal average concentrations of dust in the outdoors atmosphere and in those workplace environments where no dust exhausting equipment has been fitted. Table 4 outlines the approximate dust concentrations present in some typical industrial waste gases. All these figures are intended for a rough and ready orientation only. They are subject to frequent and substantial fluctuations, caused for instance by variations in the process parameters or processed materials, and are also dependent on the design, layout and performance of the exhausting equipment. Still, they are indicative of the magnitude of the problems that face us once we set out to control dust pollution.

TABLE 2

Average numbers of dust particles in 1 m^3 of air

Location	Number of particles larger than 0.5 microns
Industrial areas	30 to 60 millions
Large cities	15 to 50 millions
Clean rooms with normal (turbulent) ventilation	2 to 5 millions
Class 100 clean rooms with laminar flow pressure ventilation	up to 3500

TABLE 3

Average size distribution and content of dust in normal outdoors atmospheres

Particle size range (microns)	Mean particle size (microns)	Number of particles (p.p.m.)	Volume % (or weight % at equal specific gravity)
0 to 0.5	0.25	917,000	1
0.5 to 1	0.75	67,700	2
1 to 3	2	10,800	6
3 to 5	4	2,500	11
5 to 10	7.5	1,750	52
10 to 30	20	50	28
0 to 30	15	1,000,000	100

1.2 Shapes of solid particles

Dust particles may assume any shape whatsoever, but are generally classified by their shapes into three categories:

a) Isometric particles, where the three major axes, perpendicular to each other, are of at least roughly similar lengths.

TABLE 4

Approximate dust concentrations in some industrial waste gases and fumes

Source	Concentration ($g\ m_n^{-3}$)
Boilers firing stone coal:	
Bottom-blown mechanical grate,	
normally	0.5 to 10
exceptionally	up to 30
Slag type boilers,	
normally	up to 25
exceptionally	up to 70
Slag tap boilers,	
normally	25 to 30
exceptionally	up to 70
Boilers firing brown coal:	
Bottom-blown mechanical grate	2 to 10
Slag type boilers	2 to 30
Slag tap boilers	1.5 to 3.5
Fuel beneficiation:	
Producer gas,	
stage I (80—100 °C)	10 to 25
stage II (25—35 °C)	10 to 20
Water gas,	
moist	approx. 1
oil carburized	approx. 10
Iron and steel production:	
Pyrites roasting	2 to 5
Sintering strands:	
with approx. 75% of minus 60 micron fractions	up to 15
with extra fine materials	up to 28
Blast-furnace top gases,	
normally	10 to 50
exceptionally	up to 200
Oxygen convertor fumes (containing approx. 90% of sub-micron fines during oxygen blow)	9 to 40
Thomas convertors during first 5 minutes of run	20 to 60
Open-hearth furnace fumes	0.5 to 10
225-ton furnace emitting 35.2 $m^3\ s^{-1}$ at 343 °C	1.2
Oxygen-blown open-hearth furnaces	2 to 15

(Table 4 continued)

Source	Concentration ($g\ m_n^{-3}$)
Acid open-hearth furnaces (Strauss process):	
— Charging	0.2 to 0.8
— Melt-down	up to 0.25
— Oxygen blowing	up to 6
— After end of blowing	up to 0.5
— Ore charging	approx. 1.75
— Boil	1.8
— FeSi addition	up to 0.7
— Tapping	0.5
Ferromanganese production:	
Furnaces yielding 0.1 to 1 micron particles, with a maximum of 0.3 micron particles, with a bulk weight of 192.2 $kg\ m^{-3}$ before and 480 $kg\ m^{-3}$ after firing, and containing 21% Mn, 12% of alkalis, and 2% of water	up to 18
Arc-furnace fumes (containing 119 $g\ m_n^{-3}$ CO)	2.3 to 20
Cupolas:	
Normal range (depending on cupola design)	2 to 8
Peak concentrations	up to 30
Cobalt oxide calcination	5 to 15
Quarries, ore mines, and production of building materials:	
Materials handling (when half the particles are in the 2 to 15 microns range)	6 to 10
Grinding of pulverous raw materials	6 to 10
Crushers and mills (when half the particles are in the 2 to 15 microns range)	over 10
Drying facilities (when half the particles are in the minus 7 microns range)	over 10
Slag drying facilities	5 to 10
Limestone drying drums	20 to 100
Rapid drying plants with material turning devices	30 to 250
Shaft furnaces,	
normal	1.5 to 3.5
large-capacity	2 to 8
Clinker firing grate strands	5 to 15
Lepol furnaces with double gas mains	1.5 to 5
Long wet-process rotary furnaces with internal fittings,	
normally	2 to 5
peak level	up to 25

(Table 4 continued)

Source	Concentration ($g\ m_n^{-3}$)
Sludge-fed rotary furnaces,	
normally	30 to 150
calcining process	20 to 80
Rotary drying furnaces	3 to 30
Counterflow drying furnaces	40 to 70
Ceramics industry:	
Materials handling (when half the particles are in the 2 to 7 microns range)	1 to 4
Grinding (when half the particles are in the 2 to 15 microns range)	4 to 10
Glazing (when half the particles are in the 7 to 15 microns range)	1 to 6
Aluminium production:	
Bauxite grinding,	
at Loesche mills	10 to 15
at ball mills	15 to 30
Calcining furnace fumes	300 to 400
Electrolytic furnace fumes	0.03 to 0.15
Lead production:	
Sintering strands	2 to 15
Reverbatory furnaces	1 to 10
Rotary reverbatory furnaces	3 to 20
Shaft furnaces	5 to 15
Copper production:	
Chloration roasting	1 to 5
Sintering strands	5 to 7
Reverbatory reducing furnaces	4
Shaft furnaces	up to 15
Copper matte convertors	12
Refining furnaces	2
Zinc production:	
Zinc blende grate kilns	5 to 15
Sintering strands	10
Zinc oxide calcining	5 to 20
Waelz process (ZnO production)	20 to 50
Muffle kilns,	
horizontal	up to 0.3
upright	up to 2

b) Flat (or planar) particles, where two of these axes are substantially longer than the third, as is the case in platelets, lamellae, flakes, etc.

c) Fibrous particles, which are elongated along one of these axes only, forming rods, needles, or fibriform shapes.

Non-isometric particles, whether flat or fibrous, present a special problem. Their behaviour when moving in a gas is usually impossible to define unambiguously, because it depends primarily on their orientation relative to the direction in which they are moving.

Isometric particles are commonly represented, in calculations and design work, by some equivalent shape; the usual practice is to represent them by fictitious spherical particles. The diameter of these fictitious spheres can be determined in any of several ways, each based on a different set of physical assumptions or criteria. One approach is to process the several (generally three) major diameters of the real particle into an arithmetical, geometrical or harmonic mean diameter. A common alternative is to establish the diameter of a spherical particle that will have the same surface area, or the same volume, as the real particle. These are the most widespread, but by no means the only ways of working out the equivalent particle diameter.

The trend nowadays is to select an equivalent particle diameter which will characterize the behaviour of the particle as it moves in a stream of gas. This diameter is reached by a comparison of the free falling velocities of the real and of the fictitious spherical particles in a stationary gas.

Another quantity often used in calculations is the reduced equivalent diameter of a fictitious spherical particle, which we designate a_1. In this case, the equivalent diameter is recalculated so as to apply to an assumed common specific gravity, which is mostly chosen as unity, i.e. $1\,\mathrm{g\,cm^{-3}}$ ($1000\,\mathrm{kg\,m^{-3}}$). The advantage of this quantity is that it covers the way the specific gravity of a particle affects its ease of separation and behaviour in separating equipment. In other words, it covers the fact that particles with a higher specific gravity are more readily separated than those with a lower one. The a_1 figure is worked out as

$$a_1 = a\sqrt{\frac{\varrho_\mathrm{m}}{1000}} \tag{8}$$

Sometimes, however, this figure is similarly reduced to apply to a specific gravity of $2\,\mathrm{g\,cm^{-3}}$, i.e. $2000\,\mathrm{kg\,m^{-3}}$.

1.3 Specific gravity of solid particles

The specific gravity of a solid particle is defined as

$$\varrho_m = \frac{G_m}{V_m} \tag{9}$$

Since dust particles are too small to be weighed and measured individually, a large number of them have to be processed together. The result we then obtain is an average specific gravity of the given set of particles, and rests on the implicit assumption that the specific gravities of all those particles will be identical. Moreover, it is not at all easy to determine the actual volume of that set of particles, especially, when we are dealing with a highly porous material. The two techniques generally employed to ascertain the mean specific gravity of a set of particles are the pycnometric and the volumetric methods.

In the *pycnometric method,* a pycnometer of a known weight and closely defined volume is filled with a certain amount of dust and then weighed. Next, it is topped up with some liquid of a known specific gravity, which is known to wet the dust in question efficiently; the contents are mixed into a thick homogeneous suspension; and the pycnometer is weighed once more. We now have the weights of the empty pycnometer G_0, of the dust in it G_m, and of the pycnometer complete with the dust and the liquid in it, $G_0 + G_m + G_k$. Since we know the specific gravity of the liquid, ϱ_k, we can deduce the volume of liquid V_k that is sharing the pycnometer space with the dust, and since we know the internal volume V_0 of the pycnometer too, we can establish the volume of dust in it, V_m, as

$$V_m = V_0 - V_k = V_0 - \frac{(G_0 + G_m + G_k) - (G_0 + G_m)}{\varrho_k} \tag{10}$$

This procedure yields the weight of the dust, G_m, and the volume of the dust, V_m, which we need to calculate the specific gravity by means of equation (9).

This method is based on two questionable assumptions: that there will be no air bubbles or pockets left within the pycnometer, and that the liquid will penetrate every last pore in the solid particles. More often than not, neither of these conditions can be fully met even by intensive shaking and heating; and the resultant figure is then naturally on the low side.

Dust particles which tend to swell when wetted are obviously not amenable to pycnometric examination, and must therefore be treated by the volumetric method, where they remain in a gaseous medium throughout the procedure. Because of the narrow pores commonly found in the individual particles, however, this technique is likewise subject to errors in the resultant specific gravity data.

1.4 Bulk density and tapped density

The bulk density is the weight of a certain volume of a loose granular material, in the freely deposited state, under closely specified conditions; the tapped density is the corresponding figure when the material has been compacted by vibration. The bulk density can be ascertained by any of a number of methods, which differ from each other essentially in the way the particulate material is filled into the measuring vessel and levelled there so as to fill the defined volume. A procedure often used is to pour a certain weight (say 100 g) of dust into a calibrated vessel, and read off its volume V_s, whereupon the bulk density can be determined as

$$\varrho_s = \frac{0.1}{V_s} \tag{11}$$

A crucial factor in any such test is the way the dust is filled into the vessel. The lower the height from which it is poured, and the slower the filling rate, the less will the dust be compacted by impact, and the lower the resultant bulk density value. Consequently, fixtures are sometimes used to drop the dust into the vessel from a constant height. Mostly the bulk density figure has to be supplemented by a corresponding figure for the tapped density, obtained when the volume of the specimen has been reduced by manual shaking or by mechanical or electromagnetic vibration. The final volume, and hence the resultant figure, will obviously again be strongly dependent on the mode and intensity of compacting.

The bulk density and tapped density figures we gain depend largely on the shape of the dust particles, diminishing the more, the more this shape differs from a truly spherical one. Another vital factor is the particle size distribution, but (especially in the case of very fine dusts) the size issue is often confused by the agglomeration of the individual particles. There is no precise relationship between the particle size distribution and the bulk density, and never can be, because we can never expect the dust to fill the available space in its entirety. That could happen only if the smaller particles conveniently decided to fill all the interstices between the larger ones. In practice, however, we are confronted with random clusters of particles, where the individual clusters or particles very often jam to form arches or cavities. As a rule of thumb, however, the bulk and tapped densities are generally higher for coarse than for very fine dusts.

Only a very approximate relationship exists between the bulk density and the tapped density of a material. Usually, the tapped density lies within the bracket of ϱ_{st} = (1.2 to 1.5) ϱ_s, but the values for very fine dusts are mostly lower.

1.5 Particle size distribution of dust

If all the dust particles of a given set are of one and the same size, that set is said to be monodisperse. Monodisperse dusts hardly ever occur naturally, but can be prepared artificially. A polydisperse set, conversely, is one which contains particles of various sizes. In such a set, the characteristic dimension of the individual particles, for instance the reduced equivalent diameter of a fictitious spherical particle (see Section 1.2), turns into a random variable.

Thus, while in a monodisperse dust the particle size is fully defined by a single figure, for a polydisperse dust we need to know not only the range of sizes that

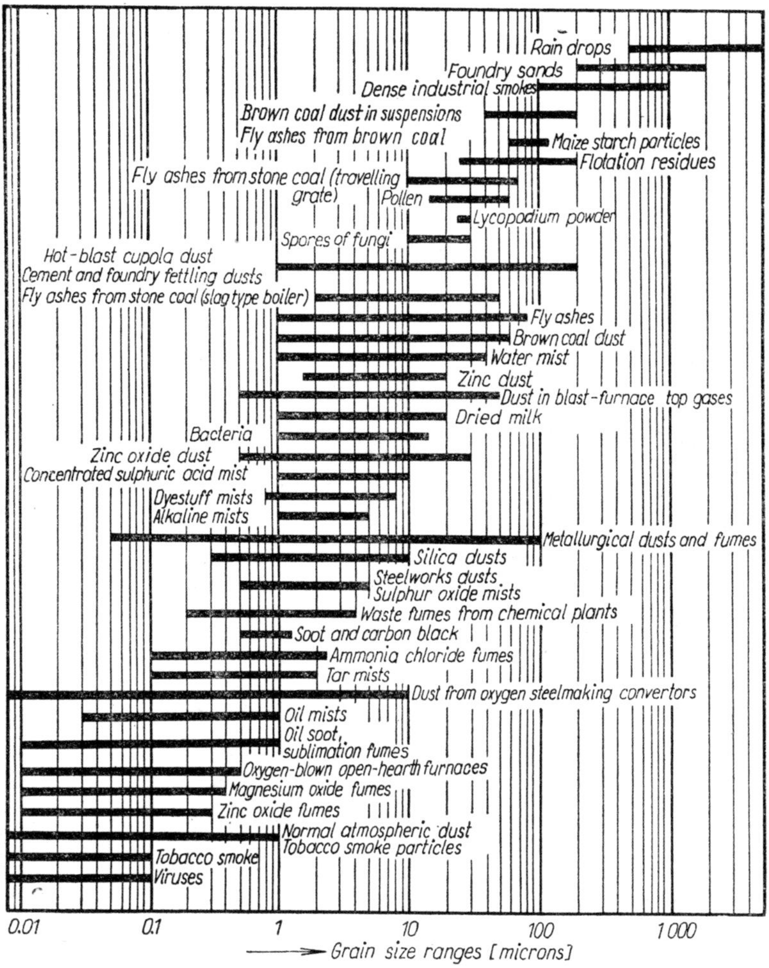

Fig. 3 Ranges of particle sizes found in various kinds of dust.

occur in it, but also the frequency of incidence of the various particle sizes. There are many ways of assessing the grain size distribution. For a first rough orientation, however, we generally classify dusts only into three broad categories – coarse, medium and fine. This classification is based only on the spectrum of particle sizes represented in the specimen, and takes no account of the relative frequency of incidence of each of them. The ranges of particle sizes found in various kinds of dust are plotted in Fig. 3.

The grain size distribution of a dust specimen, covering both the overall range of particle sizes encountered in it and the proportion of this total formed by each of the sizes, can be represented either graphically or by tabulated data. In the former case, we plot a curve of the sieve analysis, which can show either the retained (oversize) or the passed (undersize) fractions, or else a full frequency of incidence curve. The tabulated presentation has the advantage of providing numerical data, but the drawback of applying only to some selected points of one of the three curves listed above.

On a retained fractions curve, each point shows how many per cent (by weight) of the screened set of dust particles are larger than the size corresponding to that point. That size is usually expressed in microns, but is sometimes instead related to the free falling velocity of the particles. As the name implies, these are the fractions retained on a given filter mesh, or at the bottom of a cyclone at a given velocity of the carrier gas, because they exceed the size of the particles

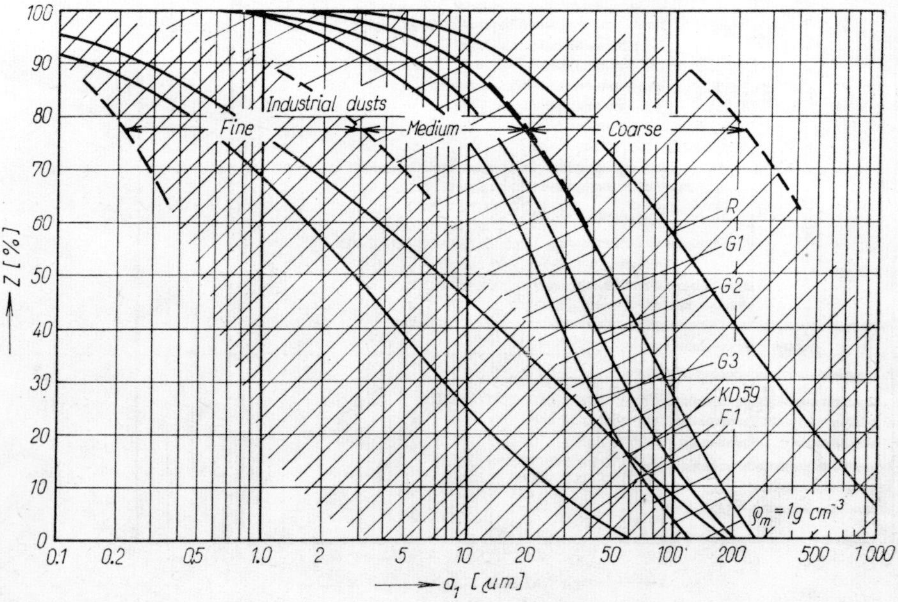

Fig. 4 Retained fractions curves of the six standard reference dusts.

which can drop through or escape. In the literature, and especially in British and American literature, we often come up against an inversion of this curve, known as an undersize fractions curve. This plot represents the proportion of particles smaller than the boundary size, i.e. the proportion of particles that have passed through a given mesh size. In effect, this is the retained fractions curve rotated around its horizontal axis, starting with the smallest particles at the bottom of the chart and ending with the largest at the 100 per cent mark (since all the particles of the examined set are smaller than the largest size still represented in it).

When we are out for a first rough comparison, or for an approximate assessment of the separating efficiencies of various types of equipment, the usual practice nowadays is to use the retained fractions curves of six different representative dusts. These standard reference dusts are designated R, G I, G II, G III, KD 59, and F 1 respectively; their retained fractions curves are shown in Fig. 4.

A frequency of incidence curve can be derived from the retained fractions curve by differentiation. This curve is known under a variety of names, such as the grain size distribution curve, grain size scatter curve, etc. To indicate the differences in presentation between curves of these two types, Fig. 5 shows the frequency of incidence curves and the retained fractions curves for six dust specimens with widely differing grain size distributions.

The grain size distribution of a dust, as represented by the retained fractions curve, is ascertained by the analysis of a specimen by any of several methods:

1. Direct classification methods involve the measurement of the actual particle dimensions, and render size figures which represent the actual lengths of the particles in one direction. The methods in question are as follows:

a) Optical microscopy is applied mainly to non-isometric particles like needles, fibres, platelets or flakes, where any other mode of examination would reveal only one not necessarily representative dimension, in dependence on the way the particles happened to be oriented at the time of their screening.

b) Electron microscopy is employed only on very fine dusts, generally with grain sizes of less than one micron, though its scope ranges from 0.05 up to 5 microns.

c) Mesh screening is utilized only when the dust is coarse, usually with a grain size exceeding 60 microns; yet wet screening can separate particles down to 35 microns in size, and air jet screening (with the Alpine apparatus) can cope with particles as small as 15 microns across. Screening is rarely used on its own in work on dust trapping techniques, because for these purposes we mostly need a detailed breakdown of the grain size distribution in the 0 to 15 or 0 to 30 microns interval. Sometimes dusts are screened to separate the coarse fractions before the remnant is classified by some settling technique. But since these two procedures classify the dust by two mutually unrelated sets of dimensions, this approach leaves us with two retained fractions curves, one for the screened and the other for the settled portion, and the two curves are generally too widely separated to be readily linked into a single, continuous resultant curve.

2. Settling techniques classify the particles by their free falling velocities under the effects of gravity or of centrifugal force. The dimension calculated from this velocity is the equivalent diameter of a fictitious spherical particle. A diagram of free falling velocities is presented in Fig. 6. There are two broad categories of these methods:

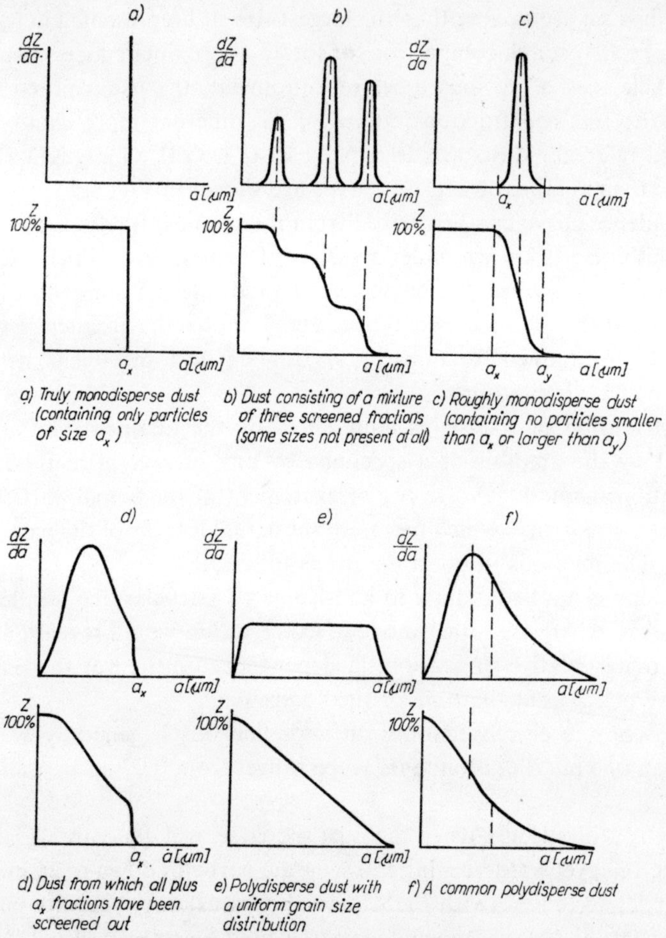

Fig. 5 How various grain size distributions are represented by the frequency of incidence curves (top) and by the retained fractions curves (bottom):
a) Truly monodisperse dust (containing only particles of size a_x); b) Dust consisting of a mixture of three screened fractions (some sizes not present at all); c) Roughly monodisperse dust (containing no particles smaller than a_x or larger than a_y); d) Dust from which all plus a_x fractions have been screened out; e) Polydisperse dust with a uniform grain size distribution; f) A common polydisperse dust.

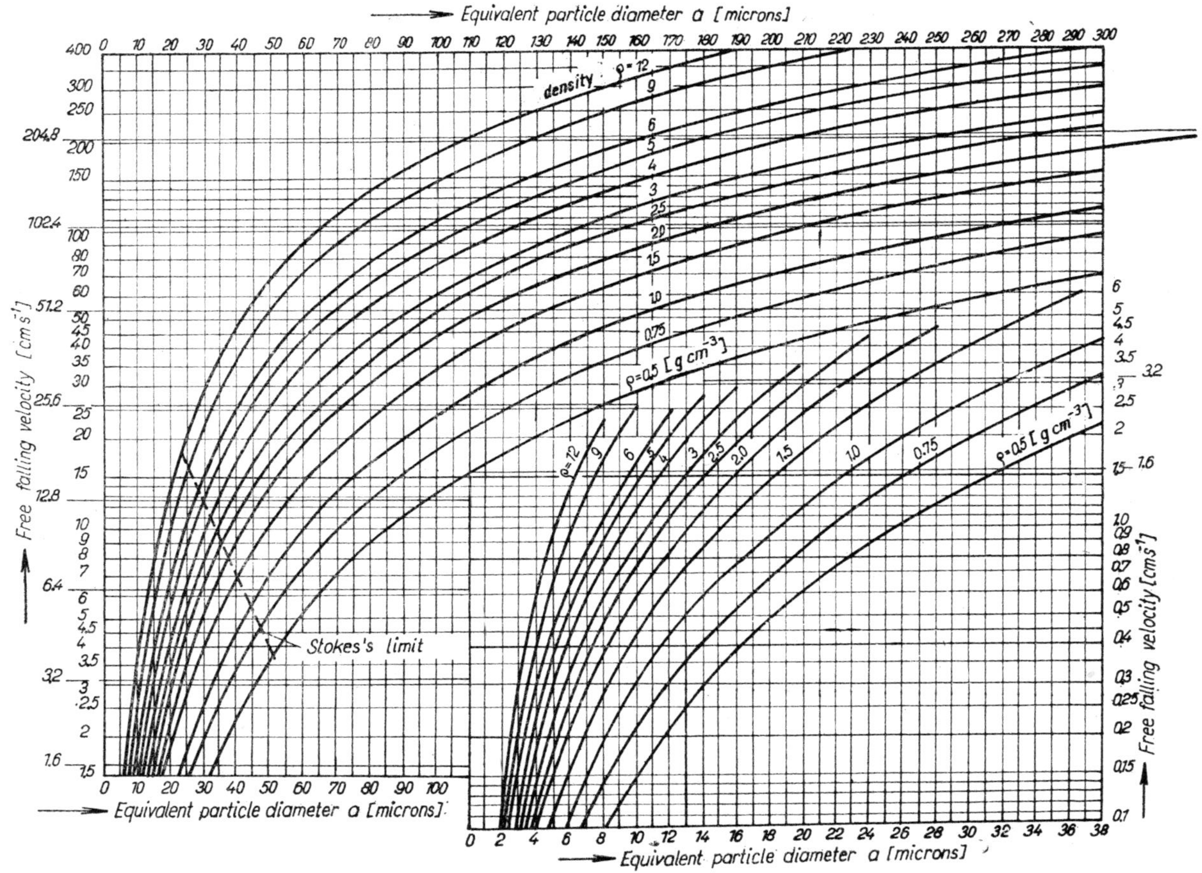

Fig. 6 Free falling velocities of particles in air (at 101,325 Pa and 20 °C) in dependence on their size a (microns) and specific gravity ϱ (g cm^{-3}).

a) Those which classify the particles in a stream of gas, either under the effects of gravity alone (for instance on the Gonell, Alpine or Fritsch devices), or under the action of centrifugal and Coriolis forces (for instance by the Bahco or Fritsch apparatuses).

b) Those which classify the particles in a liquid. These sedimentation techniques are nowadays the most widely employed of all methods for ascertaining the grain size distributions of dust specimens. Andreasen's original pipette method has since its inception in 1928 undergone numerous modifications, and has recently been joined by more sophisticated techniques. One of them is the automated procedure employing the Sartorius sedimentation balance, another is the photo-sedimentometric method, for which special instruments have been devised by Leitz, by Evans, and by some others.

The sedimentation of dust in a liquid is also exploited in various other methods for determining the grain size distribution of a dust sample, such as the displacement or the areometric techniques. Just as in procedures where the dust is classified in a stream of gas, the analysis can be speeded up, and/or the scope of the investigation can be extended towards the smallest particle sizes, by running the sedimentation in a liquid under the action of a centrifugal field, for instance in the apparatus made by MSA to Whitby's design.

3. Indirect methods establish the grain size distribution of a dust sample in terms of other properties of the set of dust particles, which are only indirectly related to the actual particle size. The best known of these instruments is the Coulter Counter, made by Coulter Electronics Ltd. of Great Britain. What this device actually measures are the electrical properties of a solution of the dust in a liquid; these data are then processed to yield the number and sizes of the dust particles suspended in the electrically conductive solution.

Several other methods are used, on a smaller scale, for investigating the grain size distribution of dust samples. One approach is to examine the permeability of a layer of dust particles, another is to classify the particles by their flotation in a stream of liquid; but none of these alternative methods of assessment need to be described in the context of this book.

1.6 Specific surface area of dust

The specific surface area is defined as the aggregate surface area of a certain unit by weight of dust, usually in $cm^2\ g^{-1}$. This definition is not as unambiguous as it might seem, because we have to distinguish between the area of the macroscopic external surface and the total surface area. The former takes no account of any internal pores within the dust particles, while the latter covers both the macroscopic and the internal surface areas (i.e. includes the surface area of pores within the dust particles which is accessible to gases or liquids). The difference between these two surface area figures, however, is not truly indicative of the porosity of the particle. The internal surfaces we are considering here are only those which are freely accessible from outside, not those of any fully enclosed cavities in the interior of the particles.

For spherical particles with a diameter of a (in microns), the specific surface area P'_{m} is established as

$$P'_{\mathrm{m}} = \frac{P_{\mathrm{m}}}{G_{\mathrm{m}}} = \frac{6 . 10^{-8}\pi a^2}{10^{-12}\pi a^3 \varrho_{\mathrm{m}}} = \frac{6 . 10^4}{a\varrho_{\mathrm{m}}} \quad (12)$$

When the particles are non-spherical, the result has to be corrected by a shape factor. The value of that factor, however, depends upon the size as well as on the shape of the dust particles.

1.7 Permittivity

The permittivity of the material of solid or liquid particles is a key quantity, since it has a qualitative effect on the electrical phenomena, and particularly on the magnitude of the particle charges, in an electrostatic precipitator. Expressed in the units of the SI system, the permittivity for a vacuum works out as

$$\varepsilon_0 = \frac{1}{4\pi c^2} . 10^7 = 8.85416 . 10^{-12}\ \mathrm{C\,V^{-1}\,m^{-1}} = 8.85416 . 10^{-12}\ \mathrm{F\,m^{-1}} \quad (13)$$

where c is the speed of light, i.e. $2.99776 . 10^8\ \mathrm{m\,s^{-1}}$.

For any other medium than a vacuum, the permittivity is determined as

$$\varepsilon = \varepsilon_r \varepsilon_0 \quad (14)$$

where ε_r is a dimensionless number, termed the relative permittivity, which indicates how many times greater is the permittivity of this medium than that of a vacuum. For a vacuum, obviously, $\varepsilon = \varepsilon_0$, which means that the relative permittivity of a vacuum, equals unity.

Since electrostatic precipitation processes are substantially dependent on the permittivity of the medium in which the particles are borne, it will be worth examining the permittivities of gases. The relative permittivity of a gas differs only very slightly from unity; like that of a solid, it depends on the specific gravity in a way governed by the *Clausius-Mossotti* relation,

$$\frac{\varepsilon_r - 1}{\varepsilon_r + 2} = A\varrho \quad (15)$$

where A is a constant dependent on the particle material or gas in question, in $m^3\,kg^{-1}$.

The value of A for a vacuum is zero, and the A values for gases are very low; air, for instance, has an A value of $1.5312 \,.\, 10^{-4}\ m^3\,kg^{-1}$.

Relation (15) yields the value of ε_r as

$$\varepsilon_r = \frac{1 + 2A\varrho}{1 - A\varrho} \tag{16}$$

When the product of $A\varrho$ is negligible, as compared to unity, the way the relative permittivity depends upon the specific gravity can be expressed approximately as

$$\varepsilon_r = 1 + 3A\varrho \tag{17}$$

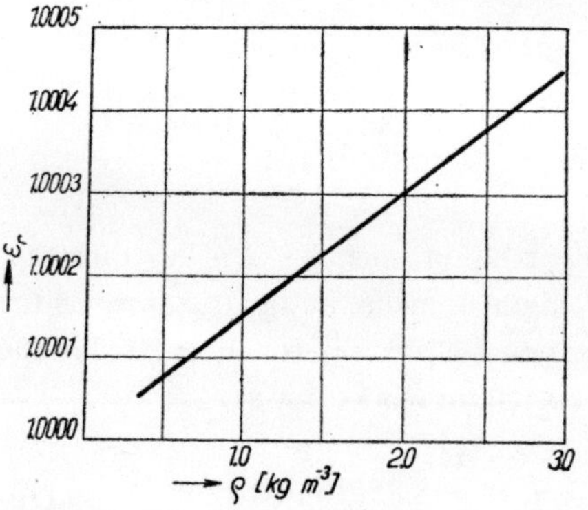

Fig. 7 How the relative permittivity ε_r of air varies with its specific gravity ϱ.

In gases, the value of $A\varrho$ is in fact insignificant relative to unity. Hence, their relative permittivity can be considered to be virtually independent of their temperature and pressure, at least under the conditions prevalent in electrostatic precipitation equipment. Under these conditions, the relative permittivity of a gas is close to unity. As an example, Fig. 7 shows how the relative permittivity of air varies with its specific gravity.

In solids, the specific gravity alters very little in response to temperature and pressure changes, and consequently the relative permittivity can again be taken to remain constant within the range of conditions normally encountered in electrostatic precipitation equipment. Table 5 lists the relative permittivity values of some gases, liquids and solids which are of particular interest in this field.

The relative permittivity of a mixture is governed by the relative permittivities of the materials which constitute that mixture. One point to note, however, is that the relative permittivity of water differs sharply from those of solids, so that humidity will always strongly affect the resultant value: the higher the moisture content of a solid mass, the higher will be its relative permittivity.

TABLE 5

Relative permittivities ε_r of some substances

Gases	ε_r	Liquids	ε_r	Solids	ε_r
N_2	1.000 61	Sulphuric acid	84	Rubber	2.5—6.8
He	1.000 07	(concentrated)		Crude rubber	2.0—3.5
N_2O	1.001 14			Fluorspar	6.8
SO_2	1.009 50	Ethyl alcohol	2.7	Quartz	4—4.7
CO	1.000 69			Metal oxides	12—18
CO_2	1.000 96	Kerosene	2.1	Talc	4.1—6.4
O_2	1.000 55			Marble	8.3—8.5
H_2	1.000 26	Water	81	Pertinax	4.8—5.1
Air	1.000 59			Fly ashes	approx. 3
				Porcelain	5—6.7
				Gypsum	5
				Sulphur	3.6
				Glass	5—10
				Silica Glass	3.7—4.2
				Mica	4—8
				Rock salt	5.8
				Coal	4
				Limestone	8

1.8 Electrical charges of particles

Solid and liquid particles are apt to acquire electrical charges both during the processes by which they arise, and in the course of their motion within the confined spaces of channels or pipings. In the latter case the charges are generated by friction, and their magnitude and polarity will depend both on the chemical composition of the particle itself and on the material of the walls at which the friction occurs. In process equipment, these charges can cause all sorts of trouble. A common danger is sparking, which upon reaching a certain intensity can lead to explosions. Another risk is that the forces of attraction exerted by charged particles will cause deposits to build up on or within the equipment. Therefore, special precautions must be taken to avert these consequences; the most important of these measures is efficient grounding (earthing) of the equipment, and the electrically conductive interconnection of all its components.

In electrostatic precipitation processes, a crucial factor is the charging capacity of particles dispersed in an ionized gas (i.e. a gas containing ions, or electrically charged gas particles). Gas can be ionized by any of a variety of external influences such as heat or radiation; in electrostatic precipitators, it is commonly ionized by a special kind of electrical discharge, known as a corona discharge. Within the precipitator, the particles move in a gas made up of ions of mainly one and the same polarity, as well as in an electrical field. The gas ions follow the lines of force of this field, collide with the particles, and on contact transfer their charges to them. This charging process continues until an equilibrium is established between the electrical field which the accumulated charges create around the particles and the external field that arises between the electrodes of the precipitator. At first, a particle accepts charges from the gas ions very rapidly; later, as the field around the particle itself builds up, the charging rate gradually diminishes; and finally, when the two fields are in equilibrium, the particle is no longer capable of augmenting its charge any further. This maximum charge which the particle can hold is termed its saturated charge.

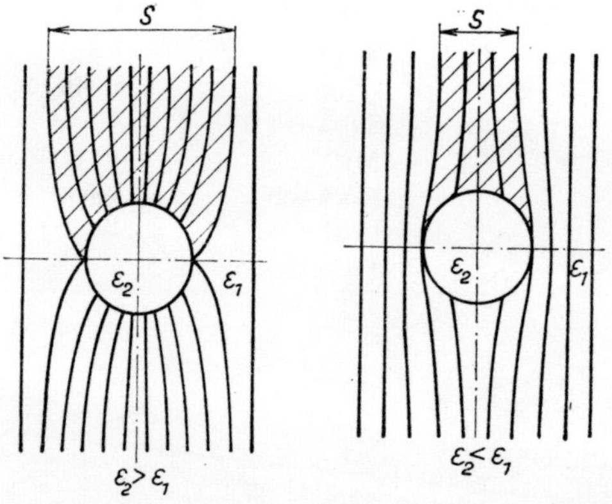

Fig. 8 How electrical lines of force respond to the difference between the field permittivity ε_1 and the particle permittivity ε_2.

There are two different mechanisms by which solid or liquid particles can pick up their charges. Large particles are charged predominantly by direct impact of the gas ions; the latter move along the lines of force of the electrical field, which are significantly distorted by the presence of a large foreign particle. Small particles, on the other hand, fail to distort the electrical field in any significant way, and are therefore charged mainly by diffusion. The diffusion phenomenon is explained by the heat-induced motions of the gas ion molecules, which cause the ions to oscillate in a plane perpendicular to their principal motion along the lines of force. These secondary ion motions are what governs the charging of small solid or liquid particles.

Fig. 8 shows how the lines of force of a homogeneous electrical field react to the presence of an uncharged particle with a permittivity of ε_2 that differs from the permittivity ε_1 of the surrounding medium. The resultant distortion depends on which of these two permittivity values is higher.

The intensity of charging by direct contact, or impact, depends on the cross-sectional area filled by the lines of force which actually touch the particle; in Fig. 8, this area is marked *S*. The larger this area, the greater is the probability of the particle acquiring a substantial charge. However, the size of area *S* depends both on the permittivity and on the size of the particle. Consequently, the saturated charge Q_n of any larger particle depends, among other factors, on the permittivity of the particle material and on the square of the particle diameter,

$$Q_n = \pi\varepsilon_0\left(1 + 2\frac{\varepsilon_r - 1}{\varepsilon_r + 2}\right)Fa^2 \tag{18}$$

The saturated charge further grows with the electrical field intensity, with the particle size, and with the value of the charging coefficient $\varkappa$:

$$\varkappa = 1 + 2\frac{\varepsilon_r - 1}{\varepsilon_r + 2} \tag{19}$$

This coefficient is obviously governed by the relative permittivity of the particle material; its values for various magnitudes of ε_r are set out in Table 6. Most of the solid and liquid particles we encounter in industrial practice have relative permittivity values in excess of unity, which, as Table 6 indicates, yields charging coefficients in the range from 1 to 3. The higher the relative permittivity, the greater are both the charging coefficient and the saturated charge of the particle in question. This again brings us up against the fact that permittivity varies with the humidity of the particle material: the saturated charge of a moist particle is higher than that of a dry one. Water and metals are notable for their high relative permittivities; so, in any given electrical field, water droplets and metal particles will tend to accept higher saturated charges than equally large particles of other materials.

TABLE 6

How the charging coefficient $\varkappa$ varies with the relative permittivity ε_r

ε_r	$\varkappa$	ε_r	$\varkappa$	ε_r	$\varkappa$	ε_r	$\varkappa$
1	1.000	3.5	1.909	7.0	2.333	50.0	2.885
1.5	1.285	4.0	2.000	8.0	2.400	100.0	2.941
2.0	1.500	4.5	2.077	9.0	2.455	500.0	2.990
2.5	1.667	5.0	2.143	10.0	2.500	∞	3.000
3.0	1.800	6.0	2.250	20.0	2.727		

The saturated charge of small particles (with diameters smaller than 10^{-6} m), at the usual processing temperatures of some 130 to 330 °C, at relative permittivities between 2 and 3, and at a space charge of the ionized gas equal to $\varrho_i = 1.6 \times 10^{-5}$ C m^{-3}, is very approximately expressed as

$$Q_n = 1.6 \,.\, 10^{-11} a \tag{20}$$

Fig. 9 shows how the saturated particle charge varies with the particle diameter, in the transition region between the domains of charging by impact and charging by diffusion. In reality, of course, there is no sharp boundary between the two modes of charging, but a broad zone within which either of these modes gradually predominates over the other, as marked by the broken line in the diagram.

The time dependence of the particle charging process is described by the relation

$$Q_\tau = Q_n \frac{\varrho_i u \tau}{4\varepsilon_0 + \varrho_i u \tau} \tag{21}$$

where u is the ion mobility.

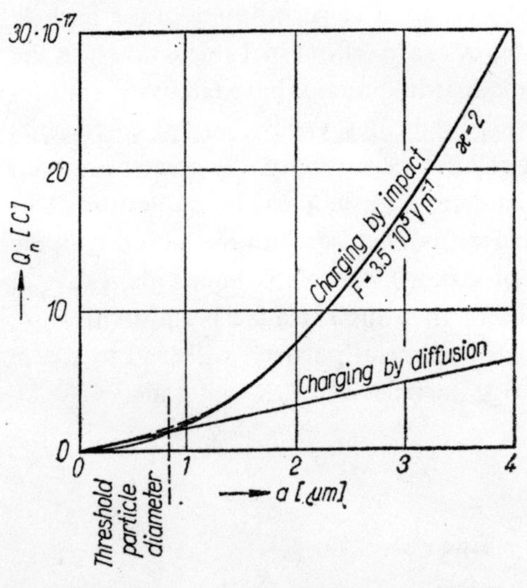

Fig. 9 Dependences of the saturated particle charge Q_n on the particle diameter a.

The value of the fraction in this expression is time-dependent: at $\tau = 0$ it is zero, at $\tau = \infty$ it equals unity. This means that, theoretically, the particle charge can attain its saturation level in an infinite time span. Of more direct interest, therefore, is the magnitude of the relative particle charge β, defined as $\beta = Q_\tau/Q_n$. Table 7 outlines the time dependence of β at various magnitudes of the space charge ϱ_i of the ions, for a constant ion mobility of $u = 2 \,.\, 10^{-4}$ m^2 V^{-1} s^{-1}. The Table further

TABLE 7

Dependence of the relative particle charging coefficient β on the residence time τ at various space charges ϱ_i of the ions

	ϱ_i (C m^{-3})		
τ (sec)	10^{-5}	10^{-6}	10^{-7}
		β	
0.5	0.967	0.745	0.226
1.0	0.983	0.854	0.369
2.0	0.992	0.921	0.539
3.0	0.994	0.946	0.637
5.0	0.997	0.967	0.745
8.0	0.998	0.979	0.824
10.0	0.998	0.983	0.854
τ_{mean} (sec)		β_{mean}	
4.0	0.977	0.863	0.485
10.0	0.989	0.930	0.671

lists the mean relative charging coefficients for charging times τ of 4 and 10 seconds. The relative charging coefficient β is defined as

$$\beta = \frac{Q_\tau}{Q_n} = \frac{\varrho_i u \tau}{4\varepsilon_0 + \varrho_i u \tau} \tag{22}$$

By the same token, the mean relative charging coefficient will be

$$\beta_{mean} = \frac{Q_{mean}}{Q_n} = \frac{1}{\tau}\int_0^\tau \frac{\varrho_i u \tau}{4\varepsilon_0 + \varrho_i u \tau}\,d\tau \tag{23}$$

where τ is the time span over which the charge in question has been building up. Integration then yields

$$\beta_{mean} = \frac{Q_{mean}}{Q_n} = 1 - \frac{4\varepsilon_0}{\varrho_i u \tau}\ln\frac{4\varepsilon_0 + \varrho_i u \tau}{4\varepsilon_0} \tag{24}$$

Reference to Table 7 shows that if an isolated particle dwells for say four seconds in the active space of an electrostatic precipitator, where the space charge of the ions is uniform throughout the length of that space, the mean charge which that particle will acquire will vary widely in dependence on the magnitude of that space charge, ϱ_i. When the active space is free of any dust, the ϱ_i values in it are commonly of the order of 10^{-6} C m^{-3}. As dust particles enter this space, however, the charges they pick up gradually create an opposing field which greatly reduces these ϱ_i values. That prolongs the particle charging time, and reduces the mean relative charging coefficient of the particles. Under these circumstances, then, the ϱ_i values that appear in relations (21) to (24) will be variable in time.

1.9 Ohmic resistance

The ohmic resistance of a layer of solid particles can be a key factor in electrostatic precipitation equipment, where it can often substantially impair the efficacy of the process. The resultant resistance R (in ohms) is largely dependent on the specific resistance ϱ (in ohmmetres) of the dust layer, which is the resistance offered by a layer one metre thick that has a cross-sectional area of 1 m^2. For any other layer thickness l and cross-sectional area S, the resistance R of the layer is established as

$$R = \varrho \frac{l}{S} \tag{25}$$

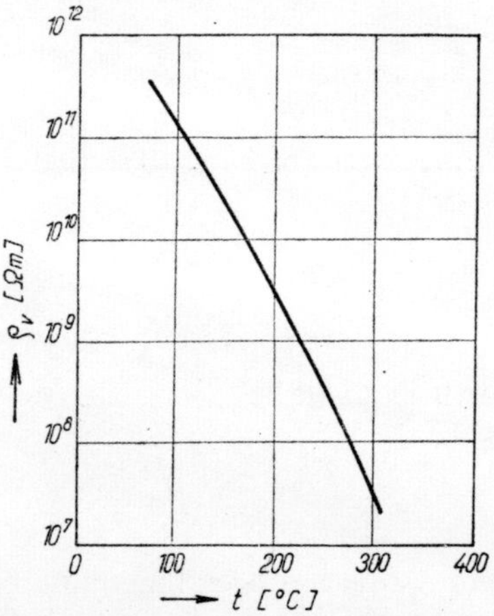

Fig. 10 Temperature dependence of the electrical resistance of minerals and of carbon.

The specific resistance of a layer of solid particles is made up of three distinct components: the specific resistance of the particle material itself; the surface resistance of the individual particles; and the contact resistances between the various particles as well as between the particle layer and the electrode surface. What appears as the resultant specific resistance of the layer is in reality the product of the whole of this system of variously combined and interlinked resistance components. The effect of each of these components upon the outcome is variable. In fact, a change in the physical conditions may alter the aggregate resistance of one and the same layer by as much as several orders of magnitude. Consequently, it is obvious that we shall get no further until we examine the conditions that govern the magnitudes of the various resistance components, and the way the resultant resistance value responds to any change in these conditions.

The specific resistance of the particle material depends primarily on its chemical composition. Dust particles are mostly of mineral origin, which means that their specific resistance values are extremely high, of the order of 10^4 to 10^{11} ohmmetres.

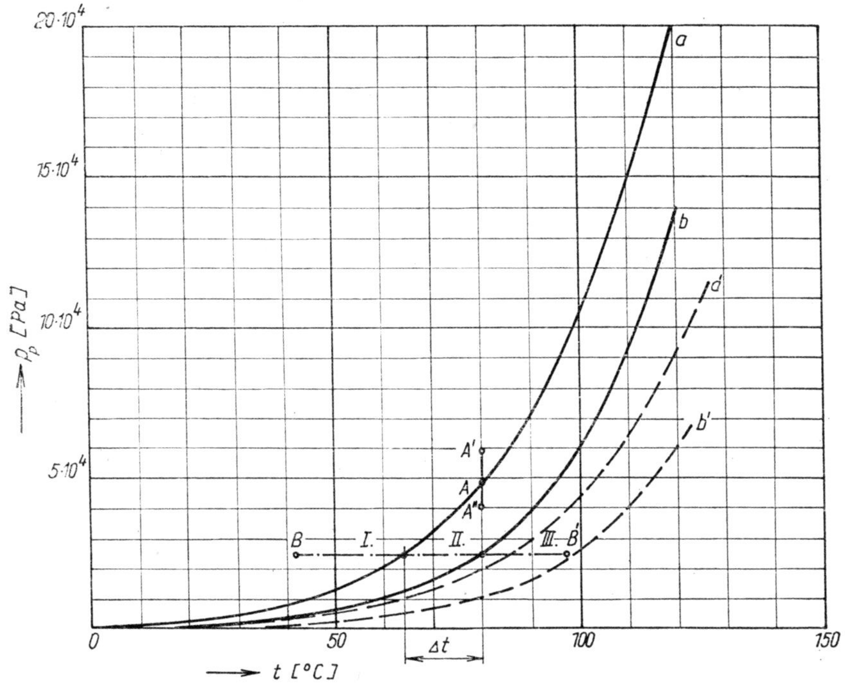

Fig. 11 Graphic representation of the gradual moistening of a particle surface layer.

In other words, these materials are practically insulators. The main exception are the carbon particles which appear as uncombusted fuel residues in fly ashes; their specific resistance ranges from 4 to $6 \cdot 10^{-1}$ ohmmetres. The internal specific resistance of a particle is also temperature dependent. As evident from Fig. 10, in mineral substances and carbon this resistance drops off as the temperature rises, often so sharply that even a slight temperature increase will reduce the resistance by several orders of magnitude.

The surface resistance depends on the various processes which take place at the particle surfaces, and which may extend down to the sub-surface layers. The most important of them are the condensation of water vapour in the pores and fissures of the particles (known as capillary condensation), and the adsorption of water vapour at the surfaces, particularly of hygroscopic particles. The point is that normally clean water, with only a low content of soluble admixtures, has a specific resistance in the region of 0.5 to 1.10^4 ohmmetres, so that any moistening of the particle surface layers tends to reduce the aggregate resistance of the dust layer. The actual degree of moistening depends on the temperature and relative humidity or dew point of the environment around the particles, but dust can be humidified even at temperatures in excess of the dew point.

The moistening process will best be understood by reference to Fig. 11. Curve *a* on this diagram indicates how the partial pressure of saturated water vapour alters with the ambient temperature. Now let us assume that the gas under consideration contains enough water vapour for the partial pressure of this vapour, at the given temperature, to correspond to point A' on the diagram, i.e. to be higher than that at point A on curve *a*. The amount of vapour corresponding to the difference between these two partial pressures will then turn liquid, i.e. form a mist, leaving the gas fully saturated with water vapour. If, on the other hand, the partial pressure of water vapour in the gas is represented by point A'', the gas will remain unsaturated. It will then be capable of accepting a further quantity of water vapour, proportional to the difference between the partial pressures at points A and A'', before it reaches its saturation point. In other words, in the region above curve *a* on the diagram we shall always be dealing with a two-phase system, consisting of liquid water and fully saturated gas, while in the region below that curve we shall always be faced with unsaturated gas. The excess water will commonly be distributed throughout the volume of the gas in the form of droplets, which will tend to settle on any surface they contact. On readily wetted surfaces, this moisture will gradually build up a continuous film.

Any solid particles in the gas stream will generally have their surfaces broken by capillary pores and cracks with highly concave surfaces. There, as the *Gibbs-Thompson* relation tells us, the water vapour pressure will drop off as shown by curve *b* in Fig. 11. Curve *b* resembles curve *a* in that it represents a boundary between two distinct regions: in the region above it, the pores in the particle surfaces will contain water and fully saturated gas, in the region below it they will hold only unsaturated gas.

Now let us examine what happens when we gradually heat a gas that contains as much water vapour as corresponds to the temperature and partial pressure at point B in Fig. 11. The moist gas, with the solid particles distributed throughout it, will pass essentially through three successive phases. The first will lie in the region above curves *a* and *b*. In this phase, the excess water in the gas will condense into a mist; the droplets will tend to attach to the dust particles; and the pores will gradually fill up with water, so that the dust particles will be wet and sticky and will tend to agglomerate into clusters. Further heating will lead to phase two, where the state of the water vapour will be somewhere between curves *a* and *b*. In this phase all the moisture in the gas and on the particle surfaces will evaporate, but the pores in the particles will still contain some water and saturated gas. The particles will seem dry, but there will still be enough moisture in their surface layers to keep their ohmic surface resistance down to low values. A further temperature rise will then reduce this moisture content in the surface layers until, at the intersection with curve *b*, it will vanish altogether. That will be the onset of phase three, where the state of the water vapour lies below both curves *a* and *b* on the diagram. In this state, the unsaturated carrier gas penetrates even the pores and crevices; the particles are dry both externally and internally; and their ohmic resistance equals that of the particle material itself, i.e. the surface resistance component disappears entirely.

As will be explained later on, in the section on electrostatic precipitation processes, the vital phase is the second one. In that phase the surface layers of the particles can be moistened, even with unsaturated carrier gas, without any actual wetting of the particle surfaces themselves. A humid surface layer is all we need to reduce substantially the overall resistance of any dust deposits. And this can be achieved, at any given moisture content of the gas, simply by keeping the gas temperature within the interval marked Δt in Fig. 11. The closer the gas temperature approaches to the dew point, the more thoroughly will the particles be moistened, and the lower will be the apparent specific resistance of the dust layers. Naturally, the dust particles borne by the gas stream will have pores of the most diverse sizes, and there will be a different curve *b* for every pore size. That is also the reason why the onset of phase three is gradual rather than abrupt: condensation may still be taking place in the smaller pores at a time when it has long ceased in the larger ones.

In practice, matters are usually rather more complicated than would appear from the above. The carrier gas is apt to contain substances (mainly oxides, and especially SO_3) which react with moisture to form very dilute acids, or react with the particle material to form soluble salts. Moreover, the particle material itself may contain soluble components which will dissolve in the moisture within the surface layer. These dissolved salts and dilute acids in the water will form second-class conductors, which will markedly reduce the specific resistance of the particle surface layers. Furthermore, they will greatly lower the vapour tension, thereby transposing curves *a* and *b* in Fig. 11 into the curves marked *a'* and *b'* respectively, and shifting the Δt interval towards the high-temperature end of the chart. As a result, the dew point of the damp gas will rise, sometimes to temperatures higher than 100 °C; this point will be the higher, the greater the ratio between the partial pressures of H_2SO_4 and of H_2O in the gas.

The third component which affects the specific resistance of a dust layer is the contact resistance. This resistance is the lesser, the greater the area of mutual contact between the particles, but its actual magnitude depends mainly on whether or not the particle layer is compressed. The more firmly compacted the layer, the smaller will be the contact resistance between its individual particles. With this much said, however, it must be admitted that much less is known about the laws which govern this resistance than is the case with the other resistance components. Consequently, the contact resistance is all too often the factor which hampers the application of laboratory results to full-scale equipment. The specific resistance ascertained in laboratory work quite commonly differs widely from the resistance encountered in the dust deposits on the electrodes of electrostatic precipitators, even given the same temperature and other environmental factors in both cases.

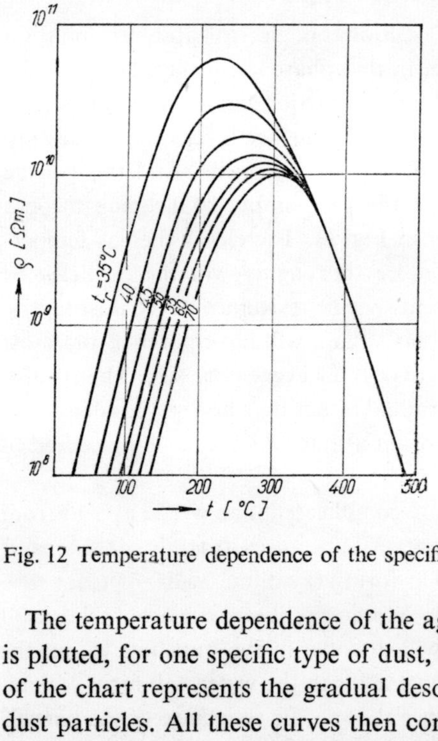

Fig. 12 Temperature dependence of the specific electrical resistance of a dust layer.

The temperature dependence of the aggregate specific resistance of a dust layer is plotted, for one specific type of dust, in Fig. 12. The array of curves on the left of the chart represents the gradual desorption of moisture from the pores of the dust particles. All these curves then converge into a single descending leg, which traces the decline in the specific resistance of the particle material itself. The seven curves in the left-hand part of the diagram are marked with the dew points of the gases on which they were established. This shows how the moisture content of the gas affects the dependence of the specific resistance upon the gas temperature. As another example, Fig. 13 indicates how the temperature dependence of the specific resistance, in one particular type of dust, responds to changes in the H_2SO_4 content of the gas. As has been stated before, the H_2SO_4 affects the outcome by reducing the surface resistance of the particles. Similarly, Fig. 14 shows how the temperature dependence of the specific resistance varies with the amount of uncombusted fuel remnants among the particles of a certain type of fly ash.

Since the specific resistance can strongly affect the particle trapping process in electrostatic precipitators, it has to be kept within certain limits in these installations. For a first rough approach, the specific resistance of the dust layers should be somewhere between 10^2 and 10^9 ohmmetres. That means either taking special measures to keep the value within those limits, or else designing the precipitator from the outset in a way which will effectively suppress the detrimental effects of an inadequate or excessive specific resistance value.

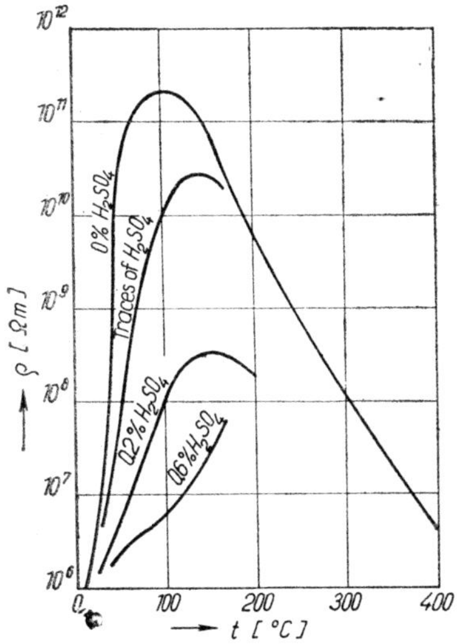

Fig. 13 How the temperature dependence of the specific electrical resistance vartes wiih the H_2SO_4 content of the gas.

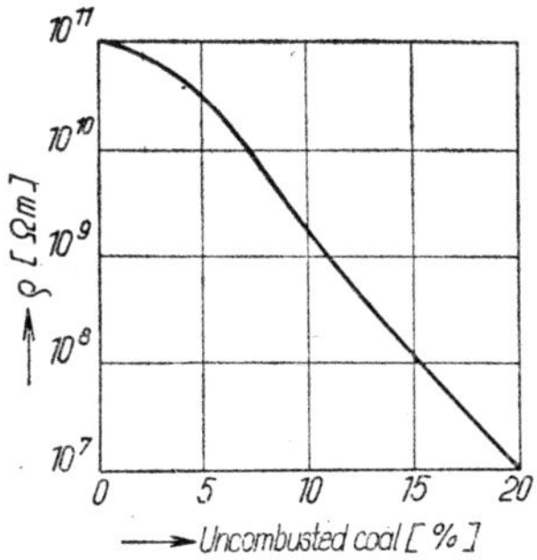

Fig. 14 How the specific electrical resistance of fly ashes depends upon their contents of uncombusted coal residues.

1.10 Angle of repose

The angle of repose is the angle included between the flank of a cone of freely deposited granular material and the horizontal. It is ascertained by means of a special instrument, consisting of a base, a pan, a protractor, and a guide rig with a funnel or hopper. The base plate bears a circular table of 50 mm diameter, which forms the base of the cone. The outlet of the funnel is provided with orifice plates of various aperture diameters for various types of granular material. While the funnel is delivering, it is raised on its guideways, by means of a screw, so as to keep its outlet constantly just above the tip of the cone of material as the latter builds up. Any surplus dust drops off the table into the pan beneath it. Once the cone has attained its final size, so that it covers the whole of the table, the slope angle is measured by means of the protractor.

The actual shape of the cone is governed mainly by the friction between the individual particles, but is also affected by some other factors. The chief among them is the shape of the particles: the more regular and rounded they are, the lesser will be the resultant slope angle. Another influence is the particle size; in most types of dust, the angle of repose grows as the mean grain size diminishes, but this does not apply to extremely fine or adhesive dusts. In these materials, we are rarely dealing with individual particles, but more usually with cohesive clusters or agglomerations which behave in a way unrelated to the original grain size distribution and particle shapes of the dust. The angle of repose further depends on the specific gravity and humidity of the dust. Finally, the results are apt to vary with the technique by which the cone has been built up, and, as in the case of specific gravity measurements, with the size of the cone itself. The slope angles ascertained in practice, on large piles, are mostly somewhat lower than those determined by laboratory tests.

1.11 Angle of slide

The angle of slide is the angle which an inclined plane must assume to dislodge the loose granular material on it, i.e. the angle at which this material begins to slip and slides down the plane. This angle is measured on a device consisting of a base, a brass plate which can be tilted through 90 degrees, and a protractor. The tilting plate is bounded on three of its sides by sidewalls 2 mm high. The dust specimen is poured onto the plate from the least possible height; surplus dust is wiped off with a template (care must be taken not to compact the dust remaining on the plate); the plate is then slowly raised, and its angle of inclination is read off at the moment when all the dust on it, or a substantial part of that dust, is set into motion.

The angle thus ascertained depends primarily on the friction between the dust particles and the inclined plane, in other words, on the material and state of the surface of the tilting plate. However, the angle also depends on the properties of the dust itself, in much the same way as the angle of repose. Which means that, in experiments performed on one and the same plate, the angle of slide will usually tend to be the greater, the finer the dust and the less the particles resemble spheres. Again, this rule fails to apply to extremely fine and/or adhesive dusts.

1.12 Adhesive properties

Although in the sphere of dust handling we commonly speak of adhesivity, this term does not represent any single, well-defined physical property. Rather, the adhesive tendency of a dust is the outcome of a number of diverse phenomena which between them cause dust particles both to adhere, e.g. to the walls of dust trapping equipment, and to agglomerate and cohere in clusters. Usually, the term adhesivity is applied only to the former case, of adhesion proper, while the tendency of the particles to form cohesive clusters is described as their agglomeration (see Section 2.6).

The dust adhering to equipment walls gradually clogs up the flow channels or active surfaces with deposits which then prevent the equipment from functioning as intended. Next to the abrasive effect of a dust, its adhesivity is thus the property most liable to impair the proper working of dust trapping devices.

The tendency of dust particles to adhere to equipment walls is governed not only by the adhesive properties of the dust itself, but equally by the process conditions within the equipment, such as the humidity, temperature, the dew-point of the carrier gas, etc. There is consequently no generally valid formula for judging to what extent a certain type of dust is likely to clog up the dust trapping facilities. Each case must be assessed separately, with due regard for the specific conditions encountered in it.

The properties of the dust itself which decide how strongly adhesive it will prove are mainly its moisture content, and the degree to which it is hygroscopic. Other important factors are its electrical properties and its grain size distribution, or the effects of previous screening upon this distribution. Especially in multi-stage dust separation processes, the extent to which the dust has already been classified by size must be taken into account when we examine its adhesive properties and their likely consequences.

Any attempt at a physical explanation of the mechanism which causes dust particles to adhere to equipment walls must needs start out from an analysis of the adhesive forces that act between the particle and the wall. Unfortunately, our knowledge of the physical conditions at the points of contact between individual particles, and between particles and the wall, are still inadequate for anything like a comprehensive investigation. Physical chemistry tells us that the forces of attraction between molecules act only down to a certain distance; once that distance is understepped, they are overcome by forces of repulsion which tend to prevent any actual contact between the molecules. Once we simplify our physical assumptions about the shapes of the contacting surfaces, we can distinguish between three broad categories of adhesive forces which will come into play mainly at the assumed point of contact:

1. *Van der Waals forces*: These molecular forces, of an electrostatic nature, can

be traced down to the fact that atoms and molecules are made up of positively charged nuclei and negatively charged electrons. These forces decay rapidly as the distance between the two opposing surfaces increases, and can safely be neglected once that gap exceeds a few tenths of a micron.

2. *Electrostatic forces*: These forces arise when the charges of the dust particles and of the equipment walls are of opposite polarities. In electrostatic precipitators these forces are induced deliberately, to attract the dust to the collecting electrodes. The same forces, however, are also generated inadvertently, e.g. when a dust particle is charged by friction against a wall. Two non-conductive particles with charges of opposite polarities are attracted towards each other by these forces, but only at extremely close ranges. The attraction virtually ceases when the gap between the particles attains half the particle diameter.

3. *Capillary forces*: The presence of a liquid layer between two dust particles, or between a particle and the wall surface, will cause capillary effects to impel the two surfaces towards each other with a good deal of force. This explains the well-known fact that humidity is one of the worst aggravating factors which intensify the tendency of dust particles to adhere and build up deposits within the equipment.

It follows from the above that the tendency of dust particles to adhere to equipment surfaces must always be considered with all of the following criteria in mind: the size of the particles; their distribution in space; the gas velocity; the intensity of turbulence in the gas stream; the velocity gradient across the boundary layer; the material and surface finish of the wall which constrains the gas stream (since there are large differences between say cast iron, stainless steel, aluminium, and brass, and between paint-coated, metallized, machined, smoothened and rough surfaces); the particle material, as regards both the chemical and the physical properties of the particle surfaces; and the type and state of the gas, mainly its temperature, moisture content, and chemical composition.

This list of criteria is obviously far too extensive and complex to permit any exact evaluation of the adhesive properties of a dust sample. In practice, we generally resort to rough and ready classifications of adhesivity, such as the tabulated survey, first quoted in Soviet literature, which underlies Table 8. In this Table, that survey has been modified and supplemented to cover the findings made since the original paper was published. In referring to this Table, it must be borne in mind that it makes no allowance for the process conditions, especially for the humidity level, and that it does not even set out to present any absolute classification of various dust types by their degree of adhesivity. All it is intended to serve for is a relative comparison as between one kind of dust and another. Some authors recommend, as a rule of thumb, that when the dust is moist, it should be shifted one step further down, to the next highest adhesivity range, than the one in which it is listed in this Table.

TABLE 8

Rough classification of dusts by their adhesivities

Category	Dust types
1. Non-adhesive	Dry pulverous slag; some dry loams; etc.
2. Slightly adhesive	Fly ashes with substantial contents of uncombusted fuel residues; coke dust; dry magnesite dust; slate ashes; top-gas dust; coarse coal dust; etc.
3. Fairly adhesive	Fly ashes free of fuel residues; peat dust and ashes; moist magnesite dust; metal powders; pyrites; oxides of lead, zinc and tin; dry cement; dust from shaft and rotary cement-making furnaces; soot and carbon black; skimmed milk powder; ore sinter dust; coal fines; and all fully minus 25 microns dusts
4. Highly adhesive	Moist cement, gypsum, limestone, kaolin, clays; flour; fibres of asbestos, cotton and wool; rich milk powder; cocoa; and all fully minus 10 microns dusts

1.13 Abrasive properties

One of the major headaches in all gas cleaning techniques is the erosion of the equipment by the dust particles carried in the gas stream. The intensity of this wear depends mainly on the type and number of these particles and on the angle under which they impinge on the exposed equipment surfaces. It further depends on the properties of the individual particles, chiefly their specific gravity, shape, size, and hardness. All other factors being equal, the wear rate will largely be governed by the character of the gas flow, which, for lack of other criteria, is usually described in terms of its flow velocity.

If we study how the wear rate of metals depends on the angle of impact α of the abrasive particles, we find that there are two quite distinct zones on the diagram, representing two different wear processes. One is wear caused by more or less perpendicular impacts, the other is wear caused by friction. The latter attains its maximum at impact angles around 30 degrees, and drops off to zero somewhere between 75 and 85 degrees. The wear caused by impact attains its peak at an angle α of 90 degrees, and is more intensive in hard metals than in soft ones. Given the same metal, the same dust, and the same gas velocity, the intensity of both these types of wear will depend essentially on the angle of impact, but this dependence is far more marked in friction-induced than in impact-induced wear. Harder metals resist wear by friction better than softer ones, especially at small α angles, but softer metals tend to withstand wear by impact better than harder ones. Quite irrespective of the angle of impact, however, the wear rate is always directly proportional to the number of impinging particles.

The amount of material eroded from the equipment surface, in other words the depth of erosion, has after a great deal of experimental research been found to depend at least approximately on the square of the impact velocity. It has also been established that this wear is effected by two different mechanisms. At large impact angles it is caused by *gradual deformation*, while at small impact angles a mechanism which involves a *microscopic cutting action* comes into play. That helps to explain why the ideal wear-resistant material for large impact angles should be relatively soft and ductile, while that for small angles should be hard and may be brittle. A particle hitting the soft, ductile material at an angle close to the perpendicular will have to deform a certain depth of the plastic metal, and will be gradually retarded over the whole of this path length. Unless the metal has actually been cut, either by a sharp edge on the particle or by an impact at high velocity, it will tend to revert to its original shape, expelling the particle in the process. Given a sufficiently resilient material and a large α angle, there should theoretically be next to no erosive wear.

The reason why hard materials wear most rapidly at large impact angles is simply that most of these materials are brittle. When exposed to an endless succession of particle impacts, they are liable to form cracks, which then propagate until microparticles are detached from the surface. At small impact angles, however, these very hard materials are highly wear resistant, because the particles fail to penetrate their surfaces. Substantial erosive wear at small impact angles is encountered only when the impinging particles are a good deal harder than the material exposed to them.

Erosive wear will obviously be at its lowest when the particles are much softer than the material they are striking. It intensifies abruptly when the particles are 1.1 to 1.6 times harder than that material. Furthermore, the wear rate grows more or less linearly with the particle size, which is attributable to the greater mass and hence greater impact energy of the larger particles. It has been observed that fly ashes cause the least wear at impact angles around 45 degrees; that is also the angle quoted in the literature as the most favourable one for sands. In the handling of metallic grit for shot-blasting purposes, the lowest wear rates have been recorded at impact angles of about 50 degrees.

The erosive effects of the dust are most apparent in cyclones, in which abrasive wear is easily the most common cause of premature failures and replacements. Wear is therefore one of the main aspects to be considered in the design of these units. This point is illustrated by the following results, gained in wear and life expectancy tests on four types of cyclones handling three different kinds of dust under a variety of service conditions. The dusts used in these trials were:

a) Corundum powder, which consists of sharp-edged cubical or pyramidal particles, with no spherical or spheroidal grains. It is so abrasive that it shows up the potential danger zones and weak spots in a very short time. The powder employed in this work was gained by the thorough mixing of dosed amounts of various size fractions, and had the retained fractions curve plotted in Fig. 15, curve *a*.

b) Dust taken from clinker coolers, which resembled the corundum in general shape but had more rounded edges. Its retained fractions curve is presented in Fig. 15, curve *b*.

c) Fly ashes from a boiler fired with coal fines. These particles were mostly spherical, with an admixture of cubical or otherwise rectangular grains; their retained fractions curve is charted in Fig. 15 as curve *c*.

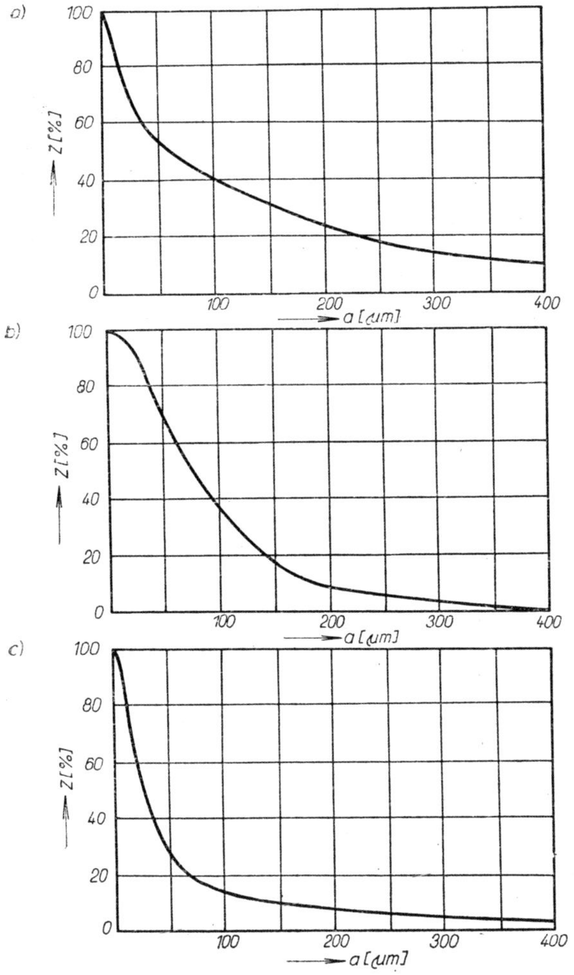

Fig. 15 Retained fractions curves of the dusts used in cyclone wear rate trials:
a) Corundum dust ($\varrho = 4091$ kg m^{-3}); b) Clinker (3178 kg m^{-3}); c) Fly ashes (2178 kg m^{-3}).

These three dusts were supplied to the cyclones under investigation at an inlet concentration of $20\,g\,m^{-3}$, and usually at three different pressure drops across the cyclone, in the range of 600 to 3000 Pa, to produce three substantially different wear rates. All the tests were conducted at 20 °C. In some of them, the initial dust concentration was deliberately varied to examine how the wear rate would respond to these changes. Some of the work was performed on scaled-down models of the cyclones. All the cyclones consisted of sheets 1 mm thick, of the same steel grade which is normally used in the routine production of these units. They were made in two exactly coaxial halves each, and the halves were then butt joined by soldering with tin. After the completion of the test runs, the two halves of each unit were separated from each other, carefully cleaned, and next had a grid marked out in soft graphite on their internal surfaces. The amounts of wear were then measured at all the intersections of this grid.

The results thus obtained on the four selected types of cyclones, designated A to D, are summarized in Figs. 16 to 19. These diagrams show the maximum wear found along the cylindrical and conical portions of the cyclones after two tons of dust had passed through each of them. The wear induced by each of the three dusts is plotted for a gas inlet velocity corresponding to a clean-air pressure drop of 1000 Pa across the cyclone. The second curve for the corundum dust applies to a gas inlet velocity which, with clean air, would have produced a pressure drop of 2000 Pa across the unit.

The four types of cyclones selected for these trials can be briefly characterized as follows:

Type A: A good general-purpose design, considered suitable for a basic standard production model.

Type B: A cyclone with an enhanced throughflow capacity, i.e. a lesser pressure drop between its inlet and outlet, but with a slightly inferior trapping performance.

Type C: A model which has its inlet inclined downwards, and has a long conical section; it combines a high separating efficiency with a fairly large throughflow capacity.

Type D: This cyclone has a constricted inlet section and a long cylindrical portion. It displays the best fractional separating efficiency, but also the lowest throughflow capacity and largest pressure drop, of the four investigated types.

The dimensions of the various types, as indicated in Figs. 16 to 19, are all converted to apply to cyclones with the same diameter of their cylindrical portions as type A. In other words, this latter type was taken as the standard reference model, and all the dimensions of the other types were related to the unit dimensions of the type A cyclone.

Apart from the maximum wear ascertained along the lengths of the cylindrical and conical portions, Figs. 16 to 19 also present the calculated life expectancies of each of these portions. These life expectancy figures were established for various gas inlet velocities (corresponding to pressure drops of 300 to 3000 Pa across the cyclone), and for each of the three types of dust tried out in this work. The life expectancy estimates are all based on a sheet metal thickness of 1 mm. A closer study of those four diagrams leads to the following conclusions:

Cyclone A (Fig. 16) always exhibited the maximum of wear at much the same places, regardless of which of the three dusts it was handling. Its cylindrical portion wore most rapidly in either of two zones: at higher inlet velocities the peak wear was concentrated in the plane of the bottom edge of the inlet, at lower inlet velocities it shifted closer to the top of the cylinder. Its conical portion displayed the most intensive wear roughly in the area where an imaginary extension of the gas outlet tube would intersect the conical surface. Only in the trapping of fly ashes did this maximum wear zone move further down the cone, probably because of the higher particle velocities, and higher axial components of those velocities, which are encountered in this application. The wear curves suggest that the effective service life of this cyclone could be significantly extended by the simple expedient of recessing roughly the last quarter of the conical section into the dust hopper. That would shift the areas of maximum wear to a zone outside the actual working space of the cyclone. Any puncture in the walls at these points

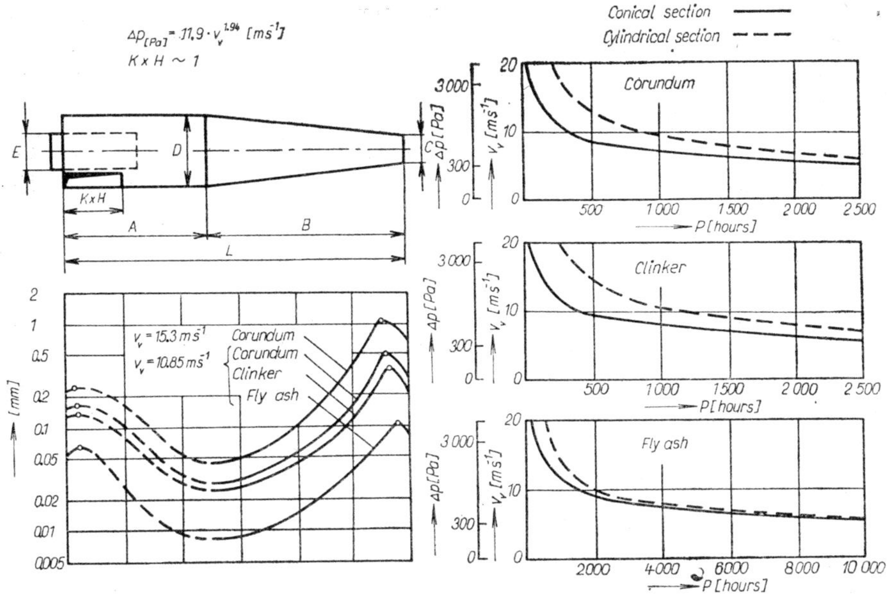

Fig. 16 Results of wear rate tests on cyclone A.

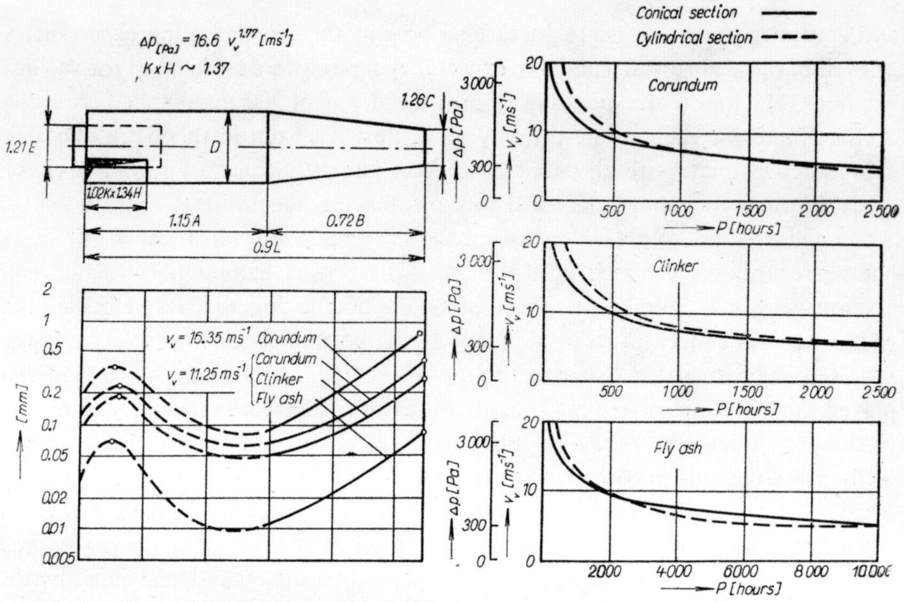

Fig. 17 Results of wear rate tests on cyclone B.

would then hardly matter, as it would not cause either the ingress of outside air or the escape of dust from the unit.

Type B (Fig. 17) always, irrespective of the kind of dust and the inlet velocity, exhibited the most intensive wear of its conical portion at the very tip of the cone. Its cylindrical portion again wore most badly around the bottom of the inlet opening when the gas velocity was high, and closer to the top of the cylinder when the inlet velocity was lower, no matter which of the three dusts was being handled. It is worth noting that even at roughly equal inlet velocities, the same amounts of dust will produce very different amounts of wear in cyclones A and B. That can be ascribed only to the basic differences between the two designs. The long conical portion of type A makes for much higher particle velocities in this section, and consequently the conical portion of type B has a roughly 25 per cent higher life expectancy than that of type A. The cylindrical portion, on the other hand, is likely to last longer in type A. That is due to the larger diameter of the outlet tube in type B, which tends to increase both the impact and friction forces acting upon the cylindrical surface as well as the number of particles impinging on it.

Type C (Fig. 18) resembles type A in the wear of its conical portion: all three types of dust produced the highest wear roughly in the area where an imaginary extension of the gas outlet tube would intersect the cone. With fly ash the maximum wear points tended to lie below this intersection, nearer to the tip of the cone; with corundum they occurred rather higher up the cone; and with clinker they were about half-way between these two points. The most likely explanation is the higher velocity attained by the fly ash particles, and especially their higher axial velocity component, because the latter governs the angle of the helix which the condensed dust layer follows in the cyclone. The cylindrical portion always wore most rapidly just beyond the area where the strongly turbulent gas stream first hits the cylindrical surface and adheres to it. In this type too, the life expectancy could be prolonged considerably by setting the bottom part of the cone into the dust receptacle beneath it.

Type C cyclones would seem to owe much of their wear resistance to their inclined inlets. These increase the radius of curvature of the particle trajectories, and thereby limit the amount of wear at higher inlet velocities to less than is common in the other three types. Moreover, this design feature restricts the centrifugal force acting on the particles, and hence the intensity of their impacts against the cyclone walls as well as the friction pressures they subsequently exert there. It also reduces the number of particles that actually contact the cyclone walls, because the centrifugal force is not sufficient to bring the particles from all over the inlet cross section up against the walls in the limited path lengths available to them. A vital factor of this design is the inclination angle of the inlet tube: the larger this angle, the lower will be the wear rate, and the pressure drop across the cyclone too. However, the overall and fractional collecting efficiencies also diminish as this angle grows, so the angle actually chosen will always have to be a compromise between the conflicting demands of efficiency and durability.

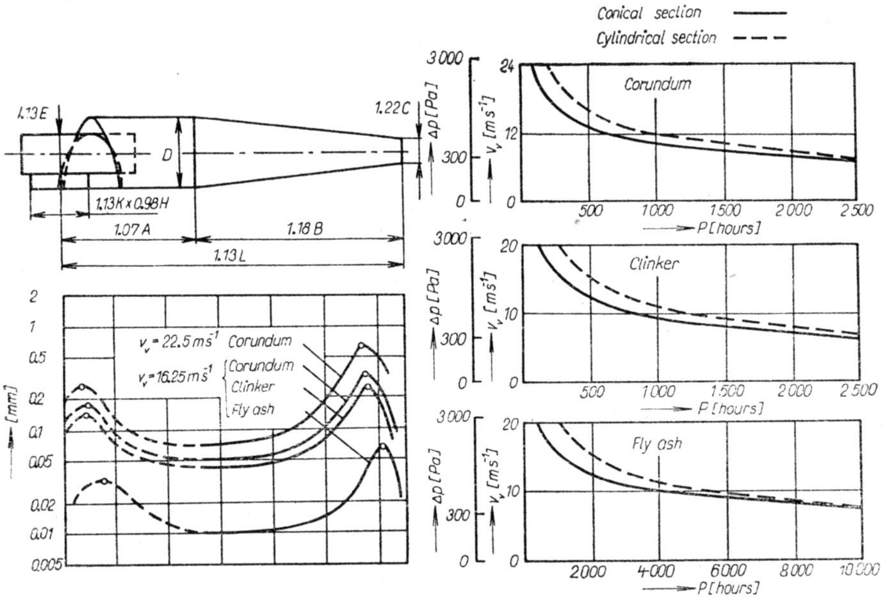

Fig. 18 Results of wear rate tests on cyclone C.

Cyclone D (Fig. 19) is normally built with twice the diameter of the other three types. It features a convergent inlet tube which helps to concentrate the dust content of the gas, but this thickening tends to accelerate the wear of the cylindrical portion. Irrespective of the type of dust handled, the cylindrical portion of this cyclone always wore fastest near its top, and the conical portion at the tip of the cone. The service life of this cyclone can likewise be extended by sinking part of its conical portion into the hopper underneath it, but it must be stressed that in type D there are limitations of scale to this expedient. The research was carried out on units with a diameter of 2*D*, or twice the diameter of the other types. When a type *D* cyclone is scaled down to *D*, or half its normal size, its cylindrical portion wears more rapidly than the cone, and there would consequently be no point in recessing the latter into the hopper, as the life of the unit would be limited by its cylindrical portion anyway.

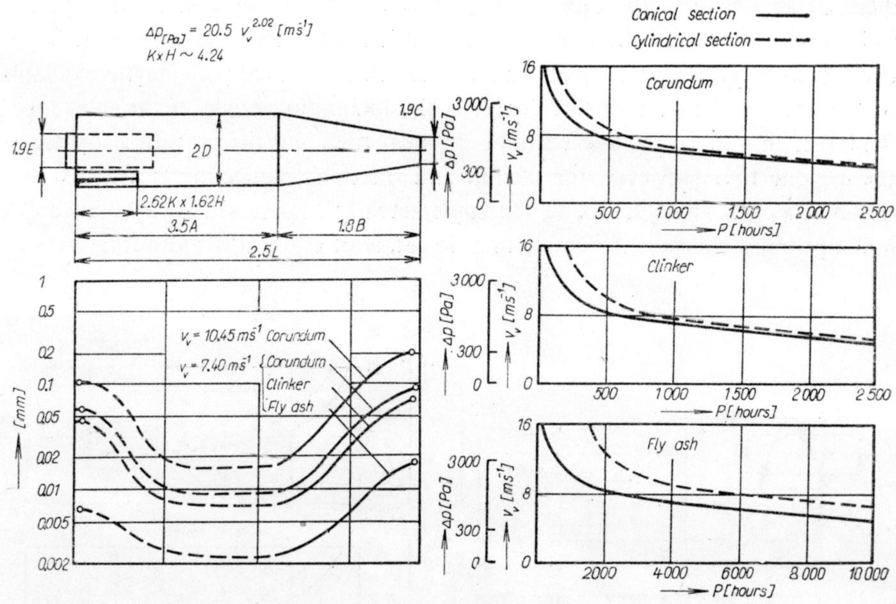

Fig. 19 Results of wear rate tests on cyclone D.

This conclusion was fully corroborated by tests on some scaled-down cyclones, in which the same inlet velocities and initial dust concentrations increased the wear rate of the cylindrical portions much more than that of the conical ones. The accelerated erosion of the cylindrical sections is obviously due to the greater centrifugal forces acting on the particles there. The relatively lower wear rate of the scaled-down conical sections can only be put down to the fact that the particle velocities are much lower there than in the full-scale cyclones. For one thing, the particles are slowed down by the increased friction during their previous passage through the cylindrical section. For another, the smaller path lengths available to them in the scaled-down conical sections simply do not permit them to build up the same speeds as in the larger units. The conclusion is that, for every given combination of inlet velocity and dust type, there is a certain closely defined cyclone diameter at which the wear rates of the cylindrical and of the conical portions will be equal. A larger or smaller cyclone diameter will produce more wear in the one than in the other of these portions.

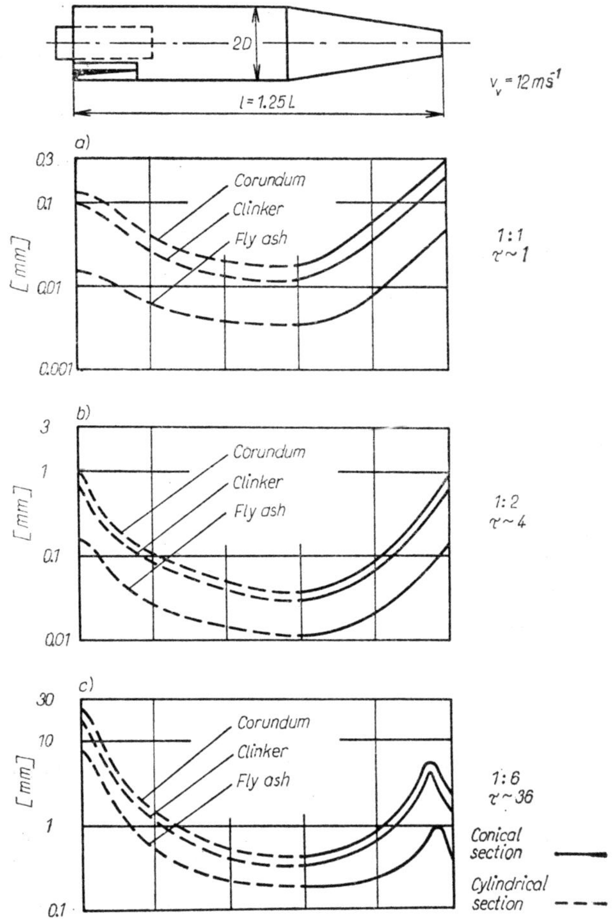

Fig. 20 Results of wear rate tests on the scaled-down cyclones D.

To illustrate this point, Fig. 20 presents the maximum wear detected along the lengths of the cylindrical and conical sections in type *D* cyclones of three different sizes. Plot *a* applies to the full-size cyclone, of diameter 2*D*; plot *b* is for a cyclone of half this size, or diameter D; and plot c represents a cyclone one sixth the size of the $2D$ unit. All the data were gained after the passage of two tons of dust through each cyclone, at an inlet velocity of 12 m s^{-1} and an inlet dust concentration of 20 g m^{-3}; naturally, the time it took to run two tons of dust through the cyclone varied with the size of the unit. As the very different lengths of the eroded surfaces in the three cyclone sizes would make any valid comparison between them rather difficult, the three charts in Fig. 20 have been drawn to one and the same standard length rather than to one and the same scale. That helps to bring out the way the relationship between the wear of the cylindrical portions and that of the conical ones varies with the cyclone diameter. It also shows that, at identical inlet velocities and concentration levels, the same amount of dust will cause almost a hundred times more wear in the one-sixth scale model than in the full-size cyclone.

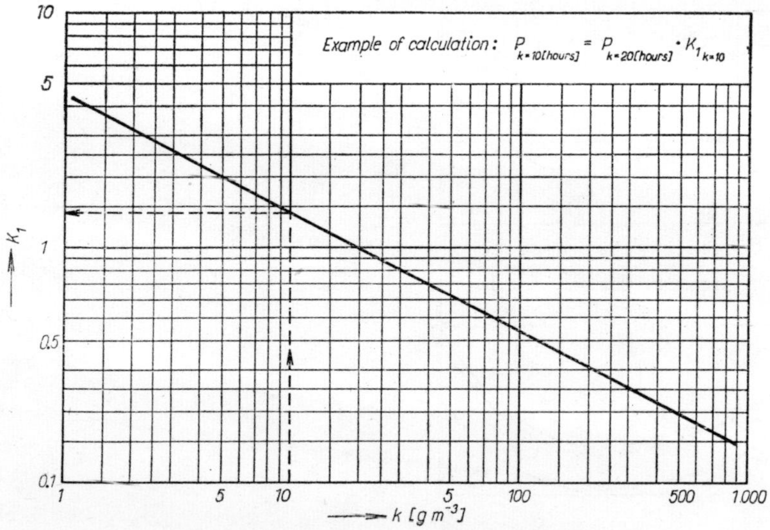

Fig. 21 Dependence of coefficient K_1 on the inlet concentration k.

All the findings charted in Fig. 20 were obtained at a dust concentration of 20 g m^{-3}. However, further trials were run at different concentration levels, to find out how these would affect the wear rate. Some of the data gained in these trials are plotted in Fig. 21, where the results at 20 g m^{-3} are taken as unity, and the effects of all other dust concentration levels are expressed by correcting coefficients denoted K_1. The life expectancy of the cyclone at any given concentration level is ascertained by multiplying the life expectancy P, established at a concentration of 20 g m^{-3}, by the appropriate coefficient K_1. These coefficient values have been found to apply to all three of the investigated dusts, and to both the cylindrical and the conical sections of the cyclones.

The obvious question, of which of these four cyclones has proved most wear resistant, is no easier to answer than the equally obvious one as to which of them offers the best dust trapping performance. There are simply too many different and often conflicting criteria to allow us to formulate any clear-cut and indisputable answer. To afford at least some basis for direct comparisons, the results recorded on those four cyclone types were converted to apply to one and the same unit diameter, and these figures were then compared for a gas inlet velocity of 10 m s^{-1} and for a pressure drop across the cyclone of 600 Pa. In this assessment, the most wear-resistant type will obviously be the one which, over a given interval of operation, displays the least pressure drop at the given inlet velocity, and the least inlet velocity at the given pressure drop. The key point about this is the constant period of operation, because, in view of the different inlet tube diameters of the various types, the amounts of dust entering each of them per unit of time will also be very different.

Of the four types under investigation, the one that came out best in this comparison was cyclone C. This type showed the least wear at the inlet velocity of 10 m s^{-1}; and when examined at a pressure drop of 600 Pa, which involved almost twice this inlet velocity, its wear rate was only slightly higher than that of cyclone D, the type which exhibited the least wear at that magnitude of the pressure drop. All these findings apply to the conical portion only, but the cylindrical part also wore more slowly in type C than in any of the other three types.

All this goes to show how complex a problem the wear of cyclones really is. The outcome depends both on the three-dimensional flow pattern within the unit, which is of course specific to every particular cyclone design and size, and on the still largely unclarified processes of erosion. In actual practice, these processes are further complicated by the concurrent effects of various other factors, such as corrosion. One thing that is clear is that cyclones wear very unevenly. That is only natural, because every least irregularity on their wetted surfaces causes intense turbulence downstream of it. In this turbulence area, the particles borne by the gas stream can easily reach and erode even the most deeply recessed parts of the wall surface, which would otherwise be shielded from the stream.

In general, the rate of erosive wear depends mainly on the velocity and direction of the particle motions, on the number of impinging particles, and on the span of time over which the individual particles are allowed to act upon the equipment surfaces. None of these factors is readily assessed, let alone evaluated numerically, and that rules out any exact calculation of the wear rates or life expectancies. Experimental findings confirm that, even given the same dust and the same inlet velocity, the wear rate will still depend on a number of other variables, chiefly those governed by the geometrical configuration and design details of the cyclone. These affect the major characteristics of the unit – the magnitude of the centrifugal forces, the degree of turbulence and intensity of the associated turbulent diffusion processes, as well as the boundary layers in the cylindrical and conical parts of the cyclone, and the behaviour of the dust particles in these layers.

Experimental work on the wear of cyclones has shown that the principal rate-governing factor is the centrifugal force, which is also the main dust separating agent in these units. Its influence is especially pronounced in small cyclones, where the radius of curvature of the particle trajectories is small and the resultant thickening of the dust stream is considerable. In a small-diameter cylinder, this high centrifugal force will result in far greater impact and friction forces than are encountered in larger units, and hence in much more intensive erosion of the walls. However, these impact and friction forces will also tend to retard the particles, and thereby save a lot of wear on the conical portion downstream of the cylindrical one, even though the centrifugal forces acting in this conical portion are generally much greater than those in the cylindrical part of the unit.

There are basically two approaches to enhancing the wear resistance of cyclones. One is to build them of some highly abrasion-resistant material, the other is to secure such service conditions which will assure the unit of an acceptable life expectancy. The former alternative nearly always entails a substantially higher capital outlay. The material we pick must combine a satisfactory wear resistance with all the technological properties required for its processing, or else must undergo a heat treatment (or any of a variety of combined chemical and heat treatments) which will render it wearproof after its fabrication has been concluded. Another course sometimes adopted is to coat or clad the basic structural material with special abrasion-resistant alloys. None of these courses, however, is really a practical proposition in the routine production of run-of-the-mill cyclone types. The materials with the right combination of properties, mainly austenitic manganese steels and various high-alloy materials, are too costly for this purpose. The less expensive alternative materials which could meet the functional requirements tend to lack the requisite technological properties. Therefore, the equipment is usually made of common structural steel grades, and extra care is devoted to the choice of its layout and operating conditions so as to ensure the longest attainable service life.

To little is so far known on these matters for anyone to judge how far the life expectancy of a cyclone can be extended by the stabilization of the flow pattern within it, by suppressing the effects of secondary vortices and excessive turbulence, or by reducing the time of residence of each particle within the unit. All that is clear from the experimental work performed so far is that cyclone manufacturers must carefully avoid any breaks, protrusions or other irregularities on the wetted surfaces, especially in their critical zones, because even the slightest unevenness can curtail the service life of the whole unit by as much as 30 per cent.

1.14 Wetting properties

Molecules of liquid are attracted towards each other by *van der Waals forces.* These forces do not normally produce any effect on molecules enclosed within a volume of liquid, because there every molecule is attracted equally in all directions, so that the resultants cancel each other out. The molecules on the surface of the liquid, on the other hand, are unilaterally attracted by the layer of molecules beneath them. Any displacement of these surface molecules towards the interior of the volume of liquid represents a certain amount of work. This is in contrast to the molecules underneath the surface, which can shift without any work being performed or consumed. This implies that the molecules of the surface layer differ from all other molecules by their ability to perform work, in other words, by having a certain reserve of energy, known as surface energy. Outwardly, this energy appears as a tension in the surface layer of the liquid, in a film with a thickness of the order of about 10^{-9} m. That surface tension acts in two distinct ways: it permits the surface film to carry or transmit a certain load or stress; and it tends to contract the surface film, as far as the applied forces (including gravity) permit, into a shape in which it will have the least possible surface area, i.e. into the state where the surface energy will be smallest.

A film with a length of ds can transmit a force defined as

$$dT = C\,ds \tag{26}$$

where C is the force, per unit length of the surface, that acts in a direction perpendicular to this length and in the plane of the surface. This quantity C, expressed in N m^{-1}, is known as the surface tension of the liquid. Table 9 lists the surface

TABLE 9

Surface tensions C of some liquids in atmospheres made up of their own vapours

Liquid	Temperature (°C)	Surface tension C (N m^{-1})
Water	18	73.5
Water	100	58
Kerosene	18	22.5
Oil	20	25 to 30
Mercury	20	472
Ether	20	16.5
Ether	150	2.9
Ethyl alcohol		22
Glycerine		62.5
Methyl alcohol		22.8

tension values of some liquids in their pure state, for the case when the space above the liquid is filled with the vapours of the same liquid.

What interests us about these forces is that they come into play whenever a liquid contacts a solid surface, and govern the extent to which the solid is wetted by the liquid.

Figs. 22a and 22b illustrate the situation when a drop of water, on contact with a solid, spreads over the solid surface and wets it. The drop will attain an equilibrium state defined by the following conditions at the interface between the liquid, solid, and surrounding gas, in the plane perpendicular to the contact surface (i.e. at point A in the diagram):

$$C_{23} \cos \vartheta + C_{12} = C_{13} \tag{27}$$

The lesser the angle ϑ, the more fully will the solid surface be wetted by the liquid. We can therefore describe the degree of wetting of the solid surface as

$$\cos \vartheta = \frac{C_{13} - C_{12}}{C_{23}} \tag{28}$$

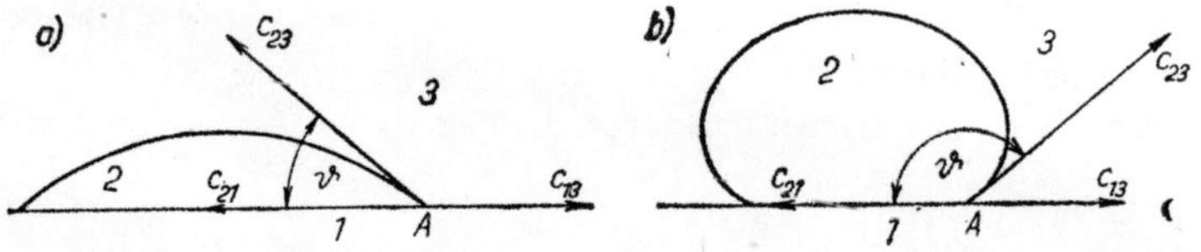

Fig. 22 Wetting of a solid surface by a liquid:
1 — solid; *2* — liquid; *3* — gas.

The value thus defined can theoretically vary between unity (at $\vartheta = 0°$, or complete wetting of the surface) and -1 (at $\vartheta = 180°$, or no wetting whatever). When ϑ lies between 90 and 180 degrees, we say that the liquid fails to wet the solid in question. That is what happens e.g. when mercury contacts a metal ($\vartheta = 145°$) or glass ($\vartheta = 140°$), or when water contacts paraffine ($\vartheta = 105°$). When ϑ lies between 0 and 90 degrees, the liquid is said to wet the solid. Full wetting, at $\vartheta = 0°$, occurs for instance when water spreads over clean glass, or oil over a metal.

Experimental evidence has shown that the wetting of solid particles also substantially depends on their shape, size, and surface properties. A jagged or pitted surface is difficult to wet, as every edge presents an obstacle to the further spread of the liquid. What in fact happens, on a broken surface, is that the abrupt changes in the value of ϑ at each edge tend to arrest the wetting process. Similarly, large particles are harder to wet than small ones. Finally, some materials, such as carbon black or textile fibres, are notoriously difficult to wet quite regardless of all the other factors involved.

Relation (28) indicates that the degree of wetting will increase as the magnitude of ϑ diminishes, i.e. the closer the value of $\cos \vartheta$ approaches to unity. It also shows that, in theory at least, there are three ways of reducing the magnitude of ϑ: by raising the value of C_{13}, or by decreasing those of C_{12} or C_{23}. In practice, of course, the only feasible approach is to lower the value of C_{23}, i.e. the surface tension of the liquid relative to the surrounding gas.

The only liquid used on any scale for the wet cleaning of gas is water. Consequently, all our efforts to improve the wetting of dust particles in a scrubber or washer, in order to facilitate their trapping, boil down to the problem of how to reduce the surface tension of water. Theoretically, we can do so in any of three ways:

a) by warming the water;
b) by dispersing fine particles in the water;
c) by adding substances which reduce the surface tension of water.

a) Heating the water is a practical proposition only when we are processing hot gases, and when the heated process water is being recirculated. Even then, the water temperature is never likely to exceed 70 °C, and charts showing the temperature dependence of the surface tension in water soon put paid to any hope of significantly improving the wetting properties this way. The surface tension drops from $73 \,.\, 10^{-3}$ N m^{-1} at 20 °C only to $63 \,.\, 10^{-3}$ N m^{-1} at 70 °C, which is obviously not enough of a reduction to have any radical effect on the degree of wetting.

This may seem out of keeping with the generally known fact that a higher water temperature often markedly improves the performance of wet scrubbing equipment. The explanation of that fact, however, lies in the *thermally induced condensation* that occurs when a relatively cold gas is moistened by hot water or steam in the course of its cleaning. When water or saturated water vapour much hotter than the gas is introduced into the gas stream, it will not only saturate a previously unsaturated gas, but also cause water vapour to condense on the surfaces of the relatively cooler particles, which in this case act as condensation nuclei. This mechanism, obviously, in no way affects the surface tension of the water.

b) The dispersion of solid particles in the water will reduce the surface tension only if the particles are extremely fine, in the micron and sub-micron size ranges, and if they are added in an adequate concentration, of say 50 g^{-1}. That is more than can generally be achieved in routine practice.

c) Substances which reduce the surface tension would have to be added in huge quantities to have any worthwhile effect, because in scrubbers we are dealing with continually disintegrating and re-forming liquid surfaces. Additions on this scale are usually ruled out both by economic considerations and by the foaming associated with them.

All this points to the conclusion that, under any ordinary circumstances, we cannot expect to lower the surface tension of water sufficiently to attain any significant improvement in the wetting of the dust particles.

However, the latest experimental findings and practical observations on the trapping of solid particles in scrubbers seem to suggest that, in most of these units, thorough wetting of the particles is not really vital for the efficiency of the separating process. It appears that all we have to ensure is physical contact of the solid particle with a droplet, because, once this contact has taken place, the forces of adhesion are strong enough on their own to prevent any subsequent separation of the two. Recent experience seems to confirm the view that in scrubbers, the wetting of the dust particles is not nearly as important as used to be thought up till a few years ago.

There is so far no generally valid definition for the absolute value of the wetting properties of solids, and consequently no accepted way of establishing any such value. The wetting properties of solids are investigated by a number of diverse methods, but none of them can do more than sort out the examined materials along a scale of ascending or descending wettability ratings. And, of course, even these relative ratings are not mutually comparable unless they have been established by one and the same method.

There are at present two prevalent techniques for assessing the wettability of a specimen of solid particles:

a) The particles are poured into a glass tube which is partly immersed in a liquid (usually water), and their wettability is judged by the height of the particle column which is wetted within a certain span of time.

b) The material under investigation is formed into a tablet. A drop of water is deposited on the tablet, the changes in the shape of this drop are filmed, and the wettability is evaluated from the film, usually in terms of the changes in time

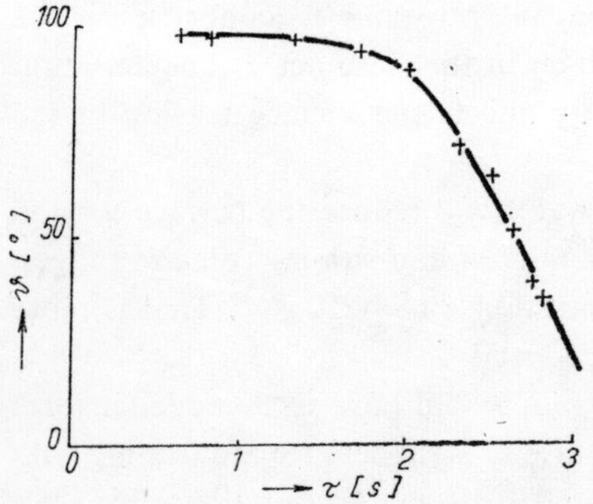

Fig. 23 Time dependence of angle ϑ (cf. Fig. 22) in the wetting of magnesite dust.

of the angle ϑ. On a plot of these changes, an easily wettable dust can be identified by the characteristic slope (in grad s^{-1}) of the line representing it, while the curve for a poorly wettable dust displays an inflection. These low-wettability dusts are then characterized by the time τ for which the drop remained virtually unaltered in shape, and by the slope of the leg of the curve adjacent to this initial interval, as in the case of readily wettable dusts. The more easily a dust is wetted, the greater is the slope of the relevant line on the chart. As an example, Fig. 23 shows the ϑ-vs.-τ plot for a magnesite dust which has an initial dwell of $\tau = 1.8$ seconds, and a subsequent gradient of 72.9 grad s^{-1}. Table 10 lists the characteristic wettability ratings of several commonly processed kinds of industrial dusts.

TABLE 10

Wetting characteristics of some industrial dusts

Dust type and origin	Slope (grad s^{-1})	Dwell time τ (sec)
Cement (Lochkov, bag filling station)	305.8	
Cement (Lochkov, crushing plant)	235	
Fly ashes (Sokolov)	254.2	
Fly ashes (Poříčí II)	178.3	
Limestone (Čertovy schody)	250	
Carbide (Nováky)	220	
Magnesite (SMZ Košice)	103.4	
Magnesite (SMZ Jelšava)	72.9	1.8
Quarry dust (Prosetín)	90	
Stone coal (Třinec coking plant)	0.63	3780
Brown coal (Poříčí II)	0.067	

1.15 Explosive properties

A mixture of dust and gas (or air) can explode only if three conditions are fulfilled simultaneously in at least some part of the space it occupies:

a) The content of inflammable dust in the gas must lie within the range in which an explosion is possible;

b) The gas must contain at least a certain minimum amount of oxygen;

c) Ignition must be initiated by some power source which has at least a certain threshold amount of energy.

There can be no explosion unless all three of these conditions are met at one and the same point and time. When considering the explosion hazards in dust trapping equipment, however, we must work on the assumption that there will always be enough energy present to ignite any explosive mixture that we allow to build up.

As for the concentration of inflammables, a mixture is dangerous whenever the content of inflammable dust in the gas is above the explosion threshold. Which does not help much, because that threshold varies widely from one dust to another. Commonly quoted values are for instance 6 to 8 $g\,m^{-3}$ for brown coal dust, 10 to 12 $g\,m^{-3}$ for stone coal dust, 12 $g\,m^{-3}$ for sawdust, 7 $g\,m^{-3}$ for aluminium powder, 12 $g\,m^{-3}$ for lignite, and 8 $g\,m^{-3}$ for pulverous rubber.

The maximum of oxygen which the gas under treatment may contain while still remaining inert, in conjunction with the kind of dust it carries, is listed for various dusts in Table 11. Once the proportion of oxygen in the gas exceeds these limits, the mixture is apt to explode as soon as the other two conditions are fulfilled.

TABLE 11

Rough classification of dusts by the maximum oxygen content at which their carrier gas can still be considered inert

Class	Dust types	Max. O_2 content (volume %)
1	Stone coal and coke dust (above 50 °C)	16
2	Brown coal and peat dusts; layers up to 2 mm thick of Class 1 to Class 4 inflammable solids deposited within separators	14
3	Fine sawdust, alcoholic resins, urethanes, phthalic anhydride, sulphur, hexamethylenetetramine	10
4	Polystyrol, pentaerythritol, methyl methacrylate, lignin resins, caseins, cotton wastes	5
5a	Aluminium powder	0
5b	Magnesium and zirconium powders	0

Geck has classified various kinds of dust, by their explosiveness, into two broad categories, and arranged the dusts of each class into a descending order of explosiveness which we quote below.

Class 1 consists of dry fine dusts with only a small content of ash matter. These materials are generally difficult to ignite when stored in bulk, but their airborne suspensions ignite easily and form intense, rapidly propagating flames. When a dust of this type is suddenly raised or swirled up, it can ignite even in the absence of any identifiable input of energy from outside. Once ignited, it is liable to explode, i.e. to bring about an abrupt and substantial pressure rise within an enclosed space. Typical dusts of this category, from the most to the least hazardous, are:

fine sawdust
cork dust
finely pulverous sugar
powdered synthetic resins
flour
malt fines
dust from sugar-beet pulp
synthetic rubber dust
starch
dextrine dust
leather and synthetic leather dusts
fibrous dust of vegetable origin
sulphurous dusts
dust from natural and synthetic fertilizers
various chemicals in the powdered state.

Class 2 comprises dry dusts with higher contents of ash matter, which are virtually impossible to ignite in bulk. Even the airborne particles will ignite only when exposed to a sufficiently high temperature for a sufficient length of time. When suddenly raised or swirled up, these dusts may ignite even without any traceable input of external energy, but under these circumstances they will burn rather than explode. Dusts in this class include the following:

synthetic cellulose dust
zinc powders
pulverous natural resins
amber dust
carbon black and soot
powdered spices
soap powders
dust from spray-applied paints
pulverous cocoa.

In the handling of inflammable dusts, fires and explosions can be averted only if a certain set of rules is observed. The nature of these precautions depends on whether only the solid particles, or the gases too, are inflammable.

A. When separating inflammable solids from non-inflammable gases:

1. The gas must be rendered inert by an addition of some non-inflammable gas, such as nitrogen, carbon dioxide, or water vapour. The object is to reduce the oxygen content below the level at which the gas will still support combustion.

The maximum residual oxygen content which can still be tolerated depends on the type of dust. For instance, for brown coal dust at temperatures in excess of 50 °C, the safe limit in the least favourable case is 14 per cent of oxygen. In flue gases, the hazards are better defined by the carbon dioxide contents, as these are easier to ascertain by quite simple instrumentation. For the above example, the carbon dioxide content should never fall below 6 per cent by volume. When sawdust or wood shavings are handled, there is no need to reduce the oxygen content below the normal atmospheric level of 20 per cent by volume, as long as the air temperature is kept to less than 100 °C.

2. The gas and equipment wall temperatures must remain below a certain limit, dependent on the type of dust involved, to prevent evaporation of the volatile constituents from the dust. The highest acceptable temperature for brown coal dust is 120 °C and for sawdust or wood shavings 100 °C.

3. The dust concentration in the gas can be kept either above or below the limits of the concentration interval within which the mixture is explosive. In either of those cases there is no risk of an explosion, no matter how fine the dust may be. The trouble is that these limits are not at all easy to determine. The Coal Industry Research Institute of Ostrava, Czechoslovakia, has published various threshold concentrations for different types of brown coal, the lowest of these figures being 6 $g\,m^{-3}$, but no such figures are available for the upper limit of the explosive interval. Moreover, even the lower threshold value rises as the dust particles grow coarser. On the other hand, an increased oxygen content is known to extend the range of dust concentrations which represent explosion hazards, both upwards and downwards.

4. Since the equipment is liable to pick up a static charge, all its parts must be interlinked by electrically conductive connecting members, and the whole system must be effectively grounded (earthed), for instance as set out in Czechoslovak Standard ČSN 34 1390.

5. To prevent the accumulation of dust deposits in the equipment, the design must avoid any horizontal surfaces over which the mixture of dust and gas would pass at a speed lower than 25 $m\,s^{-1}$. It must further avoid all corners or pockets shielded from the gas stream. The object is not so much to preclude the formation of any deposits whatever, as to keep the depth of any deposits that do build up to less than 5 mm; that minimizes both the risk of spontaneous ignition and the liberation of volatiles from the layer. This implies that the trapped dust must be continually removed from the equipment, so that the sediments present in it will always consist of freshly captured material which will dwell in the equipment only for a strictly limited time. To this end, the operating instructions must provide for the evacuation of trapped material to continue for some time after the rest of the equipment has been shut down, until the chutes, hoppers etc. are emptied.

6. In case a fire does break out, the equipment must be linked (e.g. via

a 1 1/2″ stub and a hermetic closure) to a reservoir of some extinguishing gas like compressed carbon dioxide, nitrogen, or water vapour, or some other fire extinguishing medium. The medium must of course be compatible with the gas and dust which the equipment is handling; for instance, magnesium dust would react with the oxygen contained in water vapour.

7. To prevent any fire in the equipment from spreading upstream through the feed line, the gas flow velocity must always be kept high enough for the gas in the feed piping to move at a velocity exceeding the speed of flame propagation in the mixture that is being handled. For instance, gas carrying the dusts from Czechoslovak brown coals must always be kept moving at speeds higher than 22 m s^{-1}.

8. Although no explosion can take place as long as the stipulations of the above points 2, 3, 4, 5 and 7 are fulfilled, not even when the atmosphere within the equipment is less than fully inert, that is no excuse for ignoring the possibility that a breakdown or other contingency might create conditions conducive to an explosion. Therefore, the equipment must be safeguarded by a frangible diaphragm, which in the event of an explosion will burst before the internal pressure rise can rupture the equipment itself and allow flames to escape from it. The diaphragm must on bursting release the gas into a jettisoning pipe; the latter must be large enough to offer no substantial resistance to the gas flow, not even to blast-wave surges, and must emit the gas into the atmosphere well away from any potential danger points. Equipment that is normally operated with a partial vacuum within it must further be fitted with a one-way flap which will close again once the pressure wave has been jettisoned, so as to prevent the ingress of outside air. The size of the frangible diaphragm must be selected to suit the internal volume which it is to protect, the grain size and concentration level of the dust, and the speed of flame propagation or velocity of explosion waves in the gas that is being processed. In equipment for trapping brown coal dust, for example, a diaphragm area of 1 m^2 will safeguard an internal volume of 25 m^3. The diaphragm may consist for instance of aluminium foil 0.2 mm thick, with a hairline scribed down the middle to provide a notch effect.

A special case are some of the separators of the Czechoslovak "S" system, such as the SAA and SCA units, where the frangible diaphragms do not directly adjoin the dust collecting spaces. In these units, any pressure waves generated by a fire or explosion must pass through the discharge openings and gas outlet tubes of the cyclones. The cross-sectional areas of these flow channels must therefore equal the requisite diaphragm area.

9. Special measures must be taken to prevent any ferrous objects entrained by the gas stream from entering the equipment, as their impacts against the walls might cause sparking which would aggravate the explosion hazards.

B. When the gas itself is inflammable, the requirements of points A 4, 6, 8 and 9 must be met, and moreover:

1. The temperature of the equipment walls must at all times be lower than the flash point of the gas.

2. Wherever possible, the process should be adapted so as to maintain a certain gauge pressure within the equipment. If the latter has to be run with a partial vacuum inside it, special care must be taken to ensure that it will be utterly gas-tight, and stringent leakage tests must be performed in the manufacturing and erection stages. Particularly the hopper outlet gates and seals must remain tight not only when new, but even after the normal wear and tear imposed by routine service.

C. Some kinds of admixtures cannot be separated in normal equipment at all, because they are too dangerous to handle. This applies especially to materials which contain or convey enough oxygen to support their combustion, and to materials which can be ignited e.g. by mechanical impacts or shocks.

D. No inflammable materials may be stored, and no permanently manned work-places set up, within 6 metres of the separator. This is the least safe distance in the event of a separator fire or explosion.

No matter what precautions are taken to preclude an explosion in equipment handling inflammable dusts, this is no justification for neglecting safety measures designed to mitigate the effects if an equipment failure or other unforeseen circumstance does cause an explosion. For a start, the feed and outlet pipings or ducting, like the trapping equipment itself, must be provided with frangible diaphragms which will divert the blast wave before it can burst the piping or separator casing. The blast jettisoning outlets should be sealed by non-return flaps which will prevent any air entering the equipment once the blast wave has passed out, because the ingress of air would tend to re-establish an explosive environment in the interior. The size of the diaphragms and flaps is governed by two considerations – the propagation velocity of explosion waves in the given medium, and the volume of the enclosed space in which an explosion might occur. The amount of space which a diaphragm or flap can protect depends on the kind of dust which the equipment is handling. With reference to the classification of dusts presented in Table 11, a diaphragm with an area of 1 m^2 can protect the following enclosed volumes:

Dust category	Protected space
1	30 m^3
2	25 m^3
3	20 m^3
4	10 m^3
5a	5 m^3
5b	2 m^3

These figures are naturally subject to variations. In particular, they are influenced by the way the explosiveness of the dust responds to changes in the degree of inertness of the carrier gas, i.e. in its oxygen content. A point to note is that no single diaphragm may be more than 1000 mm in diameter; where this diameter is inadequate, several diaphragms must be installed in parallel. The diaphragm must be designed to burst as soon as the static pressure is exceeded by a margin of $0.14 \cdot 10^5$ Pa, so as to damp the pressure peak at the face of the blast wave. Each diaphragm must open into a piping which will deflect the blast to some point clear of all areas accessible to personnel. When the equipment is installed indoors, the piping should end above roof level.

Another problem is to stop a fire in the dust trapping equipment from spreading upstream, through the feed piping, to the process equipment ahead of the dust trapping stage. As has been stated previously, the best countermeasure is to keep the flow velocity in the feed piping consistently higher than the velocity of flame propagation in the medium flowing through those pipes. There are two ways of ensuring the requisite flow velocity: either the feed piping diameter must be kept down so that the given volumetric flow rate will necessarily entail the desired speed of flow, or else the pressure drop across the separators, i.e. the size of that equipment, must be such as to maintain an adequate flow velocity on its own.

This is all very well in theory, but unfortunately there is still an almost total lack of data on the flow velocities actually needed to produce this isolating effect. The only figure that has been dependably established applies to equipment for trapping brown coal dust, where the flow velocity in the feed lines must never be allowed to drop below 25 m s^{-1}. In the absence of hard and fast figures, it seems reasonable to assume that this velocity will also be sufficient for all the other dusts in Geck's class 2, while the dusts which Geck has included in hazard class 1 are likely to require a flow velocity of some 35 m s^{-1}. These, however, are only tentative estimates; much work remains to be done before it will become possible to design dust trapping installations in full compliance with the safety requirements set out above.

2. PHYSICAL PHENOMENA EXPLOITED IN SEPARATION PROCESSES

Ing. J. Albrecht, CSc., Ing. J. Kurfürst, CSc., Ing. J. Urban

2.1 Effects of gravity

Solid particles borne by a horizontal gas stream are exposed to two forces – the inertia imparted by the gas flow, and gravity. If we neglect the diffusion effects of the gas flow, the vector angle of the resultant particle motions is defined, with reference to Fig. 24, as

$$\mathrm{tg}\,\beta = \frac{v_v}{v_p} \tag{29}$$

The diffusion effects, however, are not always negligible. Small solid particles are subject to diffusion caused by turbulence in the carrier gas, and to diffusion caused by the thermally induced Brownian motion of the gas molecules. The resultant forces affect the motion of the particles to such an extent that in these cases, we cannot rely on gravity to assist the separation of the solids from the gas stream.

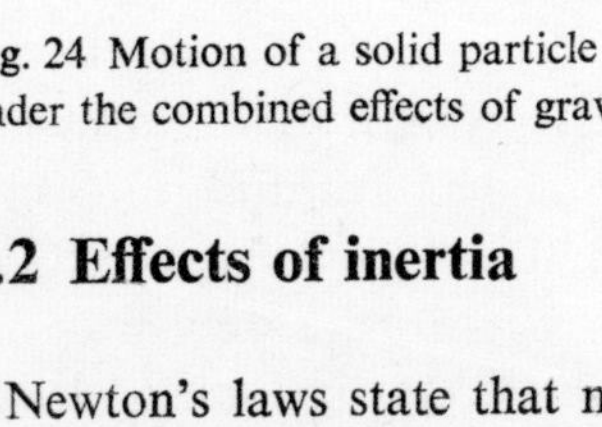

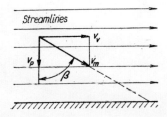

Fig. 24 Motion of a solid particle in a gas stream under the combined effects of gravity and of entrainment by the gas.

2.2 Effects of inertia

Newton's laws state that matter, as long as it is not exposed to any external forces, will tend either to remain stationary or to follow a straight path at a constant velocity. External forces will alter this state of motion, accelerating or retarding the matter in dependence on the direction in which they are acting. In the trapping of solid or liquid particles from a gas stream, these laws can be exploited in any of several different ways.

2.2.1 Trajectory deflection by stationary baffles

Let us consider a gas stream impinging perpendicularly on a flat surface. If we reduce the problem to a two-dimensional one, we obtain the pattern of resultant lines of flow shown in Fig. 25, where the velocity and static pressure distributions are represented by velocity profiles and isobaric lines.

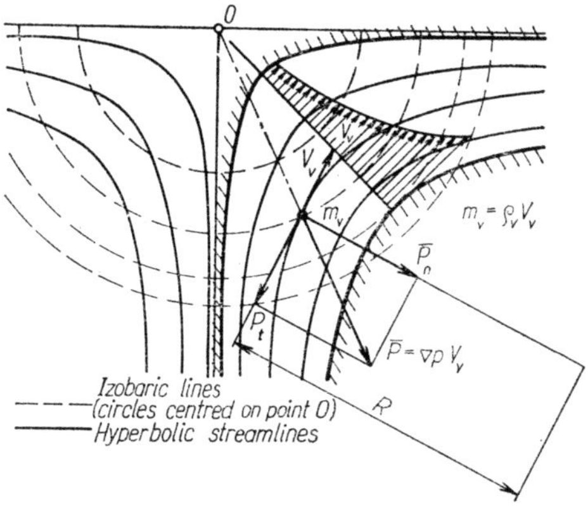

Fig. 25 Streamlines of an ideal gas in a two-dimensional flow deflected by a baffle, showing the way the buoyant force of the gas affects its macroparticles.

The Bernoulli relation indicates that in an incompressible liquid flowing in the horizontal plane, and given a constant potential energy, the product of the kinetic energy and pressure energy likewise remains constant. This implies that points with identical hydrostatic pressures will also share one and the same flow velocity. The pressure gradient of the static pressure field denotes the direction of the force which the environment at that point applies to a macroparticle of the flowing gas. This picture is of course grossly simplified, because it fails to take into account the other forces acting in the plane of motion, such as friction, the external forces exerted by particulate matter, or the effects of gravity acting perpendicularly to the plane under consideration. Still, confining our investigations to this simplified representation, we can split the vector of the pressure gradient $\bar{P} = \bar{V}r$ into two components: $\bar{P}_n$, which is perpendicular to the lines of flow, and $\bar{P}_t$, tangential to these lines. The resultant force opposes the motion of the macroparticles of gas, in other words retards them, in a way described by the relation

$$m_v \frac{dv_v}{d\tau} = P_t \tag{30}$$

Force P_n acts perpendicularly to the gas flow direction, and deflects the gas flow lines in a manner definable as

$$m_v \frac{v_v^2}{r_v} = P_n \tag{31}$$

where r_v is the radius of curvature of the lines of gas flow at the point under examination.

The solid particles entrained by the gas flow display a higher specific gravity than the gas itself, and consequently cannot be expected to follow the curvature of the gas flow lines. The static pressure gradient of the gas cannot control the motion of a solid particle as it would affect the motion of an equally large particle of gas occupying the same location in the stream. Hence, the retardation of a solid particle is described by the expression

$$P_t = m_m \frac{dv_m}{d\tau} \tag{32}$$

It follows that

$$\frac{dv_m}{d\tau} = \frac{P_t}{m_m} \tag{33}$$

Since m_m is necessarily larger than m_v, the above relations imply that

$$\frac{dv_m}{d\tau} < \frac{dv_v}{d\tau} \tag{34}$$

In other words, a solid particle with a mass of m_m will not be retarded as strongly as a macroparticle of gas with a mass of m_v.

A similar relation can be established for the direction perpendicular to the flow lines too. Since

$$\frac{v_v^2}{r_v} = \frac{P_n}{m_v} \tag{35}$$

we can similarly state that

$$\frac{v_m^2}{r_m} = \frac{P_n}{m_m} \tag{36}$$

and because m_m is greater than m_v, it follows that

$$\frac{v_v^2}{r_v} > \frac{v_m^2}{r_m}, \tag{37}$$

In most practical applications, there is no appreciable difference between the absolute velocities of the solid particles and the gas stream. We can therefore base our first qualitative approach on the assumption that $v_m = v_v$. In that case, it is clear from the above relations that the curvature r_v of the lines of gas flow will differ from the curvature r_m of the solid particle trajectories. Because

$$\frac{v_v^2}{r_v} > \frac{v_m^2}{r_m} = \frac{v_v^2}{r_m} \tag{38}$$

we can obviously infer that

$$r_v < r_m \tag{39}$$

In other words, the curvature r_m of the solid particle trajectories will be gentler than the curvature r_v of the lines of gas flow.

All this, of course, applies only to a qualitative examination of the problem. Any quantitative evaluation of the above expressions would be misleading, primarily because they neglect the consequences of the fact that the velocity of the gas stream differs from that of the solid particles entrained in it. Nevertheless, the fact remains that a stationary baffle in the flow path will produce different degrees of deflection in the solid particle trajectories and in the lines of gas flow. And this fact is exploited to separate out the particles, by inertia, in a variety of mechanical dust trapping devices that range from dust arresters and baffle-flight thickeners to cyclones.

2.2.2 Trajectory deflection by moving baffles

The inertia of solid particles can equally well be utilized to separate them from the gas stream in a flow channel which itself performs some motion, commonly a rotary one. Fig. 26 outlines the conditions in which this process takes place on a radial impeller. At any given point of the flow channel between two adjacent blades of the rotating impeller, a macroparticle of gas will have an absolute velocity vector $\bar{v}_v$ which can be determined as

$$\bar{v}_v = \bar{v}_{v\,rel} + \bar{\omega} \, . \, \bar{r} \tag{40}$$

The absolute velocity $\bar{v}_v$ further enables us to work out the spiral path, marked in Fig. 26, which the macroparticle of gas will follow during its passage through the impeller channel.

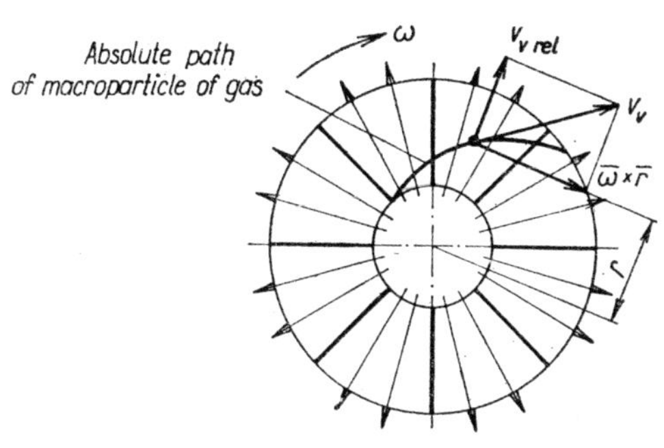

Fig. 26 Motion of a macroparticle of gas over a rotating radial impeller.

When we analyse the composite motion of a material point borne within a rotating system, we find that the overall, aggregate acceleration of that point will not be merely a sum of the accelerations imparted by the relative motion of that point and the rotary motion of the system. Any variation in the radial distance of the point from the axis of rotation will alter the velocity at which the rotating system carries the point with it, and will thereby superimpose a third acceleration component, known as *the Coriolis acceleration.* The absolute acceleration of the macroparticle will then equal the vector sum of these three components:

$$\frac{d\bar{v}_v}{d\tau} = \frac{d\bar{v}_{v_{rel}}}{d\tau} + \frac{d}{d\tau}(\bar{\omega}\,.\,\bar{r}) + 2\bar{\omega}\,.\,\bar{v}_{v_{rel}} \tag{41}$$

Overall acceleration	Acceleration of gas relative to flow channel	Acceleration of rotating flow channel	Coriolis acceleration

A curved flow channel will naturally involve a correspondingly curved path of the macroparticles traversing the impeller, but in general the relative acceleration $\frac{d\bar{v}_{v_{rel}}}{d\tau}$ will consist of two acceleration components, as evident in Fig. 29:

$$\frac{dv_{v_{rel}}}{d\tau}\,.\,\frac{\bar{v}_{v_{rel}}}{v_{v_{rel}}} = \frac{dv_{v_{rel}}}{d\tau}\,.\,\frac{\bar{v}_{v_{rel}}}{v_{v_{rel}}} + \frac{v^2_{v_{rel}}}{r_v}\,.\,\frac{\bar{r}_v}{r_v} \tag{42}$$

Overall relative acceleration	Unit vector	Tangential relative acceleration caused by variations in the absolute value of $v_{v_{rel}}$	Unit vector along tangent to path of the macroparticle	Centripetal relative acceleration caused by curvature of macroparticle path	Unit vector along radius of path curvature at examined point

The acceleration imparted to the macroparticle by the rotating system is governed by the relation

$$\frac{d}{d\tau}(\bar{\omega}\,.\,\bar{r}) = \frac{d\bar{\omega}}{d\tau}\,.\,\bar{r} + \bar{\omega}\,.\,\frac{d\bar{r}}{d\tau} = \bar{b}_w\,.\,\bar{r} + \bar{\omega}\,.\,(\bar{\omega}\,.\,\bar{r}) \tag{43}$$

Angular acceleration of rotating channel (Fig. 27)	Centripetal acceleration of rotary motion (Fig. 27)

The Coriolis acceleration, expressed as $2\bar{\omega}\,.\,\bar{v}_{v_{rel}}$, is schematically illustrated in Fig. 28.

Fig. 29 summarizes the overall acceleration of the absolute motion of a macroparticle of gas, for the general case of a curved flow channel (i.e. an impeller with blades curved so that their tips point in the sense of rotation). The diagram shows that this aggregate acceleration is the sum of the five acceleration components listed above. At the point marked *A*, the centripetal and Coriolis accelerations roughly coincide in direction, while at point *B* they oppose each other. Fig. 30 illustrates the various acceleration components, and the resultant overall acceleration, in an impeller with straight, radially aligned blades.

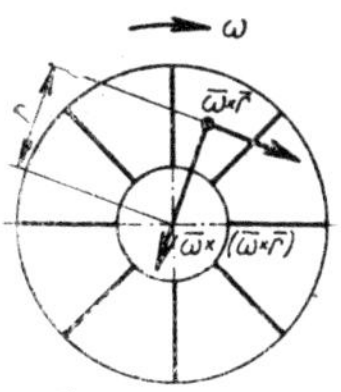

Fig. 27 Centripetal acceleration in a radial impeller.

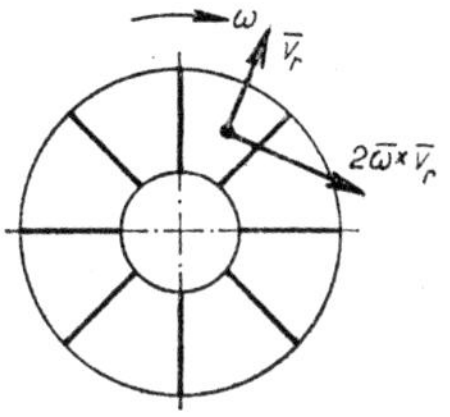

Fig. 28 Coriolis acceleration in a radial impeller.

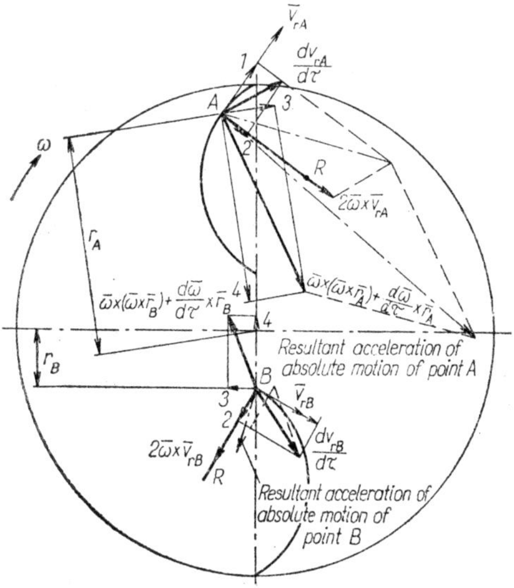

Fig. 29 Acceleration of a macroparticle of gas by an impeller with curved vanes:

1 ... $\dfrac{dv_r}{d\tau}$ — Acceleration components of motion relative to impeller,

2 ... $\dfrac{v_r^2}{R}$,

3 ... $\dfrac{d\bar{\omega}}{d\tau} \,.\, \bar{r}$ — Acceleration components imparted by impeller,

4 ... $\bar{\omega}(\bar{\omega} \,.\, \bar{r})$.

If the solid particles were to follow closely the motions of the gas around them, they would have to be accelerated in exactly the same way as macroparticles of gas occupying the same locations in the gas stream. This acceleration would have to be effected by the static pressure field (or lift) of the fluid medium. However, that lift is only just sufficient to impart the necessary acceleration to bodies with the specific gravity of the macroparticles of gas. Solid particles have much higher specific gravities, and consequently cannot follow the flow lines of the gas. Let us assume that a solid particle with a mass of m_m momentarily moves at the velocity of the surrounding gas, and is exposed to the same accelerating forces as that gas.

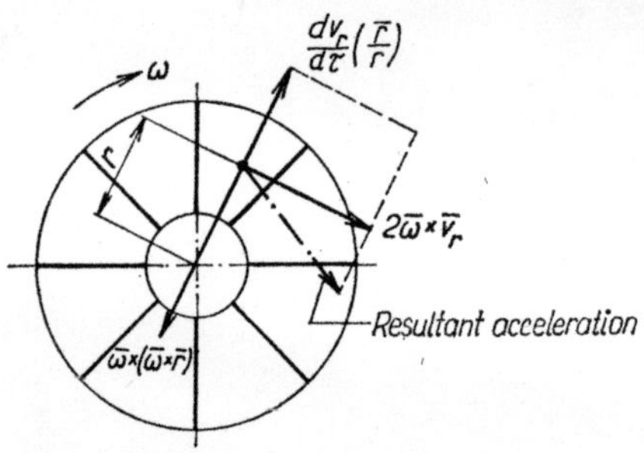

Fig. 30 Acceleration of resultant motion of gas over a radial impeller.

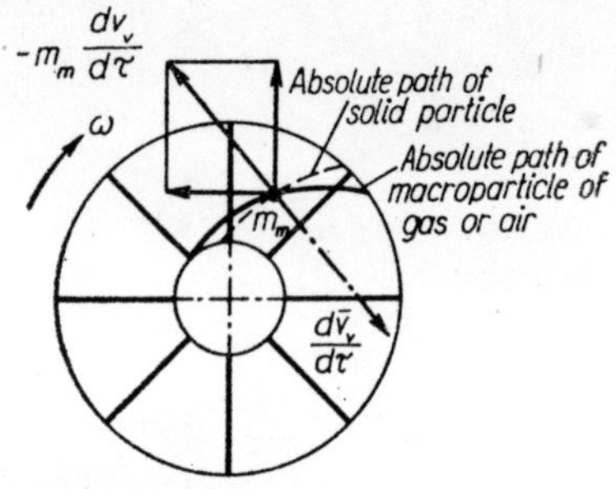

Fig. 31 Effect of the inertia of a solid particle moving across an impeller.

As evident from Fig. 31, the particle will, at the point and instant under consideration, offer a certain resistance to acceleration, and the intensity of this resistance will depend on the inertia of the particle as follows:

$$-m_m\left(\frac{d\bar{v}_{v_{rel}}}{d\tau}+\frac{d(\bar{\omega}\,.\,\bar{r})}{d\tau}+2\bar{\omega}\,.\,\bar{v}_{v_{rel}}\right) \tag{44}$$

This resistance can be broken down into two components, the one parallel to the blade and the other perpendicular to it. Fig. 31 shows that the perpendicular component accelerates the solid particle towards the blade against the resistance of the gas, while the parallel component accelerates the particle along the length of the blade. These considerations, it must be pointed out, take no account of any velocity slip between the solid particle and its gaseous environment. Fig. 31 also indicates the absolute paths followed by a solid particle and by a macroparticle of gas as they pass across the rotating impeller. The trajectory of the solid particle is the flatter and shorter of the two, simply because the inertia of that particle is greater,

$$m_m\frac{d\bar{v}_v}{d\tau} > m_v\frac{d\bar{v}_v}{d\tau} \tag{45}$$

This phenomenon is exploited in all rotary separating devices.

The preceding paragraphs, on the way the inertia of solids can be utilized for their separation from a gas stream, have ignored the fact that the particles are entrained and transported by that stream. Yet the consequences of this transporting effect limit the extent to which both gravity and inertial forces can be made to serve this purpose. It has already been stated that, in the case of gravity, certain restrictions are imposed by the diffusion of fine particles in the gas stream. Much the same applies to inertial forces too, but as these are usually a good deal greater than gravity, the constraints become apparent only at a much smaller particle size. Which means that inertia-based separation processes can cope more dependably with much smaller particles than gravity-based techniques.

2.3 Effects of electrical forces

Electrostatic precipitators separate solid or liquid particles from a gas stream by exploiting the electrical properties of these particles in an electric field. The principal phenomena utilized in this process are the ability of *corona discharges* to ionize gases; the ability of the particles to pick up electrical charges in a unipolarly ionized gas; and the various forces which act upon charged particles in an electric field. These principles are as old as the electrostatic precipitator itself, and so are the various basic arrangements which still underlie the design of these units.

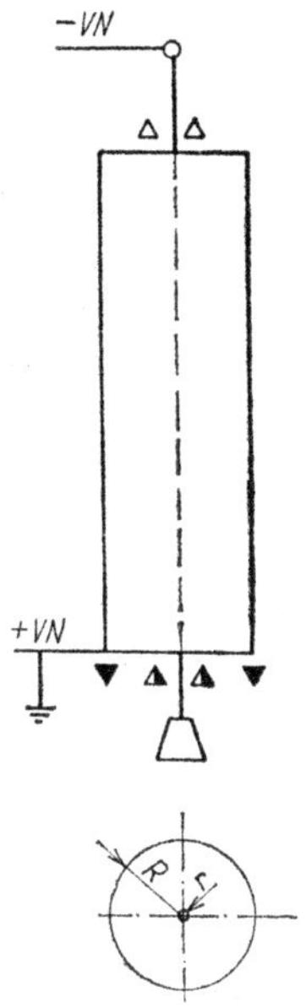

Fig. 32 Scheme of a tubular electrostatic precipitator.

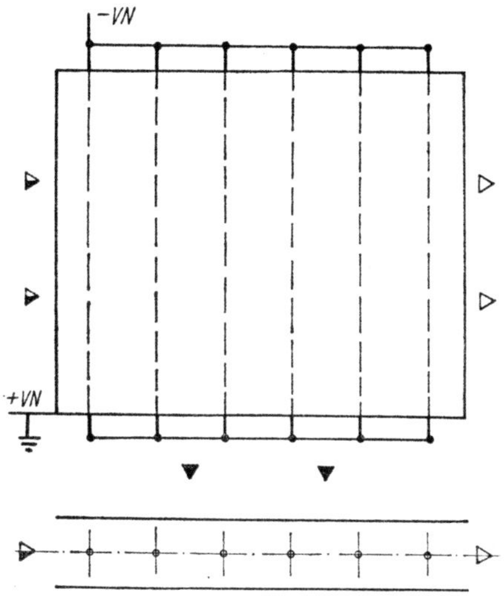

Fig. 33 Scheme of a plate-type electrostatic precipitator.

The main working parts of an electrostatic precipitator are two different electrodes. One of them, thin and small in cross section, is commonly formed by a wire or a system of points or barbs. The other is either a large-diameter tube or else a flat, corrugated or otherwise profiled plate. By the shape and layout of these latter electrodes, we distinguish between tubular precipitators (schematically illustrated in Fig. 32) and plate-type precipitators (Fig. 33). In industrial precipitators, it is customary for the wire or point electrode to be connected to the negative pole of a rectified high-voltage source, while the other electrode is both wired to the positive pole and grounded (earthed). As the particle-laden gas flows through the channels between these two electrodes or electrode systems, the particles pick up charges which impel them predominantly towards the grounded electrode, where the particles settle. However, the particles will be charged only if the electrode system consistently produces a corona discharge.

2.3.1 Corona discharges

A rectified voltage applied to the electrodes will induce an electric field in the space between them. This field can be characterized by its field intensity F (in V m^{-1}). By way of an example, Fig. 34 shows how, in one particular type of tubular precipitators, this field intensity varies with the distance from the wire electrode. It is evident from the diagram that the field intensity attains its maximum at the electrode surface, and then diminishes very sharply. The gas flowing through the gap between the electrodes consists prevalently of neutral molecules, and their motion is in no way affected by the electric field. A very small proportion of the molecules, however, will always have been ionized, i.e. electrically charged, by some extraneous factor like electromagnetic radiation or high temperatures. This process deprives the molecule of one electron, and thus leaves the remnant with

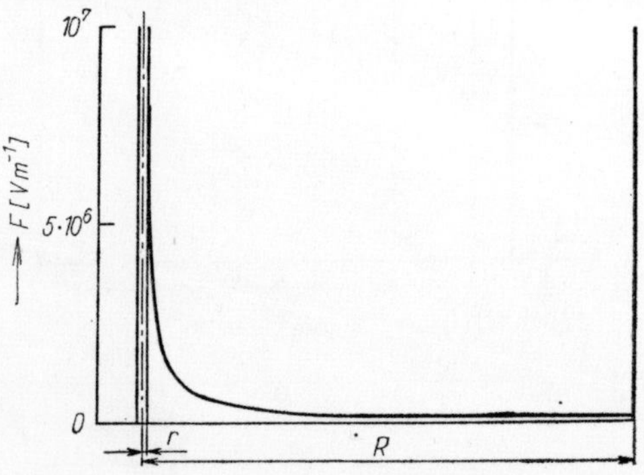

Fig. 34 How the electrical field intensity in a tubular precipitator varies with the distance from the wire electrode.

a positive charge. These ionized molecules (or gas ions), and the electrons separated from them, move towards the electrodes with opposite polarities, where they are discharged. That generates a slight current in the electrode feed circuit, but this current is normally too weak to be even detectable. A further rise in the applied voltage will increase the field intensity at the surface of the wire electrode to the point where the gas ions and electrons in the vicinity will be accelerated sufficiently (i.e. imparted enough motion energy) for them to ionize further neutral molecules by impact.

The way an ion moves in an electric field depends on several factors: the field intensity, the mass and charge of the ion, and the free path which the ion covers between two successive collisions with gas molecules. As the ion is first accelerated by the field and then retarded by impact against a gas molecule, its instantaneous speed will vary widely, but its average velocity can be established as

$$v_i = Fu \tag{46}$$

The ion mobility u depends on the mass of the ion. Consequently, the positively charged ions, which contain a massive nucleus, are less mobile. However, the ion mobility increases with the absolute temperature of the gas, and diminishes as the gas pressure rises.

It takes a certain initial critical field intensity F_0 (in V m^{-1}) for the gas in the vicinity of the wire electrode to start ionizing. For an electrode with a cross-sectional radius of r, this intensity can be determined from the empirical formula derived by *Whitehead and Brown*,

$$F_0 = K_1\delta\left(1 + \frac{K_2}{\sqrt{\delta r}}\right) \tag{47}$$

In this expression, K_1 (in V m^{-1}) and K_2 (in m$^{1/2}$) are empirical constants dependent on the geometrical configuration of the electrodes, on the kind of gas and its state, and on the polarity of the electrodes. The quantity δ in this formula is the relative weight of the gas (as compared to a certain reference state), and is calculated as

$$\delta = \frac{T_0}{T} \cdot \frac{p}{p_0} \tag{48}$$

where T_0 is the reference absolute temperature of the gas (in K), and
p_0 is the reference gas pressure (in Pa).

TABLE 12

Empirically established values of constants K_1 and K_2

Corona discharge	K_1 (Vm^{-1})	K_2 (m$^{1/2}$)
Back corona	3.102 . 10^6	3.08 . 10^{-2}
Positive corona	3.367 . 10^6	2.42 . 10^{-2}

Table 12 lists the approximate values of constants K_1 and K_2 which, given the usual electrode layout and polarity, apply to air at 25 °C and 101,323 Pa. The requisite initial critical field intensity will depend on the polarity of the corona discharge, but will always grow with the gas pressure and electrode diameter, and diminish as the gas temperature rises.

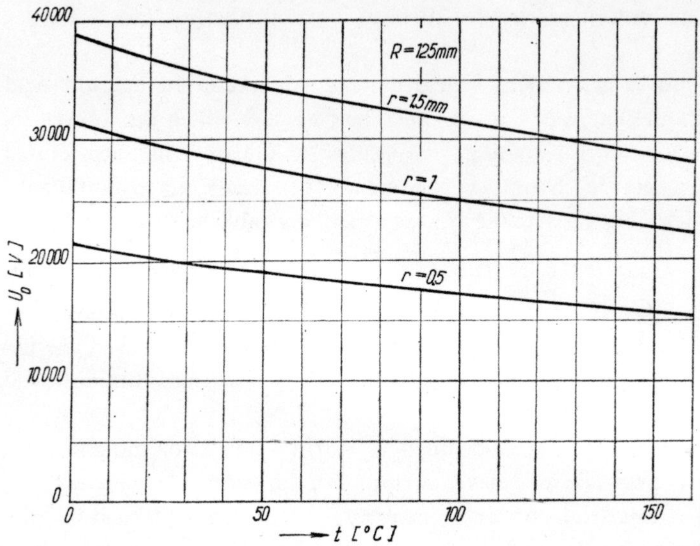

Fig. 35 Dependence of the initial critical voltage U_0 on the temperature t at various half-diameters r of the wire electrode.

At every initial critical field intensity F_0, there is a certain initial critical voltage U_0 that must be applied across the electrodes to bring about a corona discharge. For a tubular precipitator as shown in Fig. 32, this voltage is defined as

$$U_0 = F_0 r \ln \frac{R}{r} \tag{49}$$

where R (in m) is half the diameter of the tubular electrode. For a plate-type precipitator as illustrated in Fig. 33, it can at least approximately be determined as

$$U_0 = F_0 r \ln \frac{4R}{3r} \tag{50}$$

where R (in m) is the gap between the wire and the plate electrodes. Fig. 35 indicates the temperature dependence of U_0 for several different half diameters r of the wire electrode.

Any further rise in the applied voltage will enlarge the region around the wire electrode where molecules can be ionized by impact, and will increase the number of molecules so ionized per unit of time. Consequently, an ever denser stream of negative ions will leave the corona discharge region and cross the space between the electrodes towards the plate electrode; in other words, a substantial current flow will build up across this space. The intensity of this current will depend on the applied voltage in a way governed by the voltampère characteristic of the precipitator.

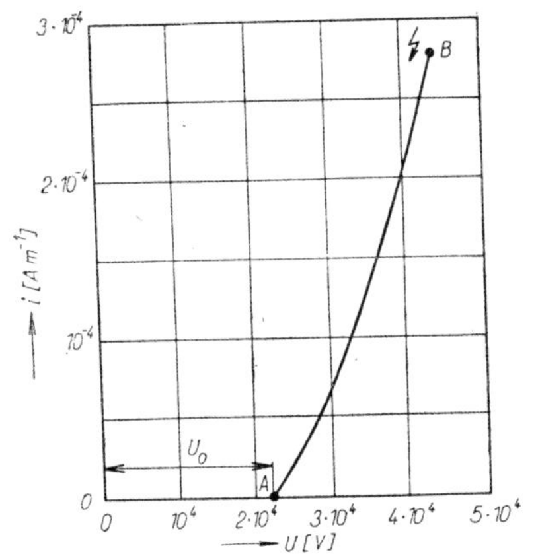

Fig. 36 Voltampère characteristic of an electrostatic precipitator.

An example of such a *V–A characteristic* is presented in Fig. 36. The current intensity plotted in such diagrams may be either the total current flow through the precipitator, or the current per unit length of the discharge electrode, or the current per unit area of the collecting electrode, and care must be taken not to confuse these quantities when comparing the characteristics of different types. It is obvious from the diagram that the voltage U across the electrodes must first build up to initial critical level U_0 before a measurable current starts flowing through the precipitator at point A. Further voltage rises then increase the current intensity, but this increase is by no means unlimited: once point B (the arc-over voltage) is reached, the corona discharge turns into an arc or spark discharge. Once the initial critical voltage has been exceeded, the current intensity depends on the product of $U(U - U_0)$ and on the ion mobility u. The voltampère characteristic can then be established from the relation

$$i = \text{const}\, uU(U - U_0) \tag{51}$$

where i is the specific current intensity (in A m^{-1}), and

const is a proportionality constant dependent on the design and layout of the electrodes.

However, the *V–A characteristic* is also affected by the properties of the gas that fills the space between the electrodes. The higher the ion mobility in this gas, the steeper will be the slope of this characteristic.

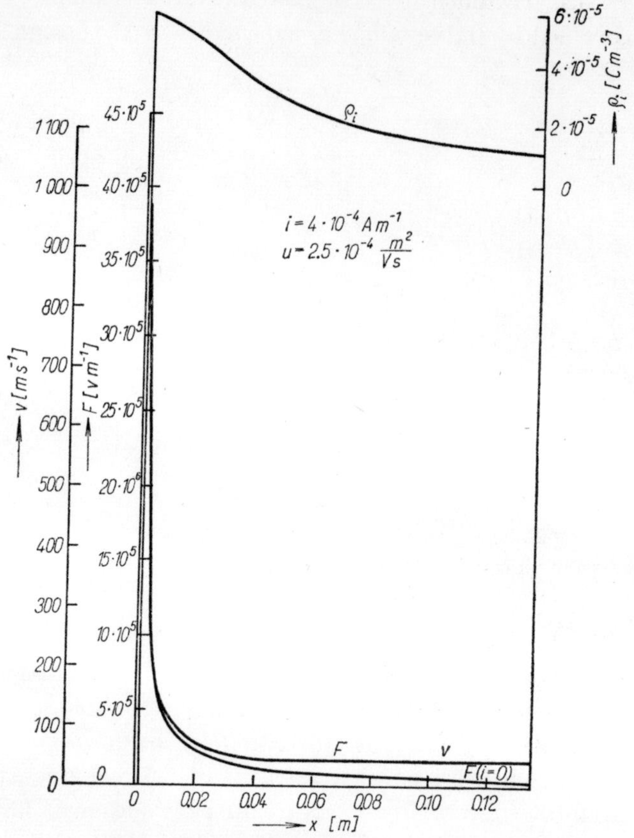

Fig. 37 Gradients of the electrical field intensity F, velocity v, and space charge ϱ_i of the moving ions.

The current flow through the precipitator naturally alters the pattern of the field intensity gradient. At the envelope of the corona discharge zone, the field will be at its initial critical intensity. In the space between this envelope surface and the collecting electrode, the field intensity gradient will depend on the magnitude of the space charge formed by the ions passing from the discharge electrode to the collecting electrode. Fig. 37 shows the relationship between the field intensity F, the ion velocity v, and the space charge ϱ_i. The area bounded by the curves represents the voltage needed to counteract the opposed field set up by the space charge of the moving gas ions. At a given current intensity, this space charge, and hence the voltage required to overcome its effects, will be the greater, the lesser the mobility of the gas ions.

The arc-over voltage depends mainly on the type of gas and the design and arrangement of the electrodes. In general, this voltage will drop off as the ion mobility increases. A good example of this trend is dry air: even the slightest admixture of some gas with a lower ion mobility (termed an electronegative gas), such as sulphur dioxide or water vapour, will raise its normally low arc-over voltage quite significantly. As for the electrodes, the arc-over voltage between say

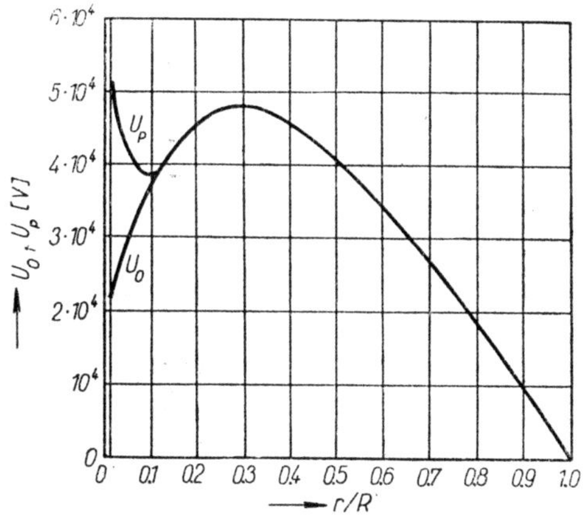

Fig. 38 How the interval between the initial critical voltage U_0 and the arc-over voltage U_p varies with the $r:R$ ratio.

a point and a plate electrode is generally lower than that between a wire and a plate electrode. Fig. 38 indicates how the arc-over voltage diminishes in response to the growing diameter of a wire discharge electrode (r being half this diameter). An interesting aspect of this chart is the way the gap between the initial critical voltage U_0 and the arc-over voltage U_p narrows as the electrode diameter increases, until, once a certain limit has been reached, the two curves meet. At and beyond this limit, we shall obtain a spark discharge right away, without any previous corona discharge.

2.3.2 Effects of the gas properties on the voltampère characteristic

A higher gas temperature increases the mobility of the gas ions and thereby reduces the initial critical voltage. Consequently, the V–A characteristic shifts towards the left side of the chart and assumes a steeper slope angle.

A rising gas pressure diminishes the ion mobility and hence raises both the initial critical and the arc-over voltages. The V–A characteristic is displaced to the right, and its slope becomes gentler.

The V–A characteristic is also affected by the *composition of the gas.* An addition of some electronegative gas, like water vapour or sulphur dioxide, will reduce the ion mobility and produce a less steeply sloping characteristic, since the gradient of this curve increases with the ion mobility. This effect of electronegative gases is due to their forming negatively charged molecular ions, which in view of their considerable mass are less mobile.

Finally, some *design parameters* of the equipment also have a marked influence on the V–A characteristic. The greater the diameter of the wire electrode, the higher will be the initial critical voltage, and the steeper the slope of the curve, because the value of the constant in expression (51) will increase. A wider gap between the electrodes (i.e. a higher value of R) will likewise increase the initial critical voltage, but will produce a gentler gradient of the V–A characteristic.

2.3.3 Charging of particles in electrostatic precipitators

The charging of a single particle in an electrostatic precipitator is described by relations (18) and (20). Given a monodisperse dust, the charge q_p per kg of dust is simply the sum of the individual particle charges q_{c1} of the n particles forming that amount of dust,

$$q_p = q_{c1} n \tag{52}$$

If the particles are spherical, their number in one kg of dust is determined as

$$n = \frac{6}{\pi a^3 \varrho_m} \tag{53}$$

so that the total charge per kg of monodisperse dust with a particle size in excess of 10^{-6} m works out as

$$q_p = \varepsilon_0 \left(1 + 2\frac{\varepsilon_r - 1}{\varepsilon_r + 2}\right) F \frac{6}{\varrho_m} \cdot \frac{1}{a} \tag{54}$$

The equivalent expression for particle sizes smaller than 10^{-6} m is

$$q_p = 1.6 \cdot 10^{-11} \frac{6}{\pi \varrho_m} \cdot \frac{1}{a^2} \tag{55}$$

Table 13 lists the q_p values for monodisperse dusts with various particle sizes, on the assumptions that $\varepsilon_r = 4$, $F = 3.5 . 10^5$ V m^{-1}, and $\varrho_m = 2000$ kg m^{-3}.

TABLE 13

The charges of 1 kg of monodisperse dust of various particle sizes

Particle size a (m)	Charge q_p (C kg^{-1}) of 1 kg of dust
$1 \,.\, 10^{-6}$	$186.0 \,.\, 10^{-4}$
$5 \,.\, 10^{-6}$	$37.2 \,.\, 10^{-4}$
$10 \,.\, 10^{-6}$	$18.6 \,.\, 10^{-4}$
$50 \,.\, 10^{-6}$	$3.7 \,.\, 10^{-4}$
$100 \,.\, 10^{-6}$	$1.9 \,.\, 10^{-4}$
$500 \,.\, 10^{-6}$	$0.4 \,.\, 10^{-4}$
$1000 \,.\, 10^{-6}$	$0.2 \,.\, 10^{-4}$

When the dust is polydisperse, i.e. described by a retained fractions curve $Z = f(a)$, we can determine its q_p figure only by resorting to the mean particle diameter a_{mean}. For particles larger than 10^{-6} m, this is

$$\frac{1}{a_{\text{mean}}} = \frac{1}{Z_1} \int_0^{Z_1} \frac{1}{a} \, \mathrm{d}Z \tag{56}$$

while for particles smaller than 10^{-6} m it works out as

$$\frac{1}{a^2_{\text{mean}}} = \frac{1}{1 - Z_1} \int_{Z_1}^{1} \frac{1}{a^2} \, \mathrm{d}Z \tag{57}$$

where Z_1 is the residue trapped on a mesh for $a = 10^{-6}$ m.

The mean particle diameter can naturally also be evaluated graphically. Once these figures are ascertained, the charge per kg of polydisperse dust is found as the sum of the two components,

$$q_{p1} = \varepsilon_0 \left(1 + 2\frac{\varepsilon_r - 1}{\varepsilon_r + 2}\right) F \frac{6}{\varrho_m} \cdot \frac{Z_1}{a_{\text{mean}}} \tag{58}$$

and

$$q_{p2} = 1.6 \,.\, 10^{-11} \frac{6}{\pi \varrho_m} \cdot \frac{1 - Z_1}{a^2_{\text{mean}}} \tag{59}$$

the total charge then being

$$q_p = q_{p1} + q_{p2} \tag{60}$$

A crucial factor in electrostatic precipitation processes is the space charge formed by the dust dispersed in the active space of the unit. This space charge ϱ_p, for a given concentration k of the dust in the gas, is defined as

$$\varrho_p = k q_p \tag{61}$$

The dust can attain a saturated space charge, ϱ_{pn}, only under certain favourable conditions. If we insert the space charge figure in place of the dust charge value, we can adapt relation (21) to express the particle charging time as

$$\tau = \frac{4\varepsilon_0}{u} \cdot \frac{\dfrac{\varrho_p}{\varrho_{pn}}}{\varrho_i\left(1 - \dfrac{\varrho_p}{\varrho_{pn}}\right)} \tag{62}$$

However, the space charge of the gas ions is far from constant, since it depends on the space charges of the dust particles, so that ϱ_i varies with ϱ_p. At any given voltage across the electrodes, the current flow through the precipitator attains its maximum (and hence induces the maximum space charge of the gas ions ϱ_{im}) when the space between the electrodes is either free of dust or when the dust particles are as yet uncharged. As the dispersed dust gradually picks up electrical charges, it forms a space charge ϱ_p which counteracts the induced field and tends to attenuate the current flow through the precipitator, in other words, to reduce the gradient of the V–A characteristic. We may assume that as long as the applied voltage remains constant, the total space charge in the space between the electrodes will also stay more or less constant. We may further assume that this total charge will equal ϱ_{im}, i.e. the maximum space charge (in C m^{-3}) of the gas ions in the state when the dust is as yet uncharged or when only clean gas is passing through the unit. In other words, as long as the voltage across the electrodes remains roughly unchanged, we may work on the assumption that

$$\varrho_i + \varrho_p = \varrho_{im} \tag{63}$$

This of course holds true only as long as the dust is not being trapped, because capture deprives the dust particles of their charges and thus affects the total space charge in the gap between the electrodes. On these assumptions, then, we can re-define the particle charging time as

$$\tau \doteq \frac{4\varepsilon_0}{u\varrho_{im}} \cdot \frac{\dfrac{\varrho_{im}}{\varrho_{pn}} \cdot \dfrac{\varrho_p}{\varrho_{pn}}}{\left(\dfrac{\varrho_p}{\varrho_{pn}}\right)^2 - \dfrac{\varrho_p}{\varrho_{pn}}\left(\dfrac{\varrho_{im}}{\varrho_{pn}} + 1\right) + \dfrac{\varrho_{im}}{\varrho_{pn}}} \tag{64}$$

Relation (64) can be converted into

$$\tau = \frac{4\varepsilon_0}{u\varrho_{im}} m \tag{65}$$

The values of m for various ratios of $\varrho_{im} : \varrho_{pn}$ are plotted in dependence on $\beta = \varrho_p : \varrho_{pn}$ in Fig. 39. This diagram shows that at a constant value of m (i.e. at a constant particle charging time) and at a constant space charge of the gas ions (i.e. at a roughly constant applied voltage), the $\varrho_p : \varrho_{pn}$ ratio drops off with the $\varrho_{im} : \varrho_{pn}$ ratio, in other words, diminishes as the maximum charge of the dust particles increases. It must be added that the above equation holds true only very approximately, because it is based on the assumption that the dust and gas ion charges are evenly distributed; and that is something we can never expect to find in practice. Moreover, this equation rests on the inherent assumption that the dust is not being trapped, so that the space charge of the dust remains constant throughout the whole length of the precipitator working space, and that each dust particle retains its initial charge indefinitely. Which means that in reality, the charges picked up by the individual particles will always be substantially greater than would follow from the above relations.

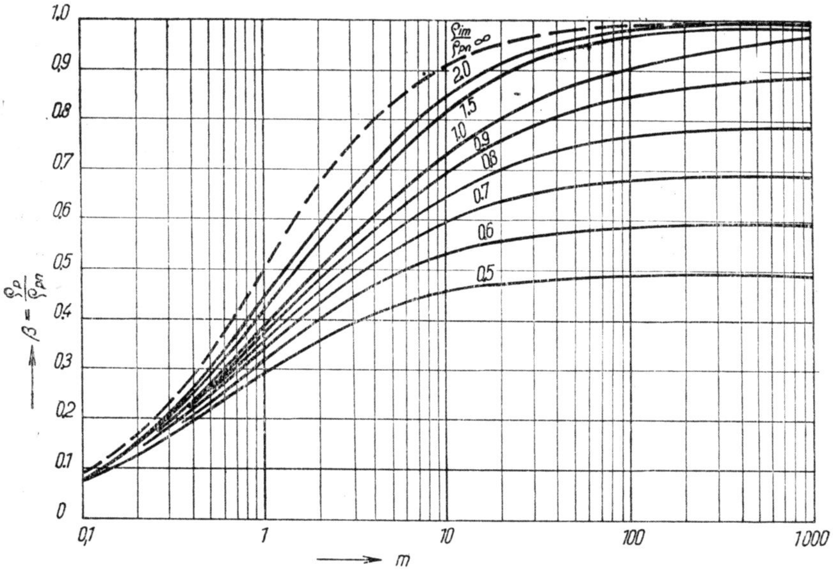

Fig. 39 The dependence of m on $\beta = \dfrac{\varrho_p}{\varrho_{pn}}$ at various values of $\dfrac{\varrho_{im}}{\varrho_{pn}}$.

2.3.4 Motion of charged particles in an electric field

In the electrical field within a precipitator, a particle bearing a charge of Q_c is exposed to the following forces:

a) Electrical forces, P:

$$P = Q_c F \tag{66}$$

P acts along the electrical lines of force, and hence towards the collecting electrodes.

b) The drag R generated by the environment:

$$R = \xi S_c \frac{v_m^2}{20} \tag{67}$$

where S_c is the wetted cross-sectional area of the particle (in m^2), and
v_m – the velocity of the particle relative to its environment (in m s^{-1}).
R always opposes the motion of the particle. The magnitude of the drag coefficient ξ depends on the shape of the particle and on the *Reynolds number*,

$$Re = \frac{v_m a}{\nu} \tag{68}$$

c) The mass G_c of the particle:

Gravity acts in a plane parallel to the collecting electrodes, and therefore in no way affects the approach of the particles towards those electrodes. Consequently, we can safely neglect it in our further considerations.

That leaves two forces which between them control the approach of the particles to the collecting electrodes. As the electrical forces begin to accelerate the particle, there is at first no drag to overcome. As the particle velocity increases, however, the drag rises, and the rate of acceleration diminishes. Finally, the drag grows sufficient to balance the electrical forces; from then on the particle moves at a constant velocity v_m, determined by the equilibrium of forces

$$P = R \tag{69}$$

as

$$v_m = \frac{Q_c F_2}{3\pi\eta a} \tag{70}$$

We can now expand the term Q_c in the above expression to gain the following relation valid for particles larger than 10^{-6} m:

$$v_m = \frac{8.855 \,.\, 10^{-12} \beta \varkappa F_1 F_2 a}{3\eta} \tag{71}$$

where F_1 is the intensity of the field in which the particle acquired its charge (in V m^{-1}),
F_2 – the field intensity at the momentary location of the particle (in V m^{-1}),
β – the coefficient of relative particle charging, i.e. the ratio between the actual and the saturated particle charges, cf. equation (22):

$$\beta = \frac{Q_c}{Q_{cn}} \tag{72}$$

The corresponding relation for particles smaller than 10^{-6} m is

$$v_m = \frac{1.6 \,.\, 10^{-11} \beta F_2}{3\pi\eta} \tag{73}$$

Velocity v_m, termed the separating brisk velocity, is a key quantity which characterizes the functioning of any given precipitator. Relations (71) and (73) indicate that its value is dependent on the particle size; and Fig. 40 shows both the theoretical dependence (curve *A*) and the relationship between the two quantities which is encountered in actual practice (curve *B*).

The greater this velocity v_m, the higher is the separating efficiency of the unit. However, as evident from relation (71), the value of v_m is not dependent on the particle size alone. It also grows with the relative charging coefficient β, with the charging coefficient $\varkappa$, and with the field intensities F in which the particle picks up its charge and in which it is trapped. Conversely, v_m is the lower, the greater the viscosity η of the carrier gas that bears the particles. Let us now examine these factors one by one.

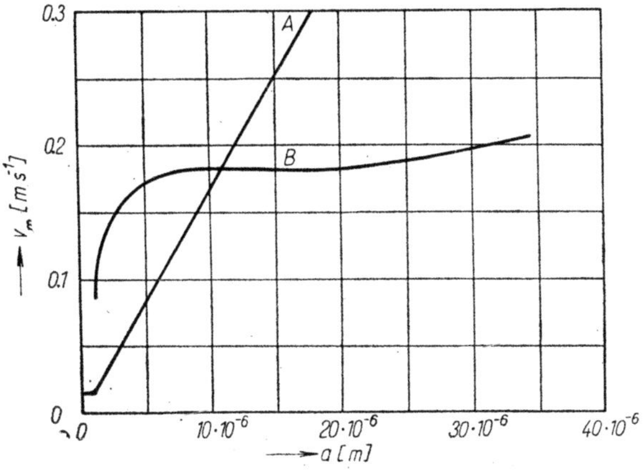

Fig. 40 The theoretical (*A*) and actual (*B*) dependence of the separating brisk velocity v_m on the particle size *a*.

For any given residence time of the particles in the electrical field, coefficient β will be the higher, the greater the ion mobility *u*, the greater the space charge of the ions formed by the corona discharge, and the greater the ratio between the space charge of the ions and the saturated space charge of the particles we intend to trap. The ion mobility *u* rises with the absolute temperature, but diminishes as the gas pressure grows. The space charge of the ions can be increased mainly by a suitable configuration of the discharge electrodes, e.g. by the use of point electrodes. This is vital particularly when the saturated space charge of the particles is large, as is the case when the initial dust concentration is high or when the dust is extremely fine. This latter relationship is evident from Table 13.

The charging coefficient $\varkappa$ grows with the relative permittivity ε_r, which is why particles with a high ε_r value are easier to capture. A case in point are moist particles, because humidity increases the relative permittivity.

The electric field intensity that we can obtain from any given electrode array will grow with the rectified voltage applied across the electrodes. Unfortunately, this voltage is limited by the arc-over point, and the latter in turn depends on the arrangement of the electrodes and the properties of the gas. The arc-over voltage can be raised, to some extent, e.g. by moistening the gas, but it will still vary with every fluctuation in the parameters of the gas entering the unit. As in practice it is usually impossible to keep the gas parameters closely constant, precipitators are commonly fitted with a variety of automatic voltage control systems. These respond to the instantaneous state of the gas, and keep the applied voltage as close to the arc-over point as is safe under the given circumstances.

The viscosity η rises with the gas temperature, which means that at higher temperatures we must needs count with a lower v_m value.

The art of designing efficient electrostatic precipitators is largely a matter of achieving the maximum feasible space charge of the ions ϱ_{im} at the maximum intensity of the field. To make the most of the specific properties of various electrode types, it is normal practice to split the precipitator lengthwise into successive compartments, each of them with its own power feed system. Generally, the first or upstream section is provided with point electrodes, and the second or downstream section with wire electrodes. The latter afford a lower ϱ_{im} value, but can handle a higher voltage and consequently provide a higher field intensity. The lower ϱ_{im} value in the downstream section hardly matters, since the particles reaching this section will already have acquired a sufficient charge in the upstream compartment; the loss certainly matters less than the higher field intensity available from wire electrodes.

2.4 Diffusion phenomena affecting solid particles in gases

In fundamental considerations on the dynamics of solid particles in gaseous media, it is customary to assume that the particles move in an ideal environment. That means neglecting the viscosity and hence the molecular structure of the gas. In a real as distinct from an ideal gas, however, the motion of the solid particles is affected both by this molecular structure of the gas, and by any turbulence of the gas stream. Both these factors can be considered as sources of diffusion phenomena which affect solid particles in a gaseous environment. In this context, the effects of the molecular structure are termed molecular diffusion, while those of turbulence are called turbulent diffusion.

Fick has formulated two laws which between them describe the diffusion of solid particles in gases, the apparent rise in the viscosity of a turbulently flowing gas, and the effects of molecular motions and flow turbulence on the processes of heat and mass transfer in a gas.

Fick's first law states that if a volume V is filled with a certain concentration n of particles and surrounded by a space where the particle concentration is nil, there will be a spontaneous migration of particles from that volume into the space with a zero concentration level. That of course implies that the concentration n in volume V will vary in time.

Fick's second law postulates that in a space where there is a concentration gradient, with particles entering that space from one side while others leave it at the other side, the actual particle concentration will be variable.

The diffusion of solid particles in a gas is characterized by the diffusion coefficient D, which is the smaller, the lower the temperature and the larger the particles. At $D = 0$, there will no longer be any spontaneous migration of the particles. D can be quantified either in $m^2\,s^{-1}$ or in $cm^2\,s^{-1}$. Rich defines the diffusion coefficient (in $cm^2\,s^{-1}$) for spherical particles of diameter a (in cm) in air, at room temperature and normal atmospheric pressure, as

$$D = \frac{24 \,.\, 10^{-12}}{a}\left(1 + \frac{18 \,.\, 10^{-6}}{a}\right) \tag{74}$$

Molecular diffusion can be regarded as the sum of the motions, performed by the molecules of a certain substance, which are induced by phenomena at the molecular level (such as thermally induced molecular motions, or partial pressure differentials). The key feature of molecular diffusion is the spontaneous motion of the molecules, which occurs as the factors that cause this process overcome the inherent resistances of the molecules. If a solid particle borne in a gas is smaller than the mean free path available to the molecules, it will collide with the molecules of gas; consequently, it will move about in a random manner, like the gas molecules themselves, following short linear paths from one collision to the next. In view of its larger mass, however, it will move more slowly than the gas molecules. This mode of motion of extremely small particles is called *Brownian motion*, after the English botanist who discovered it in 1827.

Pražák quotes *Einstein's relation*, based on the theory of *Brownian motion*, for the mean square of the path lengths ($\bar{s}^2$) which a large number of spherical particles with a diameter of a will cover per unit of time as a result of molecular collisions:

$$\bar{s}^2 = \frac{2RT\tau}{3\pi N\eta a} \tag{75}$$

where R is the gas constant,

T – the thermodynamic temperature of the gas (in °K),

N – *Loschmidt's number* (the number of molecules per mol).

When the gas stream bearing the solids is laminar, the Brownian motion will cause small solid particles to diffuse within it much as in a stationary gas. In a turbulent flow, however, the vortices which arise, and tend to persist for some time, alter the whole pattern of events. They cause both the carrier medium and the particles it conveys to move in directions which differ from the principal flow direction. Much will depend on the degree of turbulence, as defined by the turbulence intensity and the coefficient of correlation. Sufficiently small particles will be subject to Brownian motions even in a turbulent environment. In that case, the motions of the carrier medium molecules will cause the small particles to diffuse in a turbulent flow much as they would do in a laminar stream or in a stationary environment. Moreover, the circular motion within the vortices, their lateral displacements, and the resultant movements of small carrier medium particles, will all superimpose further motion components on the solid particles. Once the size of the latter particles exceeds roughly one micron, Brownian motion will rapidly dwindle away, but the vortices in a turbulent flow will still generally be powerful enough to shift even these bigger particles across the main flow direction. In other words, turbulence will still cause diffusion even when the particles are too large to display any Brownian motion. The overall outcome of all these effects is difficult to assess, but the literature presents various relations for the coefficient of turbulent diffusion, roughly analogical to the expressions for the coefficient of molecular diffusion.

2.5 The mesh effect

The mesh effect means the direct retention of particles which are larger than the openings of a given sieve or in a given filter material. It is important especially in those types of equipment which employ filter cloths, and in the trapping of pollutants by means of fibrous, granular or porous filter materials. To what extent the particles borne in a gas stream are actually trapped depends mainly on the particle size and the size of the filter openings or interstices; when the particles are irregularly shaped, it also depends on their orientation at the moment when they impinge on the filter material. As the filter gradually clogs up with captured particles, the effective size of its openings diminishes, and hence the efficacy of the mesh effect in the trapping of further particles tends to increase.

2.6 Agglomeration of solid particles

Solid particles are sometimes liable to agglomerate into clusters which exhibit a certain degree of cohesion. This process may be induced or supported by any of a variety of factors – thermal, sedimentation (gravitational or centrifugal), turbulent, electrostatic or acoustic forces. Any of them can bring two or more particles close enough together, and into such mutual positions, that adhesive forces will come into play and join and hold the particles together. This process is as a rule irreversible. It interests us mainly because it reduces the number of particles per unit volume of space, and because, once a substantial proportion of the particles have joined together into clusters, it radically alters the grain size distribution. Where agglomeration is at all likely, we must always distinguish between the particle concentration levels and particle size distributions that apply before and after this process has taken place.

3. REVIEW OF SEPARATOR TYPES

Ing. J. Albrecht, CSc., Ing. J. Hejma, CSc, Ing. O. Štorch, CSc., Ing. J. Urban

3.1 Dry mechanical separators

3.1.1 Settling chambers and inertial dust arresters

The simplest kind of dust trapping device is a settling chamber, also known as a gravity chamber. Unfortunately, the efficiency of these units is so low that they are nowadays hardly ever used for actual gas cleaning. The only purpose for which they are still employed on any scale is the screening of polydisperse dusts into two or more size fractions. A typical application of this kind is the separation of coarse particles from a gas flow upstream of a fan.

A settling chamber is essentially just a receptacle where the cross-sectional area of the flow channel is much larger than in the adjoining inlet and outlet tubing or ducting. The resultant drop in the flow velocity permits the heavier particles to fall, by gravity alone, to the bottom of the chamber, from which their deposits are then removed. The effect will be better understood if we consider that in the tubing or ducting, the entrained particles are apt to move at velocities of the order of 10 to 25 m s^{-1}, but that their free falling velocity is usually only a few centimetres or millimetres a second. Furthermore, the flow in the ducting is turbulent, and is therefore associated with an intensive mass transfer which tends to average out the local concentration levels, especially of the finer particles. In the settling chamber, where the flow velocity generally declines to something like 1 m s^{-1}, the gravity-induced velocity component has a much more marked effect on the resultant motion of the particles; and turbulent diffusion decays sufficiently to allow the particles to settle by gravity.

These phenomena, however, can produce a significant effect only on particles with sizes of the order of a few hundred microns or more. A settling chamber capable of capturing finer particles, in the 10 to 100 micron size range, would have to be impracticably large; and fines less than 10 microns across, i.e. the fractions most affected by turbulent diffusion, will evade trapping no matter how large the chamber may be.

All the above applies only to the simplest of gravity chambers, as sketched in Fig. 41. A refinement which can substantially improve the performance of these chambers is internal fittings of various types. Some of these fittings, e.g. in the form of chains, are intended to trap and retain the dust directly, others are essentially baffles which deflect the flow and thus harness the centrifugal forces generated by a curved trajectory to augment the separating action. This utilization of inertial effects can greatly enhance the efficiency of the unit. Inertial dust arresters of this kind exploit the same basic physical principles as cyclones and sometimes share their general shape too. True, the centrifugal accelerations attained in them are much smaller than those common in cyclones, and their functioning still resembles that of a settling chamber much more than that of a cyclone, but even so they represent a distinct advance over the plain gravity chamber.

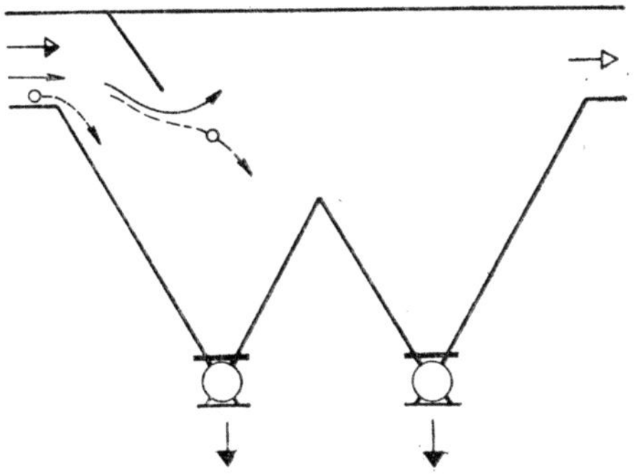

Fig. 41 Schematic representation of a settling chamber.

Another device in this general category, which is sometimes used as a coarse pre-cleaning facility, is a tube bend or elbow installed upstream of the main gas cleaning equipment. This elementary measure can often markedly prolong the service life of the cyclones downstream of the bend or elbow. The latter must naturally be made of some abrasion resistant material. The gains attainable by this device are greatest when dense concentrations of coarse, highly abrasive dusts are handled at relatively low mass flow rates.

3.1.2 Cyclones

Cyclones, or "vortex separators", are at present by far the most widely used type of mechanical separating equipment. Not because they are the most efficient type: some other and more complex kinds of purely mechanical separating equipment, which also exploit the difference between the specific gravities of the gas and of the admixture particles, can easily beat the cyclone for sheer efficiency. Most of them, however, are less attractive than the cyclone because of the extra costs and complications involved in their manufacture, operation and maintenance. Moreover, no rival type has yet undergone as long a development process as the cyclone, and some of these types (such as the Hurriclone described in Section 3.1.5) need an appreciably higher power input. In practice, these disadvantages between them are apt to outweigh any marginal gains in efficiency which the more complicated units can offer.

The cyclone is old, as the ages of present-day plant equipment go: the first patent relating to it was granted in the late nineteenth century. And since at first sight there is not much outward difference between the earliest and the latest cyclones, people are inclined to regard the whole cyclone concept as dated. There have of course been very real advances in performance over the years. In fact, the most recent cyclone designs attain a level of efficiency which is not likely to rise much further, no matter how much more development work is devoted to these units. Still, it could in all fairness be claimed that the cyclone is in such widespread use, and likely to remain so for many more years to come, chiefly because of the inherent limitations of its main rivals. Electrostatic precipitators, wet scrubbers and cloth filters are all restricted in their scope by factors which make little difference to a cyclone. In the latter, the main constraint is its poor performance in the trapping of very fine particles. At the present state of the art, cyclones are little use for particles less than some 10 microns in diameter, at a ϱ_m value of 1000 kg m^{-3}.

Electrostatic precipitators are unlikely to replace cyclones in the foreseeable future because of their size and cost, which make them applicable only where the throughflow is really large. At present, the minimum acceptable throughflow would seem to be somewhere between 20,000 and 50,000 $m^3\ h^{-1}$, and there would be little point in attempting to build a precipitator for smaller flow rates.

Wet scrubbers are still confined to only certain kinds of dust and certain processing technologies. For instance, they are obviously not suitable in cases when the trapped dust is to be further processed in the dry state. They cannot handle dusts which are prone to bond or cement when wet, and are both difficult and costly to run when the dust and/or the carrier gas contains a high proportion of aggressive, corrosion-inducing constituents. Furthermore, scrubbers are liable to freeze up in winter; need costly sludge handling ancillaries; and often create serious sludge disposal problems. Finally, their power requirements are as a rule higher than those of cyclones.

Cloth filters would seem to stand the best chance of ultimately replacing the cyclone in at least some applications. However, the range of their potential usefulness, and the rate at which they are adopted, will depend heavily on the evolution of new filter materials capable of withstanding higher temperatures, and on radical improvements in the reliability of these units. Even then, given the best of filter materials and an utterly dependable design, the cloth filter will always be more costly, need more maintenance, use up more power, and take up more space than an equivalent cyclone unit.

It thus appears that at least in some fields, the cyclone is likely to stay with us for as far ahead as we can see. One of these fields is power generation, though the types which seem to have the best prospects in this branch differ from those in use there at present. One promising type would be a simple, virtually non-selective, low-efficiency multiple cyclone unit, with a collecting efficiency of only about 60 to 70 per cent, for the primary cleaning stages ahead of electrostatic precipitators. Another is a multiple cyclone installation with the highest attainable efficiency, to serve small boilerhouses (e.g. in district heating plants) which would not justify the costs of electrostatic precipitating equipment.

The food industry is another domain where cyclones are apparently here to stay, especially in processes which involve drying. This field will continue to call for a variety of cyclone types, and not all of them will need to be particularly efficient: in some instances the cyclones will act mainly as classifiers, in others they will form a pre-cleaning stage upstream of the main filtering equipment. One special requirement will be a high-efficiency cyclone to handle large throughflows in the region of 10 to 15 $m^3 s^{-1}$, probably something like the RS unit shown in Fig. 43. Similar installations are likely to come into use in the timber and woodworking industries too, although these are recently beginning to introduce cloth filters as well.

Other spheres where cyclones will evidently remain in use for a very long time to come, either on their own or as a preliminary stage upstream of other separating equipment, include large parts of the chemical industry, and the production of some building materials. Cyclones will also find it easy to hold their own in all technologies which entail subsequent drying, or pneumatic conveying, and will be very difficult to replace in locations where the equipment is exposed to high service temperatures. In short, cyclones, though often considered obsolescent, still obviously have a future ahead of them, and the diversity of cyclone types in demand is actually more likely to grow than to dwindle.

For all the decades of their evolution and for all the variety of types on the market, all cyclones work on much the same basic lines. The dust-laden gas entering the unit is forced to follow a helical path (or, as the common but quite erroneous definition goes, to "perform a rotary motion" – only in plan view does the path look circular). The helix angle varies with the diameter of the flow channel. When examined from the potential flow aspect, and viewed in the cross section perpendicular to the centre line of the cyclone, the flow looks rather like a combination of a potential vortex and a sink. Still, this picture is presented only by way of an illustration, because cyclones are not amenable to investigation by the potential flow approach.

With this much said, we can classify cyclones by the ways the helical motion is induced in them. One category are "*tangential*" *cyclones*, where the dust-laden gas *enters* a cylindrical chamber *tangentially*. Another are "*axial*" *cyclones*, where the gas *is fed in axially* and then deflected by baffles or vanes. Axial cyclones further fall into two categories: *reverse flow types*, where the gas enters at the top, performs a 180-degree U-bend at the bottom, and re-emerges at the top; and *straight-through types*, where the gas enters at the top and leaves the unit at the bottom.

Fig. 42 shows a variety of cyclones which have been evolved or manufactured in Czechoslovakia by the ČSVZ works (Československé vzduchotechnické závody). To facilitate comparisons, all these types are drawn with the same basic diameter. The T 1 to T 5 types have perpendicularly arranged tangential inlets; the 300 V type, derived from the Soviet-built 400 N unit, features an inclined gas inlet. Fig. 43 illustrates some larger-diameter designs: the Alden, made in the U.S.A.; the RS and VC units by ČSVZ of Czechoslovakia; the ASH type made by Keller of West Germany; and the Soviet Liot type with helical guides and a flow regulat-

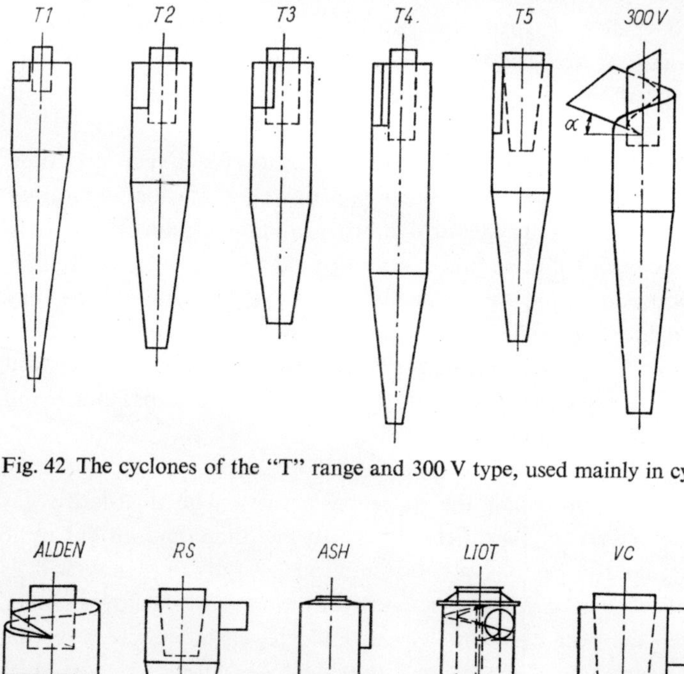

Fig. 42 The cyclones of the "T" range and 300 V type, used mainly in cyclone batteries.

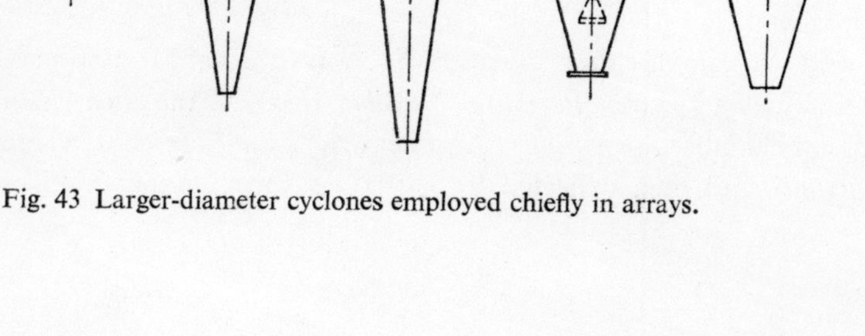

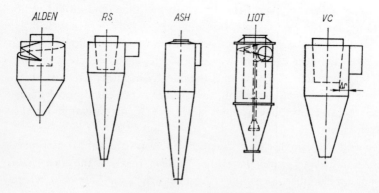

Fig. 43 Larger-diameter cyclones employed chiefly in arrays.

ing cone inside it. The RS type is commonly used for trapping dried milk, the others are intended mainly for coarser or fibrous pollutant particles, e.g. in sawmills or flax scutching mills. Fig. 44 represents an SHA system incorporating the Davidson R cyclone; this is an axial straight-through unit typical of the cyclones designed for pre-cleaning applications, which makes up for its relatively low efficiency by taking up very little space.

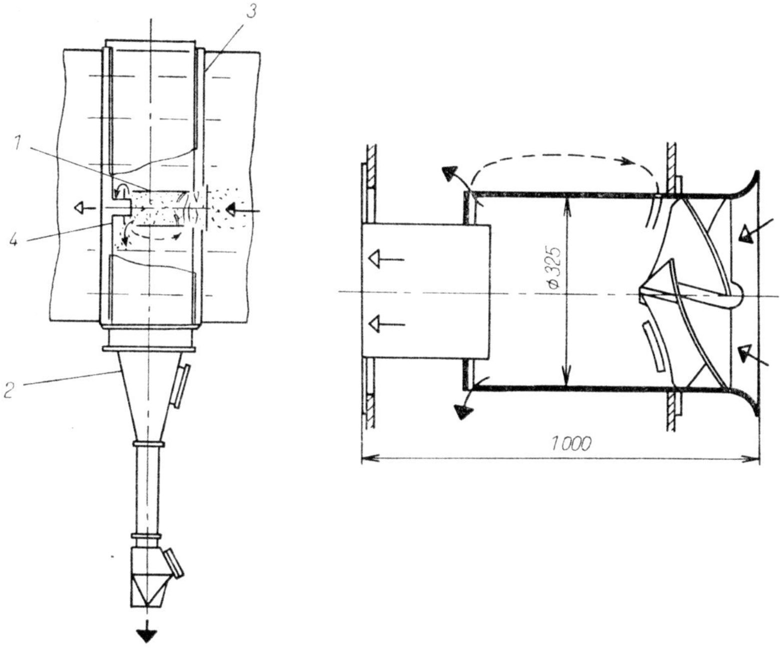

Fig. 44 The Davidson R cyclone and its installation in the SHA units:
1 — cyclone; *2* — hopper; *3* — inlet flange; *4* — gas outlet flange.

3.1.2.1 Key parameters of cyclones

The most important parameters of a cyclone, as of any other piece of separating equipment, are its *collecting efficiency* and *the pressure drop* across the unit. The former describes the ability of the cyclone to capture and retain dust particles, the latter indicates how much power the unit will need to do so. Another vital parameter is the life expectancy of the unit under average service conditions.

By and large, the pressure drop tends to rise with the collecting efficiency; and although there are numerous exceptions to this rule, the mechanism behind it is worth explaining. A cyclone works by altering the flow direction of the dust-laden gas that enters it, thereby subjecting the mixture of gas and solids to intense centrifugal forces, and then exploits the higher inertia of the solid particles to divert their trajectories from the flow path of the gas. It follows that the efficiency of the unit will grow with the number and sharpness of the bends in the flow path, but a long and tortuous flow route will obviously entail a larger pressure drop than a short and straight one. That does not imply that a high pressure drop must always be conducive to a high collecting efficiency, but it does mean that a cyclone with an extremely low pressure drop is unlikely to be particularly efficient.

This relationship can well be illustrated on the range of cyclones made at the ČSVZ works. The Davidson R axial straight-through cyclone, in which the swirling motion of the gas is induced only by an array of vanes, has the lowest pressure drop, but also the lowest efficiency of all the units in the range. The V 1/315 type, a reverse flow axial cyclone, pays for its higher efficiency by a greater pressure loss, which is inevitable when a flow reversal has to be superimposed on the swirling motion of the stream. Both the efficiency and the pressure drop attain a peak in the tangential types, where the three-dimensional path of the gas flow is more complex than in the axial units. However, it would be wrong to infer that the pressure drop must necessarily be greater in tangential than in axial cyclones, because both this drop and the efficiency are also affected by a number of design factors which are not directly related to the overall concept of the unit. These variables include the geometrical configuration of the cyclone, the cross-sectional areas of its various flow channels, etc.

The starting point for all design work on cyclones is usually *the requisite level of efficiency*. This can be assessed only if we know the grain size distribution of the dust, the dust concentration in the incoming gas, and the acceptable dust concentration in the gas leaving the unit, but then these figures are mostly known, at least approximately, in advance. These data largely determine the type of cyclone needed, while the total throughflow to be handled, and the permissible pressure drop, govern the number and size of the units.

The performance of a cyclone is best judged from the curve showing how the fractional collecting efficiency varies with the particle size of the dust. The overall collecting efficiency of that particular cyclone, in the application under consideration, can then be estimated from this plot and the retained fractions curve or sieve analysis of the dust. The fractional collecting efficiency can also be expressed analytically, in the form of an exponential function, but this approach calls for extreme caution. Given well-selected constant values, the exponential expression can describe the performance of some cyclones very closely, but can still prove totally misleading when applied to other types.

Fig. 45 shows the $O_f = f(a)$ dependences for some types of cyclones. At dust densities around 1000 kg m^{-3}, these curves will generally start at particle sizes in the vicinity of one micron, or, for the most efficient units, at $a = 0.5$ microns. But if the origins of all these curves tend to lie at much the same values, the far ends of the curves, where O_f approaches the 100 % mark, display a good deal of scatter. That is natural, because the 100 % mark will mostly lie at relatively large particle sizes, in view of the odd oversize particle which every now and then evades capture. Moreover, the inherent errors of our measurement techniques usually preclude any exact definition of the 100 % point on the chart. Fortunately, the precise location of that point does not really matter; in practice, the upper ends of the curves are generally drawn so as to yield a continuous plot rather than to end at a closely established point. Even in this conventional representation, however, it is obvious that the upper ends of the curves for various cyclone types lie much farther apart than the bottom ends.

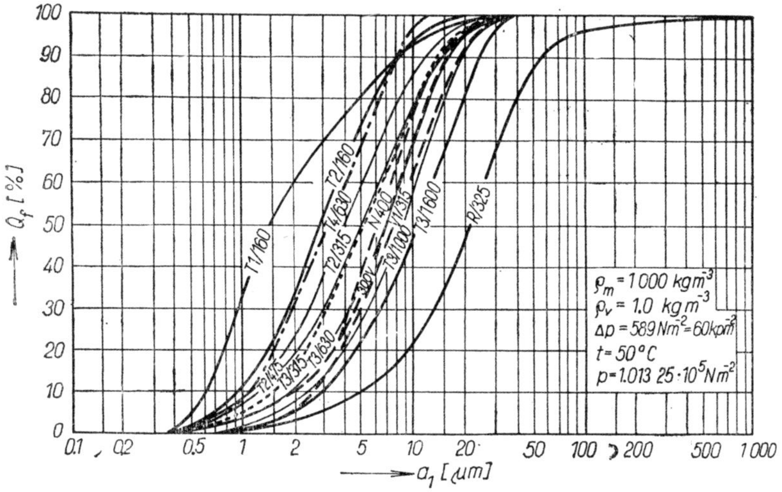

Fig. 45 Fractional collecting efficiency curves of various cyclone types.

On the charts for those cyclones which display the highest collecting efficiencies and best screening effects, the curve will approach the 100 % point at particle sizes of about 10 microns. In most cyclones, however, the curves converge to this level only around 20 microns, and the curves for low-efficiency types come near this level only in the particle size interval of 30 to 40 microns. Since all the curves originate at much the same grain size, this means that the particle size at which the O_f curve nears the 100 % level is a good yardstick of the selectivity of a cyclone: the smaller this size, the higher the screening or particle classifying effect of the unit. One point worth noting in Fig. 45 is that some of the curves, e.g. those for the T 1/160 and T 4/630 types, intersect each other. This indicates that the former type does better on the finer dust fractions, while the latter type exhibits higher O_f values at particle sizes just below 10 microns.

All the above goes to show that even charts of the fractional collecting efficiency cannot provide any unambiguous criteria of the overall collecting efficiency.

Comparative trials conducted with a fairly coarse dust will often fail to reveal any appreciable difference between various cyclone types, as the O_c values of all of them will be close to 100 per cent. Again, take the case of the two mutually intersecting O_f curves in Fig. 45. If we test these two types on two dust specimens with pronounced peaks on their grain size distribution curves, the peak for one sample lying at 3 to 4 microns and that for the other at 8 to 10 microns, we shall only find what we already know – that the T 1/160 is superior on the finer and the T 4/630 on the coarser dust. So that, regrettable though the fact may be, there is no short cut to a hard and fast answer: the efficiency of a cyclone has to be evaluated from case to case.

If theory provides no formula that will yield a clear-cut figure of merit, we can at least fall back on a considerable fund of empirical knowledge about the duties to which various cyclone types are commonly assigned. This body of experience enables us to distinguish between three broad categories – *high-efficiency, medium-* and *low-efficiency types*, though especially the last of these categories will obviously span a very wide field.

Another factor which has to be borne in mind in all these considerations is that the collecting efficiency of cyclones is temperature dependent. The specific gravity of gases diminishes as the temperature rises (so does the specific gravity of dusts, but not to any comparable extent). Therefore, the difference between the specific gravities of the gas and of the solids grows with the temperature. And as all cyclones function by exploiting this specific gravity differential, one might expect them to work better at high than at low temperatures. However, a temperature rise will also increase the viscosity of the gas, which hampers the separating action, and this latter effect predominates. The overall collecting efficiency will consequently drop off as the temperature rises, but the degree to which it does so differs from one cyclone to another, and also varies with the particle size of the dust. In the handling of any given dust, the temperature-induced loss of collecting efficiency will depend both on the actual temperature and on the slope of the fractional collecting efficiency curve. For example, assume that we are trapping a dust with $a = 5$ microns, and contemplating a rise in the operating temperature from 20 to 200 °C. In the T 1/160 type, which has a fairly flat O_f curve, this will reduce the O_f value from 81 to 78 per cent, but in the T 4/630 type, which has a steepish O_f plot, the O_f will drop from 77.5 to 72 per cent. In the trapping of the standard type R and G reference dusts, temperature dependent efficiency losses normally vary between 1 and 3 per cent, which means that they are far too high to be neglected. With efficiency levels running as high as they nowadays do, an extra 3 per cent of escaping particles may significantly alter the overall pollution picture.

The other fundamental parameter of a cyclone is *the pressure drop* Δp (in Pa). It is usually defined as

$$\Delta p = \xi_D \frac{v_D^2}{2} \varrho \tag{76}$$

The key point to note about this expression is that the drag coefficient ξ_D is here related to the velocity v_D which the flow would attain in a tube of diameter D (i.e. of the same diameter as the cyclone), in other words, to a purely fictitious velocity. That, however, is unavoidable. If the drag were related to any of the flow channels in the actual cyclone, e.g. to its inlet section, the resultant figure would not be indicative of the true state, because two cyclones with fully identical characteristics can have different inlet velocities and would consequently have different ξ values. The magnitude of this coefficient depends chiefly on the type, design and geometry of the cyclone, i.e. on its dimensions and the ratios between them. In the range of types evolved at the ČSVZ works, its value ranges from a few dozen to several hundred. Another point to note about the above expression is that it does not cover the state of the cyclone wall surfaces. Still, in actual service, as long as the walls remain reasonably clean, their state has virtually no effect on the resultant pressure drop.

The relationship between the collecting efficiency and the pressure drop of a cyclone is more complex than is generally thought. There is a widely held belief that the collecting efficiency is bound to increase with the throughflow, i.e. with the pressure drop, but this is now known to hold true only within a certain interval of those parameters. The point can well be illustrated on the types of Czechoslovak design, because the key data of all these types have already been established; we know their overall collecting efficiencies, and the curves representing their fractional collecting efficiencies at throughflows corresponding to pressure drops of 400, 600 and 1000 Pa at $\varrho = 1$ kg m^{-3}. Within this interval, all the types display the highest efficiency around 1000 Pa, and the lowest around 400 Pa. However, once we plot the overall collecting efficiency of the T 4/630 type against the throughflow, it becomes evident that from a certain point onwards the efficiency tends to diminish. The rate of this decline, as well as the location of the $O_{c\,max}$ point, would seem to depend to some extent on the type of dust being handled. Thus experimental research on the T 2/160 type has demonstrated that with very fine dust, where the overall collecting efficiency was about 40 per cent, this O_c value remained practically constant throughout the pressure drop interval between 600 and 2000 Pa. Some cyclones evolved at the NIIOGAZ institute provide a reasonable efficiency at pressure drops as low as 300 Pa.

The logical conclusion from all this is that, space and other considerations permitting, we should always select a cyclone size which will afford a pressure drop near the bottom end of the recommended range, say 600 to 800 Pa for $\varrho = 1$ kg m^{-3}. We must naturally take into account that at higher temperatures the pressure drop will be reduced by the lower specific gravity of the gas; but as long as the specified flow velocities can be substantially maintained, that need not necessarily affect our choice.

3.1.2.2 *Size dependence of cyclone parameters*

The key parameters of a cyclone are governed chiefly by its dimensions, i.e. by the cross-sectional areas and lengths of the individual flow channels.

The first dimension to consider is *the diameter of the cyclone,* which strongly influences the collecting efficiency. By and large, a smaller diameter makes for a higher efficiency. True, even units more than a metre in diameter can reach high efficiency levels, provided that a high resistance coefficient can be tolerated; but the whole complex of service requirements and behaviour dictates that high-efficiency cyclones should have diameters in the 200 to 600 mm bracket.

The dust discharge opening should be as small as possible, to forestall its clogging. The ratio between the discharge port diameter d_0 and the cyclone diameter D normally ranges between 0.18 and 0.40; the higher ratios are usually found on the large-diameter cyclones intended for trapping coarser dusts.

The overall height of the cyclone affects both the resistance and the collecting efficiency, both of which become more favourable as the height increases. These heights generally vary between $2D$ and $6D$. The apex angle of the cone mostly lies between 10 and 20 degrees, the smaller angles being more usual on high-efficiency units. The gas outlet tube is commonly extended down into the cylindrical chamber so as to end at a slightly lower level than the lowermost point of the inlet tube; sinking it further down than this point is apt to reduce the collecting efficiency.

Both the resistance and the collecting efficiency of a cyclone are strongly affected by the cross-sectional areas of its *inlet* and *outlet openings.* The sum of these two areas varies from some 25 per cent of the cross-sectional area of the cyclone itself, in units with peak efficiencies and high resistances, to something like 50 per cent in general-purpose designs, and to as much as 70 per cent in special low-resistance units. The way this total area is split between the inlet and the outlet depends primarily on the kind of dust which the unit is to handle. Abrasive dusts call for lower inlet velocities and hence larger flow channels at the inlet end. Non-abrasive particles are better induced at higher velocities, though in contemporary practice these seldom exceed a limit of about 18 m s^{-1}, and in some high-efficiency types the inlet velocity is deliberately kept down to roughly 6 to 8 m s^{-1}. The ratio between the inlet and outlet cross sections is mostly 1 : 1, but various current designs employ ratios ranging all the way from 1 : 0.7 to 1 : 2.

The inlet should clearly be shaped so as to deliver the dust into the closest possible vicinity of the cyclone wall. Consequently, cyclones intended for use in arrays or batteries mostly feature long narrow inlet tubes, while types designed for use on their own, as separate units, are commonly provided with inlets that admit the flow along a spiral path. The aim in both cases is to make the distance marked Δr in Fig. 43 as long as is feasible. Another common device is to taper the clean gas outlet tube, expanding it from the bottom upwards; experience with for instance the T 5 has proved that the velocity drop in this expanding tube can slightly reduce the overall resistance of the unit.

One of the major stumbling blocks in the design of cyclones is the fact that *almost every measure which enhances the efficiency will also increase the resistance, and almost every cut in the resistance also entails a loss of efficiency.* The sole exception is increasing the height of the cyclone, which curtails the resistance while actually improving the efficiency. When it comes to picking the size of the inlet and outlet, however, we are immediately faced by conflicting requirements: high efficiency calls for small openings, a low resistance for large ones. The decision will thus obviously depend on the duties for which the cyclone is intended.

3.1.2.3 The separating process in cyclones

So far we have dealt with the key parameters of cyclones – their overall collecting efficiency, pressure drop, and the particle size dependence of their fractional collecting efficiency. These data define the performance and practical potentialities of a cyclone, but tell us nothing about what actually happens inside the unit. Various theories have been advanced about the separating process in a cyclone, but all of them rest on simplifying assumptions which also involve a gross simplification of the actual particle behaviour. Consequently, these theories have at best proved valid only for one particular cyclone type, and many of them have been found to bear no relationship whatever to subsequent observations made in actual plant practice. One common feature of most of these theories is that they take into account only two forces: centrifugal force as the factor which separates the solids from the gas stream, and the entraining effect of the centripetal velocity component as the factor which causes some particles to avoid trapping. That means they neglect a host of other factors, most of which impede the separating process, such as the induction of outside air through the dust discharge opening, or turbulent diffusion within the gas stream.

Before we attempt to describe the events in a cyclone, we must first define the quantities which affect these processes. In a stream of clean gas or air, the chief of these variables will be the pressure and the flow velocity, and, if we want to examine the character of the flow too, the intensity of turbulence.

The pressure field induced by the centrifugal force in no way affects the separation of solids. In the gas or air mass, each gaseous particle is in a state of equilibrium between the pressure and centrifugal forces (a close analogy to the relationship between the pressure and velocity in a potential vortex). The centrifugal force is proportional to the weight of the particle, but no such dependence applies to the resultant of the pressures acting on the particle surface. If we now replace the gaseous particle by a solid one (which usually has a specific gravity something like three orders of magnitude greater), the centrifugal force will grow in proportion, but the resultant of the pressures will remain unchanged. Which means that in considerations of the separating process, we are fully justified in neglecting the effect of the pressure gradient that acts from the high-pressure zones at the cyclone walls towards the low-pressure zone around the centre line of the unit.

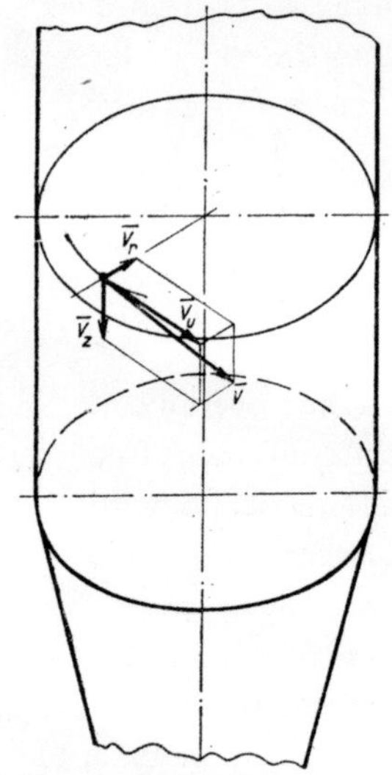

Fig. 46 The flow velocity vectors in a cyclone.

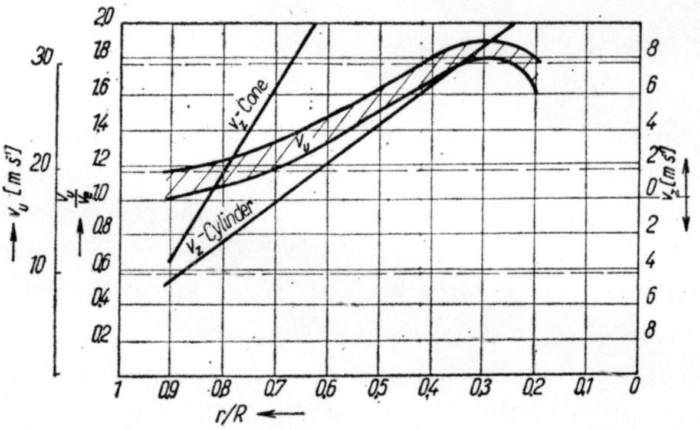

Fig. 47 Dependence of the circumferential and axial flow velocities upon the geometry of the cyclone.

The velocity field is best investigated in terms of the velocity vectors, as indicated in Fig. 46. That enables us to isolate the circumferential, axial and radial velocity components. Each of these components varies along the cross-sectional radius of the cyclone, in a way that differs from one cross section (or vertical level) to the next. Still, a closer study of these variations reveals a distinct pattern that is common to most of these vertical levels. The circumferential velocity v_u at first increases with the distance from the cyclone wall, reaches a maximum at the circumference of the vortex core, and then drops off abruptly. The axial component v_z differs in direction along the cyclone radius, since there is generally a downwards motion at the walls and an upwards one near the centre line of the unit. The centripetal velocity v_r is difficult to determine, as its absolute values are small enough to be easily lost among the random velocity fluctuations. However, it seems safe to assume that the level where v_r attains its maximum is just beneath the bottom end of the gas outlet tube; that along the cyclone walls the v_r component is nil; and that the peak v_r values, in the cross-sectional plane, will occur near the surface of an imaginary downward extension of the gas outlet tube. Note that this need not necessarily hold true in the lower part of the conical section, where the flow lines are distorted by the converging walls, and where it is usually difficult to discern any sort of pattern in the random flow.

Measurements on a cyclone of 315 mm diameter have shown that, in the cylindrical part of the unit, the v_u gradient can be adequately described by the expression

$$v_u r^{\varkappa} = C \tag{77}$$

where, for $Q = 1200\ \mathrm{m^3\ h^{-1}}$, the average values of the constants are $\varkappa = 0.5$ and $C = 150$. However, this work has also demonstrated that the gradient is largely specific to the vertical level at which it is examined, and that in the vicinity of the dust discharge opening the whole pattern is deformed beyond recognition.

The v_z component varies along the cyclone radius r almost linearly, but the slope of this dependence alters from one vertical level to the next. One interesting finding was that the $v_z = 0$ points were located, almost exactly, on the surfaces of an imaginary cylinder and cone. In other words, this $v_z = 0$ zone would seem to form a continuous boundary dividing the internal space of the cyclone into an inner and an outer region, as indicated in Fig. 47. In the outer region, next to the walls, the flow descends; in the inner region, around the vertical centre line of the cyclone, it rises. Since the same v_z profile was detected right down to the dust discharge opening, we must assume that part of the downward flow along the walls actually enters the hopper at the bottom of the unit; which obviously implies that an equal amount of air or gas must be re-induced from the hopper into the cyclone. Moreover, since the ascending flow is narrower than the descending one, the velocity of this re-induced flow must actually be higher than that of the downward one, which raises alarming implications of substantial re-entrainment.

Let us now turn to the centrifugal force field set up by inertial effects as the particles follow a curved trajectory. Odd as this might seem at first sight, the centrifugal force cannot be classed as an external force acting on the particle. The only mechanical forces which the environment can exert upon a particle are those generated by friction and pressure. In the basic equation of motion, $\Sigma P = ma$, centrifugal force has no place among the externally applied forces; only under certain simplifying assumptions can its resultant appear, on the right-hand side of this equation, as an accelerating force.

Fig. 47 gives some indication of the sort of circumferential velocities which are commonly encountered in cyclones. Once we know both this velocity v_u and the dimensions of the cyclone, we can work out the centrifugal acceleration b_v as

$$b_v = \frac{v_u^2}{r} \tag{78}$$

The result is usually in the range of several hundred to several thousand g, and proves that plain gravity, or the free fall of the particles, cannot possibly play any role in a cyclone.

Another conclusion to be drawn from the above is that the collecting efficiency of a cyclone does not depend on the centrifugal force alone. If that were so, we could achieve an almost unlimited rise in efficiency simply by reducing the cyclone diameter so as to increase the centrifugal force. It is perfectly true that smaller-diameter cyclones are as a rule more efficient than larger ones, but experience shows that the efficiency gains are only rarely proportional to the centrifugal force increment. Obviously, then, a reduction of the cyclone diameter must also produce other effects which are detrimental to the collecting efficiency.

The chief phenomenon that impairs the collecting efficiency appears to be turbulent diffusion. A turbulent flow is known to effect some mass transfer between the individual streamlines, so that the particle trajectories in a tubing will not be linear, and those in a cyclone will not follow a smooth, regular curve. The basic particle velocity has a fluctuating velocity component superimposed on it. The latter varies both in magnitude and direction, but in cyclones generally attains some 5 to 10 per cent of the basic velocity (or even more near the walls and around the discharge opening). The particles most susceptible to these velocity fluctuations are naturally those with the lowest inertia, i.e. those in the under 5 microns bracket. To illustrate the scale of these effects: in the atmosphere, the free falling velocity of a particle with $a = 5$ microns, at $\varrho_m = 1000\ \mathrm{kg\ m^{-3}}$, is only $0.75\ \mathrm{mm\ s^{-1}}$; yet measurements in a cyclone have revealed transversal velocity fluctuations, along the cross-sectional radius, as high as $2\ \mathrm{m\ s^{-1}}$. The centrifugal acceleration at the point of measurement was roughly 500 g, and therefore produced a particle velocity equal to 500 times the free falling velocity, or about $0.375\ \mathrm{m\ s^{-1}}$, which is just about an order of magnitude less than the fluctuating velocity component. And the smaller the particle, the more unfavourable will this relation become, with obvious consequences for the separating process.

The situation can be summed up as follows: the flow in a cyclone is highly turbulent, and nothing outside the cyclone itself can alter this picture. Flow deflecting baffles or vanes could reduce the turbulence intensity to some extent, but apparently not enough to make much difference to the collecting efficiency. Besides, they are not really viable in routine practice, at least for the time being, both because of the extra costs involved in their production and installation, and because they might easily jeopardize the reliability of the unit under service conditions. Turbulence is also evidently one of the factors that limit the size of the particles which the cyclone can still trap, and designers will do well to take this fact into account.

Once we abandon the flow of clean gas or air, and consider a flow containing solid admixtures, the problem becomes even more complicated. For a start, we must always count on the dust being polydisperse, and must therefore investigate events in a cyclone along two different lines: part of the dust entering the unit will generally be coarse enough to be trapped almost in its entirety, while the rest of the dust will be finer, and will be retained less efficiently.

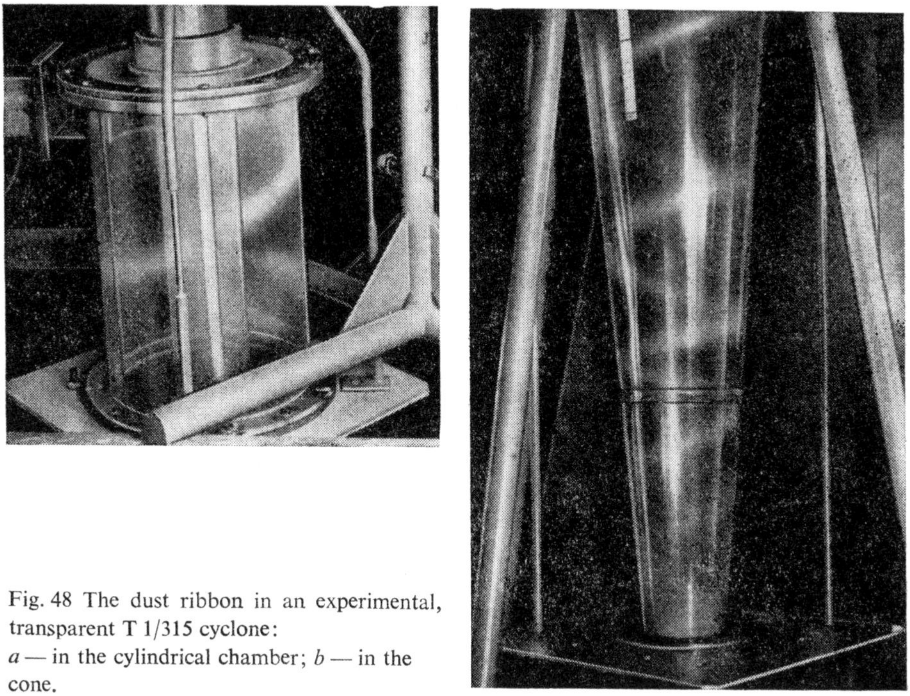

Fig. 48 The dust ribbon in an experimental, transparent T 1/315 cyclone:
a — in the cylindrical chamber; *b* — in the cone.

The coarse particles are separated out from the gas stream practically as soon as they enter the cyclone. They then form what looks like a continuous ribbon, and descend in a spiral along the walls towards the discharge opening. Their actual paths are determined by the gas streamlines in this zone. The direction of the spiral is controlled mainly by the shape of the inlet opening. In some cases the ribbon forms in the immediate vicinity of the inlet opening, and then descends at a constant helix angle. In others the particles at first descend very gently, and form into a continuous ribbon only near the boundary between the cylindrical and conical sections of the cyclone. Fig. 48a shows the ribbon in the cylindrical and

Fig. 48b in the conical section of a T 1 cyclone of 315 mm diameter, which was built for experimental purposes; these pictures were taken when the cyclone was fed with paper discs of 1.5 mm diameter, punched from a teleprinter tape. From the practical point of view, the most important aspect of this ribbon of coarse particles is the abrasion it causes on the equipment walls. Some designers, such as van Tongeren, have countered this effect by fitting the cylindrical wall with a helical channel that guides the ribbon along a predetermined route and thereby saves wear on the rest of the cyclone surfaces.

The behaviour of finer particles is quite different. These particles, which are normally trapped with an efficiency of say 50 to 70 per cent, are affected both by the centrifugal force and by turbulent diffusion. Fines less than about 5 microns in size are so prone to diffusion that they are apt to be spread almost uniformly throughout the volume of the carrier gas. As far as they are trapped at all, they seem to avoid contact with the cyclone walls, and tend to be blown out of the discharge opening into the hopper. This view was corroborated by spot concentration measurements on the trial cyclone of 315 mm diameter. In this work, fly ashes with the usual grain size distribution yielded values which conformed to the conventional concept of a marked concentration growth at the cyclone walls: the dust concentration in this layer was about twenty times that in the incoming gas. However, dust samples which had undergone previous screening failed to comply with the classical dogmas. A dust which spanned only the 12 to 40 microns size range produced inconsistent concentration levels which differed widely from any theoretical forecast, and a dust consisting entirely of under 10 micron fractions formed local concentrations which simply would not fit into any smooth and continuous curves.

That leaves us with the following main conclusions:

1. There are marked differences between the behaviour of coarse and of fine particles. The coarser fractions follow a helical path, of a shape governed by the gas streamlines along their route. The finer particles tend to diffuse throughout the internal space of the cyclone, and will be difficult to trap efficiently unless we can manage to harness such supporting factors as for instance particle agglomeration.

2. The collecting efficiency attained by the best of present-day cyclones (an O_f 100 % at a = 10 microns) is no longer likely to be surpassed. This means that future design and development work should be aimed at reducing the pressure drop and extending the life expectancy of cyclones with this efficiency level, rather than at any further improvements in efficiency.

3. There is a considerable scope for design modifications which will make some cyclone types more reliable. Abrasive wear and clogging can both be strongly influenced by control of the axial velocity component, which in turn is largely determined by the shape of the inlet section.

4. Investigations of particle behaviour in a cyclone are no easy matter. Apart from the cyclone itself, the carrier gas, and its flow pattern, there are many other variables which affect the outcome, and a good many of them so far defy all attempts at a theoretical definition. The chief of these elusive factors are the agglomeration of individual particles into clusters, sometimes complicated by the subsequent decomposition of these clusters, and the comminution of large particles within a cyclone. These effects often radically alter the grain size distribution ascertained by sampling upstream of the unit, and leave us in the dark about the size distribution of the particles actually processed in it. Other factors which tend to confuse the issue include the induction of outside air, the ingress of moisture, the chemical activity of some of the dusts and gases, etc.

Even this short list shows how complex a picture is created by the numerous interacting factors which all affect the separating process in a cyclone. The outcome is influenced by the shape of the unit and the resultant flow streamlines; the character and turbulence of the flow; the various properties of the dust, such as its tendencies to adhere, to agglomerate, to disintegrate, etc.; and by far too many other variables for any one formula or theory to encompass them all.

3.1.2.4 Cyclone types and cyclone arrays

Over the past few years, the ČSVZ works of Czechoslovakia have evolved a standard range of cyclone types designated by the prefix T. These units, available in diameters ranging from 160 to 1600 mm, are schematically outlined in Fig. 42. Table 14 lists their diameters and principal parameters, including the fractional and overall collecting efficiencies for standard reference dusts. This range is supplemented by the British Davidson R type, which is intended to serve as a coarse pre-cleaning facility upstream of electrostatic precipitators, and by three cyclones which are essentially just dimensionally modified versions of the Soviet NIIOGAZ N 400 types. The three versions differ from each other by their inlet angles (marked α in Fig. 42), which are 11, 15 and 24 degrees respectively, and by the size of their inlet openings.

This range of types can be classified by their collecting efficiencies into three broad categories: class 1, the high-efficiency types; class 2, of medium efficiency; and class 3, the low-efficiency category, which spans by far the widest range of efficiency levels of any of them. With this in mind, we can briefly review the individual types as follows:

T 1/160: A class 1 unit with the highest collecting efficiency for fines, but with so high a resistance coefficient that it is hardly ever used in multiple cyclone arrays.

T 2/160: A class 1 unit with an acceptable resistance level, which is commonly incorporated in SAA and SBA cyclone batteries. It is particularly

TABLE 14

Technical data of some current Czechoslovak cyclones

Type	Diameter (mm)	Flow rate Q* ($m^3 s^{-1}$)	V_D** ($m s^{-1}$)	ξ_D	Fractional efficiency O_f† (microns)	Overall efficiency O_c (%)†† on standard dusts			
						R (%)	G I (%)	G II (%)	G III (%)
T 1	160	0.024	1.21	807	0.4—22	99.0	97.9	95.9	93.8
T 2	160	0.042	2.13	255	0.4—18	98.9	97.8	94.5	92.7
	315	0.164	2.15	255	0.4—25	97.7	95.3	93.0	87.8
	475	0.375	2.13	255	0.4—28	97.5	94.1	91.9	85.0
T 3	315	0.255	3.28	109	0.5—28	97.3	94.1	89.3	83.7
	630	0.875	2.80	150	0.5—28	96.8	92.8	88.4	82.4
	1000	2.210	2.82	150	0.8—30	95.7	91.0	84.5	78.2
	1600	5.650	2.81	150	0.9—40	93.0	87.4	79.8	72.6
T 4	630	0.570	1.83	353	0.5—10	99.1	97.3	94.9	91.5
T 5	300	0.163	2.30	226	0.5—25	97.2	95.5	91.9	87.7
V 1	315	0.340	4.40	61	0.9—33	95.0	90.5	83.2	76.5
300 V	300	0.277	3.89	78	0.5—28	95.7	91.5	86.5	80.3
400 N (11°)	400	0.321	2.55	190	0.5—27	97.5	94.3	90.2	85.5
400 N (15°)	400	0.409	3.24	115	0.5—30	97.2	93.9	89.5	84.5
400 N (24°)	400	0.538	4.25	65	0.5—40	95.7	91.0	86.0	79.7
Davidson R	308	0.890	10.43	11	2.0—150	84.5	73.3	65.1	53.4

*Flow rates Q apply at $\Delta p = 600$ Pa and $\varrho = 1.0$ kg m^{-3}.
**The V_D figures are for a pressure drop of $\Delta p = 600$ Pa.
†These ranges indicate the initial ($O_f = 0$) and final ($O_f = 100$ %) points of the O_f curves for $\Delta p = 600$ Pa at 20 °C.
††At $\Delta p = 600$ Pa, $t_3 = 20$ °C, and $\varrho = 1.0$ kg m^{-3}.

suitable for dry, non-adhesive dusts, but is not recommended for sticky or extremely fine dusts, nor for the second stage of cleaning installations, where it is liable to clog up.

T 2/315: A class 2 unit, with much the same resistance coefficient as the types of 160 mm diameter, employed in the SDA batteries.

T 2/475: Another class 2 unit, used in type SCA batteries.

T 3/315: A class 3 cyclone with a relatively low resistance coefficient, sometimes utilized in SDB multiple cyclone arrays.

T 3/630: A class 3 type which, like the T 3/1000 and T 3/1600, is often grouped into type SEA batteries.

T 4/630: This class 1 unit is the most selective of all the types listed here, having an O_f curve that attains 100 per cent in the vicinity of 10 microns, but also has a fairly high resistance coefficient. It is used either singly or in type SEB or SGA arrays; can handle virtually any kind of dust; but is too tall to be geometrically scaled up any further.

VI/315: A class 3 unit with a low resistance, generally grouped into type SDC arrays.

300 V: This class 3 unit has a tangential inlet, but differs from the T range by the helical design of its lid; it is mostly incorporated in BMM arrays.

400 N (11°): A type that falls between classes 1 and 2; it is employed in SKA and, less frequently, in BMM arrays.

400 N (15°): Differing from the 11° version by its inlet angle and by belonging squarely to class 2, this unit is also used in the SKA and, to a lesser extent, in the BMM arrays.

400 N (24°): A class 3 unit with the same fields of application as the other two 400 N types; all three of these types have helical lids, and, when used in arrays, are mostly mounted in an inclined position.

R: A licence-produced Davidson type with an extremely low resistance and a relatively low collecting efficiency, intended only as a coarse pre-cleaning stage. This unit is produced in a single version only, is made of cast iron, stands out by its abrasion resistance and small size, and is very easy to group into arrays.

T 5/300: Somewhere between the T 2/160 and the T 2/315 in its efficiency rating, this type has a rather lower resistance coefficient than any of the T 2 units.

T 6/300: This type is slightly less efficient than the 400 N (15°) unit, but also has a lower resistance coefficient, with $\xi_D = 70$.

The T 5 and T 6 types share the same inlet and outlet dimensions and spacings, and are therefore interchangeable in one and the same casing.

Cyclone batteries can be divided into loose arrays or clusters of cyclones, and true multiple cyclone installations. The former share common dust removal facilities, but the gas inlets and outlets may be either branched to each cyclone separately, or led in and out through plenum chambers. In the multiple installations, the whole array of cyclones is enclosed in a common housing.

The cyclones of an array are mostly arranged either in a circular pattern or in rows. The circular arrays of Czechoslovak manufacture are the SEA and SEB

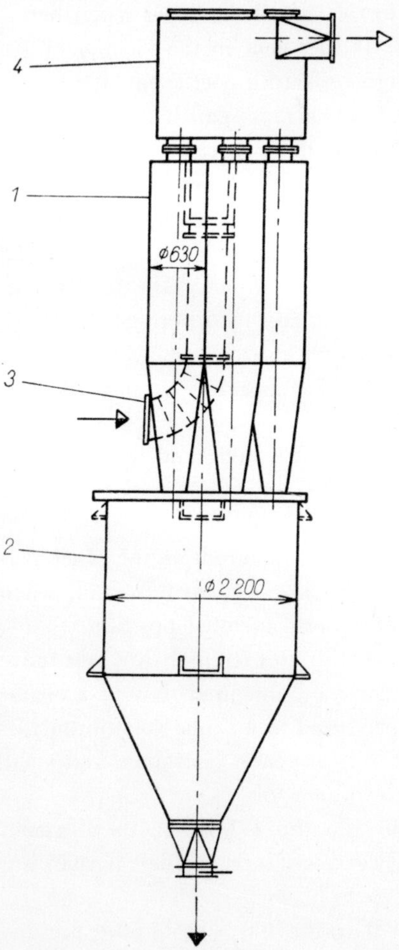

Fig. 49 A type SEB separator fitted with T 4 cyclones:
1 — cyclone; *2* — hopper; *3* — inlet stub; *4* — outlet plenum chamber.

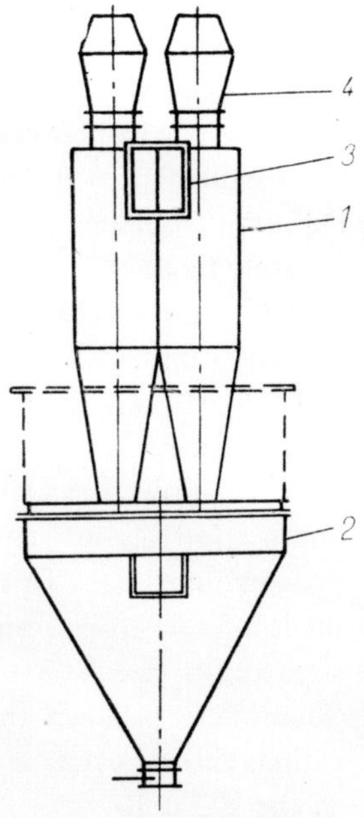

Fig. 50 An SEA separator with type T 3 cyclones:
1 — cyclone; *2* — hopper; *3* — inlet duct; *4* — gas outlet manifold.

types, illustrated in Fig. 49, and the less widely used SCA units. The SEA system mostly comprises either one or two T 3 cyclones of 630, 1000 or 1600 mm diameter; is used for separating only coarser admixtures; and is usually installed downstream of a fan, so as to exhaust into the atmosphere, as indicated in Fig. 50. The SEB system can consist of up to twelve cyclones provided with one common hopper, and is always fitted with the T 4/630 high-efficiency units; it can handle up to 30,000 m^3 per hour, and has been widely adopted for trapping sawdust,

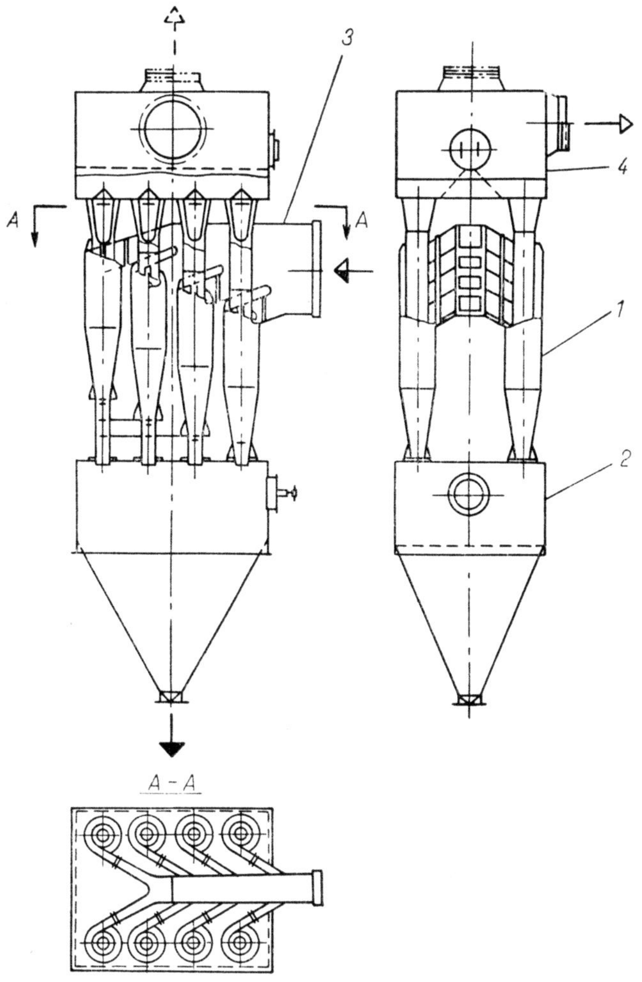

Fig. 51 Type NIIOGAZ cyclones in an in-line array:
1 — cyclone; *2* — hopper; *3* — inlet plenum chamber; *4* — collector main.

wood shavings, and other materials amenable to high-rate mechanical separation. The SCA system, shown in Fig. 53, is made up of one to twelve T 2/475 cyclones, can deal with throughflows of up to 1600 m^3 an hour, and is used mainly in the chemical industries. Other arrays in this general category are equipped with the NIIOGAZ 400 N cyclones, in circular or in-line configurations, as exemplified in Figs. 51 and 52.

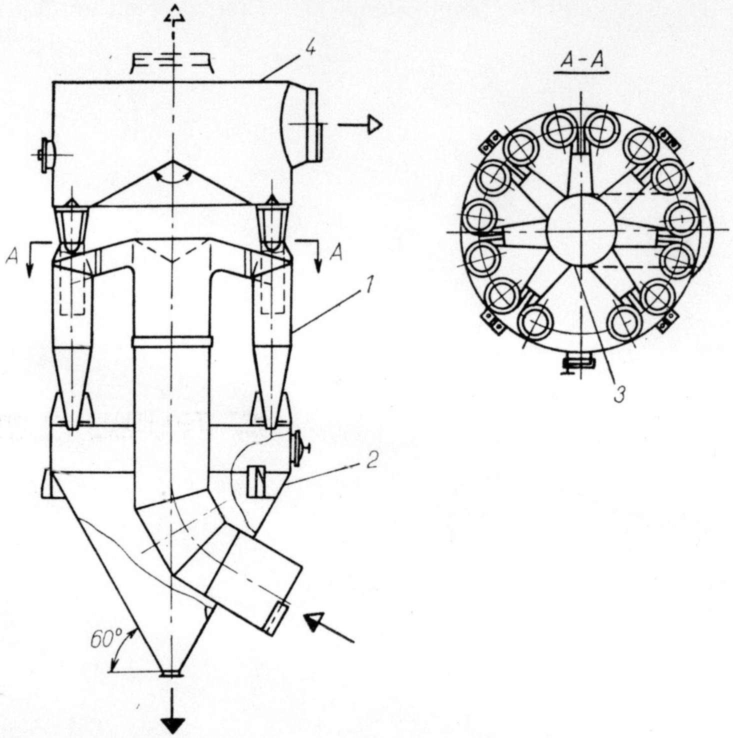

Fig. 52 Type NIIOGAZ cyclones in a circular configuration:
1 — cyclone; *2* — hopper; *3* — inlet plenum chamber; *4* — outlet plenum chamber.

The various arrays differ mainly by their inlet arrangements. The polluted gas may enter either through a plenum chamber, or through individual inlets to each of the cyclones, as in the SEA system in Fig. 50. They further differ by their circular or in-line layouts: the former yields a more convenient hopper configuration, while the latter simplifies the design of the inlet and outlet manifolds. Regardless of these differences, however, the resultant cluster is always difficult to expand by the addition of further cyclones, and also tends to present thermal insulation problems. That is why multiple cyclone units have been devised to overcome these obstacles.

In a multiple cyclone unit, the individual cyclones are linked in parallel, and housed within a common casing that forms the inlet chamber. The dust discharge and gas outlet openings of the cyclones must naturally be joined, to the hopper and to the exhaust ducting (or outlet plenum chamber) respectively, by gas-tight seals. Typical examples of such units are the BMM and SKA systems, shown in

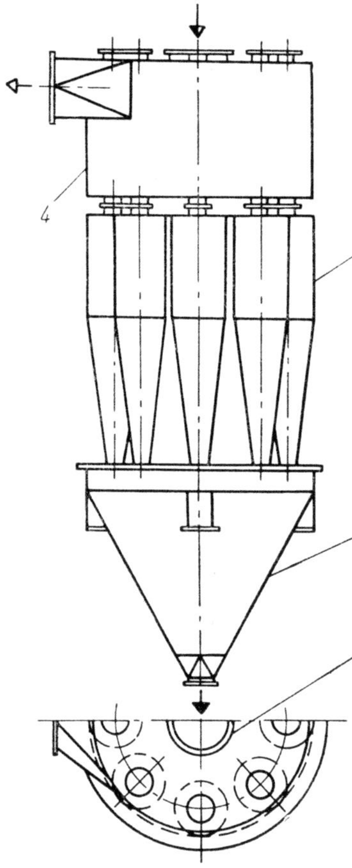

Fig. 53 An SCA separator fitted with T 2 cyclones:
1 — cyclone; *2* — hopper; *3* — inlet flange; *4* — outlet plenum chamber.

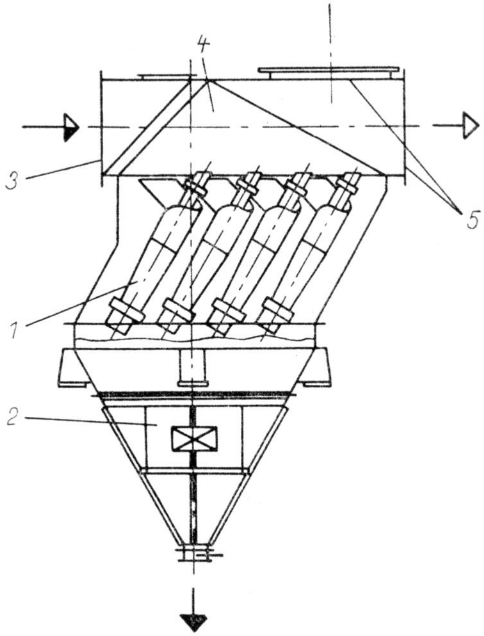

Fig. 54 A BMM separator with type 300 V cyclones:
1 — cyclone; *2* — hopper; *3* — inlet flange; *4* — outlet compartment; *5* — outlet flange.

Figs. 54 and 55 respectively. A roughly similar design, the Lurgi Multiklon, seen in Fig. 56, is notable for having the whole of the cyclone cones set into the hopper; each cyclone consists of two separate parts, the cylinder and the cone, and is sealed off from the hopper space by a sand bed. In all the systems mentioned above, the dust is trapped in the cyclones only. An alternative approach is to provide the inlet casing with dust removal facilities too, and capture some of the dust in this compartment before the gas is admitted to the cyclones proper. Two systems which incorporate such a pre-cleaning arrangement are the SGA (Fig. 57) and SDC (Fig. 58).

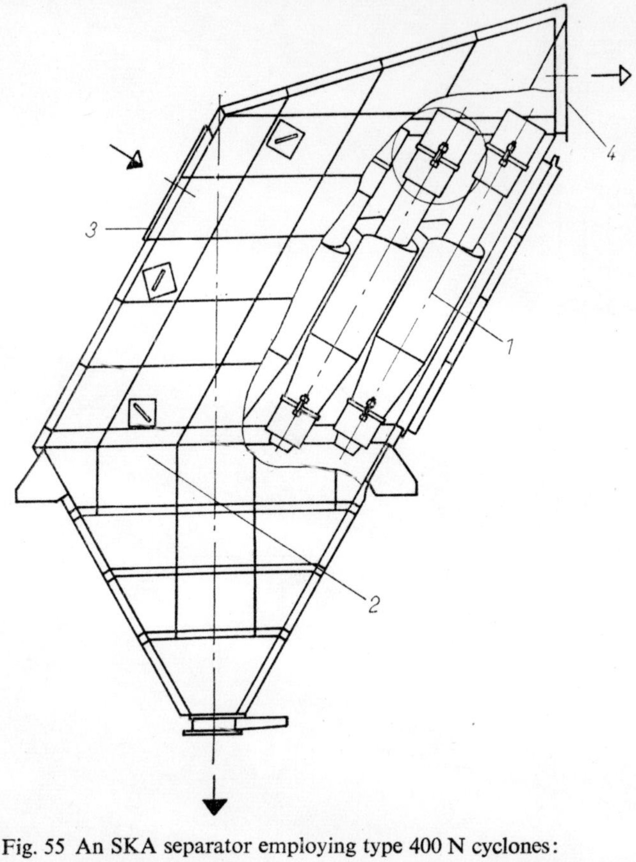

Fig. 55 An SKA separator employing type 400 N cyclones:
1 — cyclone; *2* — hopper; *3* — inlet flange; *4* — outlet flange.

The BMM and SKA systems are used predominantly in power stations, as pre-cleaning facilities ahead of electrostatic precipitators. They are at present also beginning to be installed, as the sole items of dust trapping equipment, at small boilerhouses. They are further employed downstream of high-temperature equipment, for instance in the production of cement and magnesite, where it would be difficult either to find any other heat-resistant units with a comparable efficiency, or to fit larger dust trapping installations into the limited space available for them.

The SBA system stands out by its very small floor space requirements. Fig. 59 shows such a system with 120 cyclones of the T 2/160 type; in some existing installations, four such arrays have been vertically superimposed in a single compact stack of 480 cyclones. However, these arrays of small-diameter cyclones

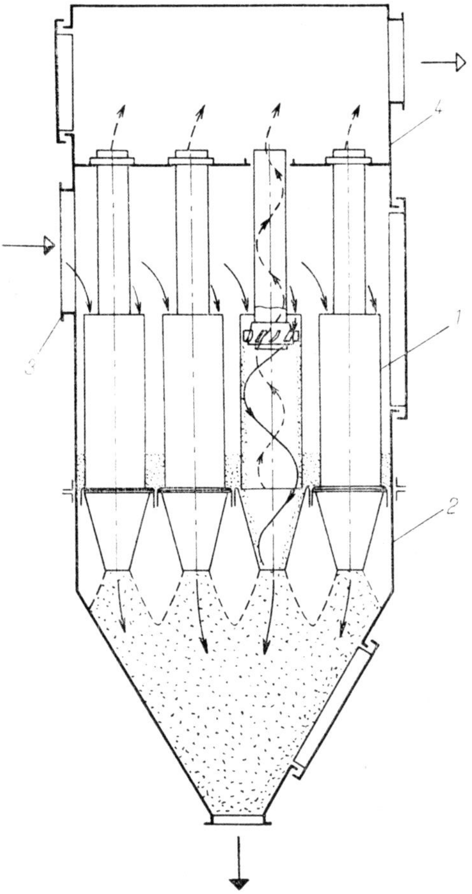

Fig. 56 The "Multiklon" made by Lurgi of West Germany:
1 — cyclone; *2* — hopper; *3* — inlet flange; *4* — outlet chamber.

mounted in an inclined position are very prone to clogging, which means that they can reliably handle only coarse and dry dusts.

The SHA battery in Fig. 44 is made up of horizontally installed Davidson R cyclones, and is often exploited as a primary cleaning stage ahead of electrostatic precipitators. In view of its inherently low resistance, it can accept throughflows as high as 500,000 m^3 per hour. One of the most successful batteries so far evolved is the SGA type, illustrated in Fig. 57, which consists of T 4/630 cyclones. It combines a high efficiency with a relatively long service life of its components, and is available in sizes capable of dealing with 20,000 to 100,000 m^3 an hour.

Multiple cyclone installations fall into two main classes which differ by their dust removal provisions. In one class, two separate hoppers are provided for the cyclones and for the casing respectively, while in the other a single hopper serves the whole of the unit. The former category is exemplified by the SD and the latter by the BMM system. The cyclones can be installed horizontally, as in the SHA type, or in a slanting position, like those of the BMM system, or vertically, as

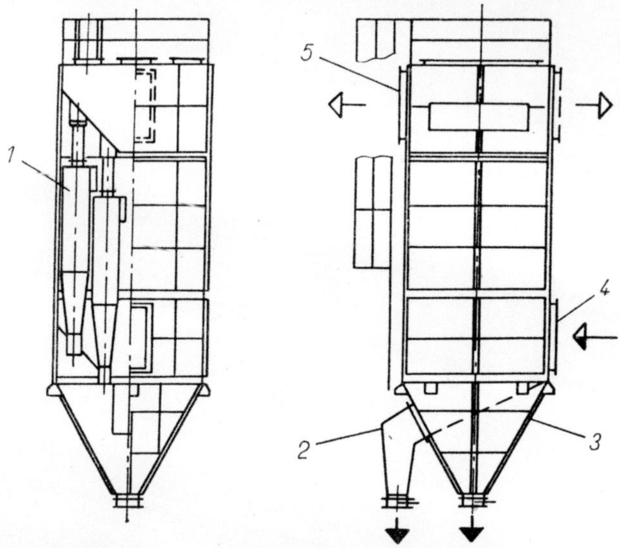

Fig. 57 An SGA separator incorporating T 4 cyclones:
1 — cyclone; *2* — chamber discharge hopper; *3* — cyclone discharge hopper; *4* — inlet flange; *5* — outlet flange.

in the SDC units. They are more often than not all mounted in one and the same horizontal plane, but can also be stacked vertically, as in the SBA type. Finally, there is a vital difference in the various ways in which the structural design has provided for future maintenance work. Some of the all-welded structures, like those of the SBA systems, do not allow any replacement of the individual cyclones. Others, like the SD types, though also fully welded, permit the replacement of the complete cyclone assembly only. In arrays such as the SKA or BMM, every cyclone can be replaced individually.

Large cyclone batteries are nowadays not nearly as common a piece of plant equipment as they used to be some years ago. The decline in their popularity is due partly to the ever more stringent requirements on air pollution control, partly to the gradual evolution of other kinds of separating equipment. Especially electrostatic precipitators and cloth filters have largely supplanted big arrays of cyclones as the main dust trapping facilities at power stations and in the production of building materials, and are now increasingly able to dispense without any mechanical pre-cleaning stages. Which means that some cyclone types are likely to drop

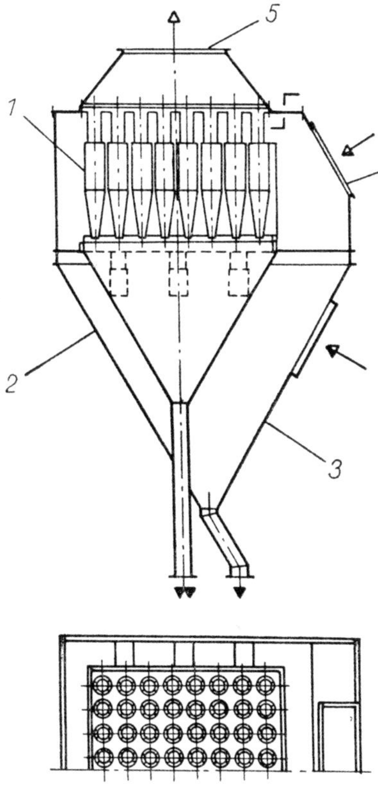

Fig. 58 An SDC separator equipped with V 1 cyclones:
1 — cyclone; *2* — casing hopper; *3* — cyclone set hopper; *4* — inlet flange; *5* — outlet flange.

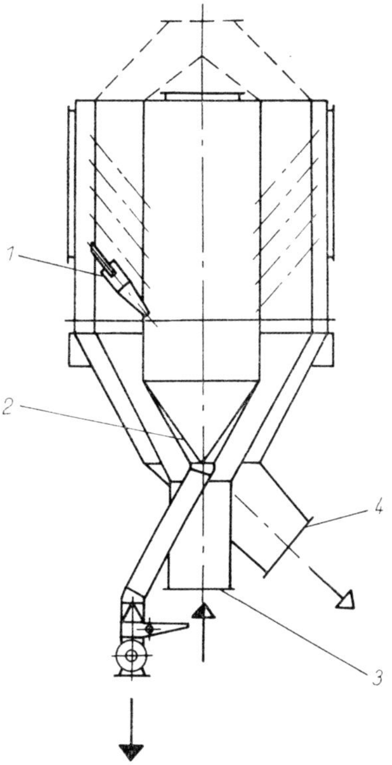

Fig. 59 An SBA separator provided with T 2 cyclones:
1 — cyclone; *2* — hopper; *3* — inlet flange; *4* — outlet flange.

out of use not on account of any inherent shortcoming, but simply because their traditional place in the arsenal of dust trapping equipment will have been taken up by other, more recently evolved devices.

Even the consistent improvement in the performance of the cyclones themselves has in some respects narrowed the scope of their application. It used to be fairly common practice to arrange two sets of cyclones in tandem, for a two-stage gas cleaning procedure. With present-day high-efficiency cyclones, however, this would almost inevitably result in the clogging of the second stage, especially if it consisted of small-diameter cyclones. The first stage would tend to screen out and retain the coarse fractions, so that the second stage would receive only the finest particles, which would promptly clog it up. Two cyclone stages in succession were all very well with the earlier low-efficiency units, which failed to capture the fine adhesive fractions in the minus 10 to minus 15 micron brackets, but are no longer a practical proposition on newly installed equipment. Yet high throughflow rates and abrasive admixtures do call for a two-stage process, if only to save wear on the cyclones. That is the whole object of multiple cyclone installations with inlet plenum chambers designed to retain the coarsest fractions.

Cyclones for some time held their own in two-stage cleaning techniques where the second stage was formed by an electrostatic precipitator, but even this arrangement is becoming less prominent in contemporary plant designs. The precipitators can now accept much higher inlet concentrations than formerly, and also need much less shutdown time for maintenance. Consequently, they can now very often get along without any pre-cleaning facility, and where one is necessary, it usually needs nothing like the level of efficiency offered by the best of present-day cyclones.

3.1.2.5 Selection of a cyclone type

Picking the right cyclone for any given application would seem to be a perfectly straightforward matter: the requisite collecting efficiency governs our choice of design configuration, and the acceptable pressure drop and desired service life between them determine the size of the units. Unfortunately, the decision is complicated by a variety of factors which tend to impair the performance of the cyclones, or reduce the amount of service we can get out of them. The reliability of a cyclone under service conditions can be regarded simply as its degree of resistance to these adverse influences. As a rough rule of thumb, these detrimental factors can be grouped into three broad categories:

1. Mechanical defects of the cyclone or its ancillaries, such as leakage through welds or manholes, failures of rotary outlet gates or seals, etc.
2. Clogging of the unit by dust deposits.
3. Excessive wear, usually by abrasion.

The mechanical defects can mostly be traced down to negligence or bad workmanship in the design, manufacture or installation of the unit. Their consequences can generally be remedied or at least mitigated by relatively simple measures, though these are sometimes costlier than their simplicity would suggest. As always, the preferable course is to prevent the incidence of any such failures. That is not so much a matter of further technical progress, as is so often assumed, but rather a matter of adhering to the well-established rules at every stage in the design, fabrication and erection procedures.

Clogging can be caused by a bewildering variety of factors, but on closer examination these break down into three major groups:

a) Mechanical defects (category 1 above) can either cause the cyclones to fill up with trapped matter (since evacuation of the hoppers need not always empty the cyclones too), or else lead to the ingress of moisture. This cause of clogging can usually be cured without any drastic changes in either the equipment or the operating procedures.

b) The chemistry of the dust can cause it to bond or cement. The normal countermeasures are improved thermal insulation, stricter observance of the specified operating conditions, and meticulous maintenance of the equipment. Even at the best, however, the trouble will be much more difficult to rectify than in case a), and of course there are some types of cyclones which cannot be cleaned out at all.

c) The physical properties of the dust, particularly its grain size distribution, may be such that some degree of clogging is inevitable. This is especially common in two-stage cleaning processes with a cyclone secondary stage, and in other such instances when high-efficiency cyclones have to deal with dust of the minus 10 to minus 15 micron fractions. This type of clogging is the most difficult of all to cure; in some cases, there is nothing for it but to replace the cyclones. The only possible remedies are to modify the processes upstream of the gas cleaning plant so as to gain a coarser dust, or else to arrange for partial or complete by-passing of the primary cleaning stage.

Excessive abrasion will ultimately wear a hole in the equipment. A hole in the hopper or in the battery casing is easy enough to repair, by fixing a sheet of metal over it, but holes in the cyclones themselves are much more of a problem. They usually occur in either of two critical zones – in the cylindrical part just beyond the tangential inlet opening, and in the conical part at its very bottom. The trouble is that no two cyclone types ever behave quite alike under wear-inducing conditions, so that the point where the puncture is likely to appear, and the running time likely to elapse before it does appear, can only be established by experiments. The least troublesome case is a hole in the lower part of the conical section when that part of the cone is recessed into the hopper; this need have no substantial effect on the functioning of the unit. A special case are multiple cyclone installations, where any hole in the inlet chamber will disrupt the whole flow pattern in the cyclones and reduce their collecting efficiency. When that happens, there is no point in replacing only those cyclones which have actually been punctured, because even the cyclones which still seem sound will probably have been worn to the point where punctures in them are imminent.

Too many people are inclined to attribute excessive wear only to the properties of the dust, overlooking the fact that the properties of the cyclone itself may also be causing, or at least contributing to, the high wear rate. A typical case is a wrongly selected cyclone size, which leads to an excessive flow velocity within the unit.

Over the past few years, the abrasive wear of cyclones has been extensively investigated both by laboratory research and on equipment in actual plant service. The main findings of the laboratory work are briefly reviewed in Section 1.13 of this book.

The applications in which cyclone walls are subject to the most severe wear are found in the fields of power generation and the production of building materials. Fly ashes and similar particles will of course always cause some abrasion, but the fact that the wear rate is so uneven, and especially the fact that in extreme cases this wear can reduce the life expectancy of the equipment to less than a year, obviously merit closer investigation. The factors that influence the wear rate are numerous. The chief among them seem to be the kinetic energy of the particles; the angle of their impact against the equipment walls; the character of the gas flow (i.e. the magnitude and orientation of the flow velocity components, and the degree of turbulence); further, the hardness and shape of the particles; the average and local particle concentration levels; the geometry and diameter of the cyclone; the material of its walls; the gas temperature; the type of boiler that generates the fly ashes, and the ash content of the fuel; various chemical interactions; and a host of lesser contributory factors. Many of these factors are interrelated. For instance, the ash content of the fuel will affect the retention of fly ashes in the steam generator, and hence the particle concentration level at the separator inlet. That leaves us with the conclusion that the wear rate is affected by everything from the properties of the fuel or raw material, and the technique employed for its combustion or comminution or other processing, down to the selection of the cyclone type and size.

Not all these factors influence the wear rate equally, so that some of them afford excellent and others only poor prospects for reducing the overall wear rate. Moreover, in many cases there are so far no quantitative data on the way and degree in which these variables affect the wear rate, and some of these factors, such as the type of fuel used at a power station, will usually be outside our powers to alter anyway.

If wear of the separator walls is unavoidable, there are at least several ways of limiting it, both in existing equipment and, more particularly, in installations which are still in the design stage. The most obvious measure is to provide the critical zones of the cyclone with thicker walls. That will extend the service life of the unit roughly in proportion to the increase in the wall thickness.

The actual service life of a cyclone depends mainly on the flow rate, i.e. on the velocity at which the dust particles enter the unit. The effect of the concentration level is usually less pronounced. Of course the dust concentrations vary within very wide limits, and extremely high concentrations can affect the outcome almost as much as a high flow velocity, but then it is mostly beyond our powers to control the incoming dust concentration. Consequently, the variable by which we can control the wear rate most effectively is the volumetric flow rate, in other words the inlet velocity of the dust-laden gas. In existing installations, there are essentially two ways of reducing this velocity:

a) The cyclones can be replaced by some type with an inherently lower inlet velocity. For example, in BMM arrays the usual 400 N (15°) cyclones can be replaced by the 400 N (24°) type. Fig. 60 indicates how the service life P depends

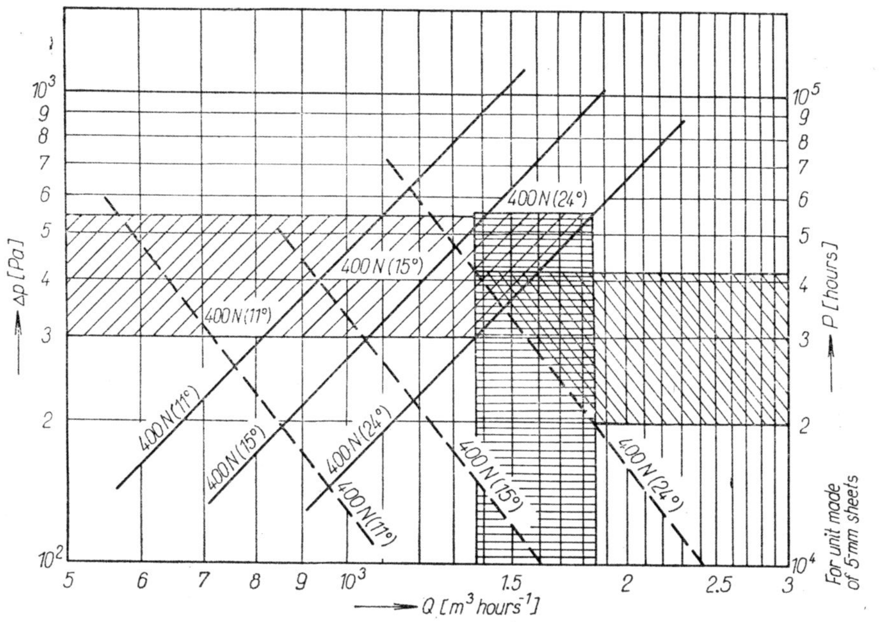

Fig. 60 Flow rate versus pressure drop and life expectancy characteristics of the three NIIOGAZ 400 N cyclones—the 11°, 15° and 24° versions.

on the volumetric flow rate Q or on the pressure drop Δp in cyclones of the 400 N range, with inlet angles of 11, 15 and 24 degrees respectively, made of sheets 5 mm thick. The diagram also shows the range of throughflows which each of these types can handle while retaining an adequate collecting efficiency (at the minimum flow rate) and a life expectancy of at least three years, or about 20,000 running hours (at the maximum flow rate of the range).

b) Part of the flow can be made to by-pass the separator either through external ducting, or by the removal of some of the cyclones from a multiple cyclone installation, or else by "shorting" each of the cyclones, for instance by a slot in its outlet branch. That will reduce the volume of dust-laden gas that enters the cyclone per unit of time, but will also leave that proportion of the gas uncleaned. Like the alternative listed under a) above, this measure inevitably lowers the overall collecting efficiency, and is acceptable only where the cyclones serve as a primary cleaning stage upstream of an electrostatic precipitator.

Fig. 61 shows how the service life P and the outlet dust concentration k_v of a type 400 N (15°) cyclone vary with the by-pass ratio $Q_0 : Q$, where Q_0 is the volume of the by-passing flow and Q the full volume of the original, non-bypassed flow. Thus at the 0.2 point of the $Q_0 : Q$ axis, 20 per cent of the original flow

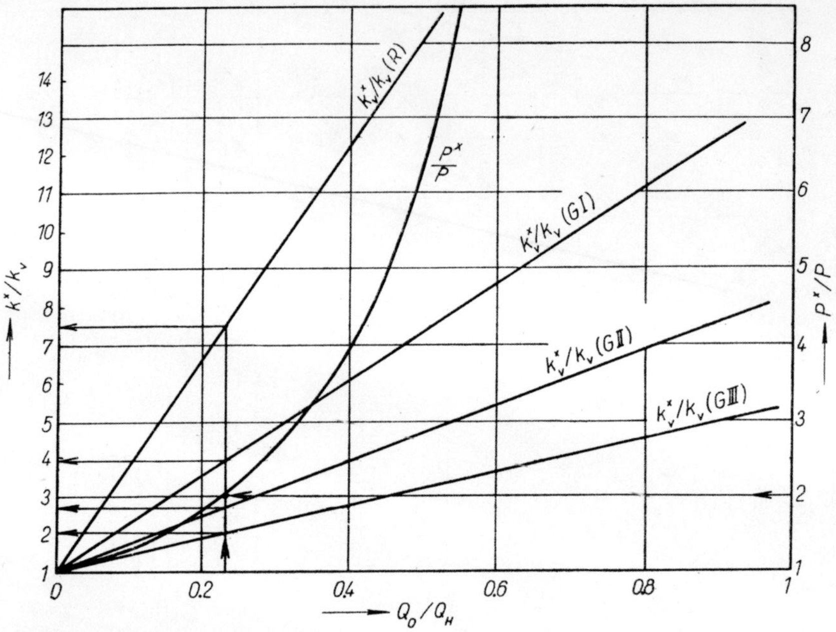

Fig. 61 An example of how the by-pass ratio affects the service life and outlet dust concentration of a separator.

is diverted to the by-pass channel and only 80 per cent of that flow still passes through the cyclone. The vertical scales of this diagram show the ratios between the outlet dust concentration k_v^x and service life P^x at the given by-pass ratio, and the initial outlet dust concentration k_v and service life P that existed before by-passing was adopted. This diagram yields some interesting items of information. For example, it indicates that the service life can be doubled, i.e. a $P^x : P$ ratio of 2 : 1 attained, for a by-pass ratio $Q_0 : Q$ of only 0.23 : 1. What this will entail in terms of increased dust emission can be read off, for four basic kinds of dust, at the relevant intersections of the $k_v^x : k_v$ lines.

Neither of the alternatives presented above is feasible when the cyclones are used on their own, with no second-stage gas cleaning facilities beyond them. In such cases, the only possible response to an insufficient service life is to replace the whole equipment with a set designed to last longer. Not all equipment selectors yet appreciate that the life expectancy is at least as important a design consideration as the collecting efficiency or power consumption of the equipment they are choosing.

3.1.3 Baffle flight separators

These units have become fairly rare in dust trapping applications, in Czechoslovakia as elsewhere, but have lately begun to be installed as thickeners ahead of electrostatic precipitators. The way they work is evident from Fig. 62. A length

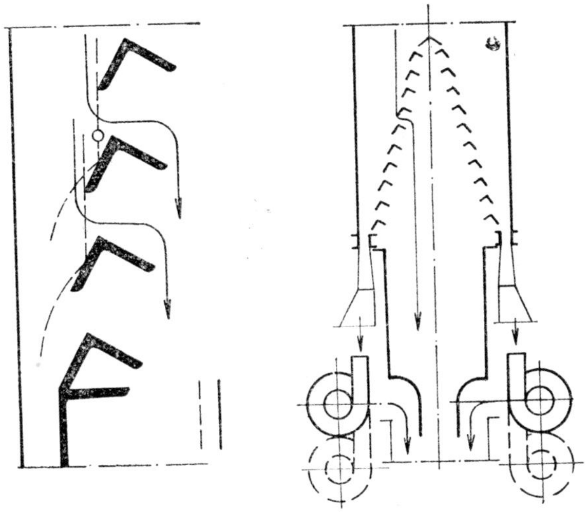

Fig. 62 Detail sketch and overall scheme of a baffle-flight separator (or flow thickener).

of ducting, usually of rectangular section, is fitted with a flight of stepped baffles or louvres, generally made of angle sections or variously shaped strips of sheet metal. These abruptly deflect the stream of dust-laden gas, causing the particles to leave the streamlines and impinge on the next baffle in their path. The particles then either bounce down the flight, or are swept along it, towards the extraction point at its end, from which part of the gas flow is diverted into the cyclones or other separators downstream of the unit. The object of the device is thus not to capture the dust, but rather to increase the particle concentration in that part of the flow which reaches the separators further downstream. The efficiency of gas cleaning plants which incorporate these units is therefore limited by that of the actual separators, and is in fact always lower than that limit, because a substantial part of the dust remains in the part of the gas stream which never reaches the cyclones or precipitators.

A point to note is that the gas delivered to the cyclones downstream of these thickeners must as a rule be forced through them, because their resistance to the flow is apt to exceed that of the baffle flight units. The requisite pressure gradient can be built up either by a fan, or by an ejector beyond the cyclones which induces the flow emitted by them.

At present, these units are employed only where high concentrations of very coarse dusts have to be pre-processed to make them more manageable by other gas cleaning techniques.

3.1.4 Rotary separators

The term "rotary separator" or "spin separator" applies to all mechanical devices in which admixture particles are separated from a gas stream by means of a rotating impeller. The only category of these devices which merits closer examination are those where the impeller is mounted on a fan shaft, and housed in the same casing as the fan. The other types are essentially just inclined rotating drums, where the particles are accelerated so as to strike the casing wall; they then slide down this wall to the dust discharge opening, while the clean gas emerges at the centre of the drum. The way particles move across the impeller of a fan-type separator has already been described in Section 2.2.2.

The units in which a fan and a separator are housed within a single casing are of two principal types. In the one, the impeller is mounted upstream of the fan, so as to clean the gas before it reaches the latter; these units are known as centrifugal separators. In the other, the impeller and fan are combined into a single component which both cleans and moves the gas stream; these units are termed fan-type separators. Fig. 63 shows a typical centrifugal separator, Fig. 64 an American fan-type design dating back to 1932.

In Czechoslovakia, rotary separators are not in routine production, and are built only to cater for special cases. It is easy enough to prove theoretically that these units subject the admixture particles to separating forces much greater than those attainable in cyclones, but hardly anything is known about the negative factors that come into play in rotary separators. Nor is it likely that very much will be discovered in the near future, because it would be very difficult to measure directly such quantities as the turbulence intensities or velocity fields in the rotating mass of gas. The flow patterns in these units differ from those in cyclones mainly by the Coriolis acceleration imparted by the rotating impeller, a point which the makers of these separators never fail to stress. Yet the practical performances of these units do not suggest that their collecting efficiencies are significantly higher than those of the best cyclones. The Soviet literature even quotes many cases when cyclones equalled or surpassed the efficiency of the most highly lauded of rotary separators.

Obviously, a well-designed and well-fabricated rotary separator should achieve a substantially higher collecting efficiency than any cyclone, and the point is easy enough to prove in the laboratory. But there is no reason for supposing that a full-scale industrial separator of this kind could reliably capture the minus 5 microns fraction, so that the fractional collecting efficiency curves for these units will apparently not be much different from those of really good cyclones. One undeniable advantage of rotary separators is their compactness: an adapted and slightly enlarged fan, with a hopper at the bottom, does not take up much

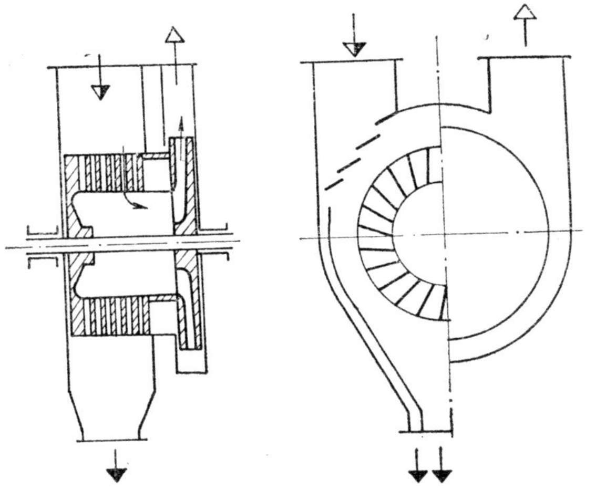

Fig. 63 A rotary separator with the fan mounted downstream of the impeller.

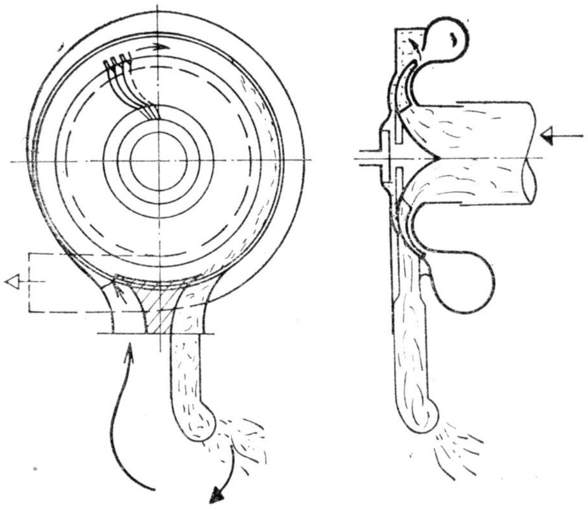

Fig. 64 A conventional fan-type rotary separator (the AAF 1932 model).

space. Another advantage is that these separators are easy to scale up, so as to handle something close to 100,000 m^3 per hour. On the other hand, a rotary separator is a much more complicated piece of hardware than any of the other kinds of mechanical separating equipment, and this tends to outweigh any advantages which it may have to offer. Hence, it seems unlikely that these units will be adopted on any major scale in the foreseeable future.

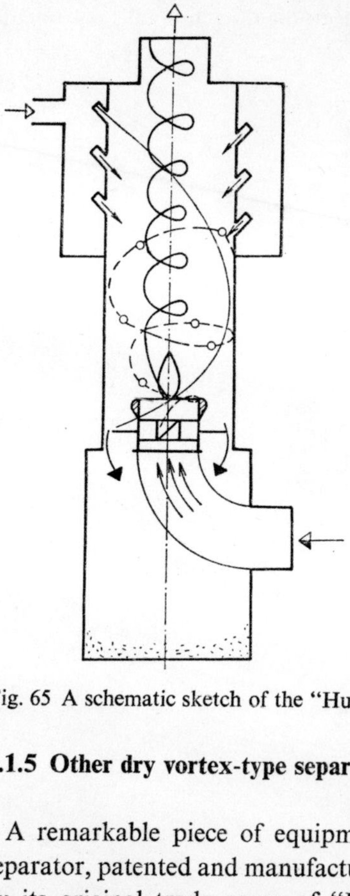

Fig. 65 A schematic sketch of the "Hurriclone".

3.1.5 Other dry vortex-type separators

A remarkable piece of equipment in this category is the counterflow vortex separator, patented and manufactured in West Germany, which is generally known by its original trade name of "Hurriclone". This is essentially a cyclone with internal fittings which impart a swirling motion to the incoming gas stream, and with nozzles set into its jacket to admit secondary air flows. The latter intensify the rotation of the mass of gas, and hence the centrifugal forces acting on the solid particles, and also help to carry the trapped particles into the hopper. Fig. 65 shows one of the many existing variations on the basic design. Hurriclones admittedly achieve higher collecting efficiencies than even the best of ordinary cyclones; but they fall short of the efficiencies common in wet scrubbers, while consuming nearly as much power as the average equivalent scrubber. Still, there is a large domain of applications where the dust must be trapped in the dry state, and where extreme efficiency is crucial. These are the fields where Hurriclones come into their own.

Another item of equipment which seems to be gaining in popularity is a combination of a cyclone with a filter, usually of the sand-bed type. The complications of these units are more than offset by their numerous advantages, especially their resistance to high service temperatures. Their collecting efficiencies are so high that the dust concentrations at their outlets are mostly lower than 100 mg m_n^{-3}.

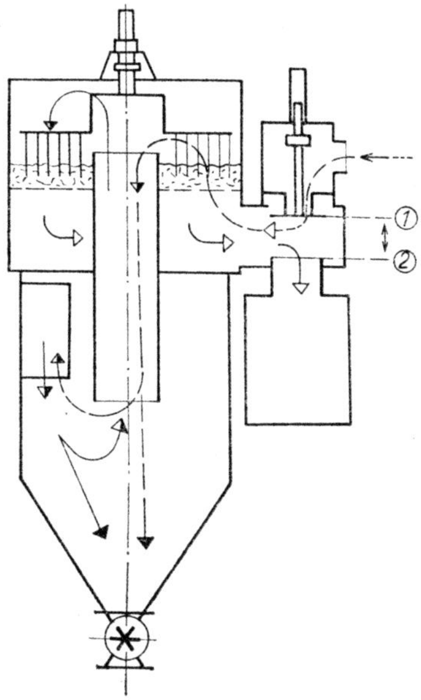

Fig. 66 Scheme of a separator made up of a cyclone and a sand-bed filter (the Lurgi "Drallschichtfilter"). ——— Filtration flow, — — — Filter cleaning flow.

The filter bed offers relatively little resistance to the flow, and therefore need be no larger than other equivalent separating facilities. The filtration velocities are much higher than in cloth filters, generally in the interval of 0.5 to 1 m s^{-1}, and the pressure drops across these units normally range from 1000 to 2000 Pa. The filter bed is commonly made up of screened sand with a grain size around 5 mm, in a layer that is typically some 50 to 150 mm deep.

The filter units vary a good deal in their design. Of those in use at present, the best performances seem to have been achieved with the "Drallschichtfilter" made by Lurgi of West Germany. However, this type is not suitable for very fine dusts, and must always be installed in sets of two or more units. The way a unit of this kind works can be followed on Fig. 66. The dust-laden gas first enters the cyclone, which retains the coarser fractions, and then passes out to the top of the circular filter compartment, where the sand layer, resting on a wire mesh or screen, traps the finer particles. When the layer clogs up with dust, which is indicated by an abrupt rise in its resistance, the valve is reset to its "*2*" position; that shuts off the clean gas outlet, and admits a blast of purging air supplied by a special fan. At the same time, a raking mechanism is started up to loosen the layer as the blast of reverse air blows it clean. The dust flushed out of the layer is entrained by the purging air and carried back into the cyclone through its gas outlet tube. Since this dust consists predominantly of large agglomerated particles, most of it is trapped and leaves the cyclone through its dust discharge gate. The purging air emerges through the cyclone inlet into what is normally the feed piping, and enters the other unit of the pair (or some other unit of the set).

So much for the basic principle, which is simple enough in itself. Complications set in mainly because this unit is intended for high service temperatures – according to its makers, it can be used at up to 500 °C. It is no easy matter to ensure that all the control gear will work reliably, and the whole unit seal tightly, at these temperatures. Moreover, the motor that powers the raking mechanism is mounted on the filter, and is consequently also exposed to the heat. Nevertheless, these units have proved valuable in applications where a high collecting efficiency is essential, and where the use of other high-efficiency separating devices is ruled out by the operating conditions, mostly by the high service temperatures. A further advantage of this design is its considerable resistance to abrasion. Both of these features make this type particularly useful in the production of building materials, where the fact that it is not suitable for trapping the minus 5 microns fraction does not make much difference. Lurgi now build these units with two filter layers on top of each other, which doubles the effective filter area without increasing the diameter of the unit.

3.2 Wet scrubbers

3.2.1 Spray towers, packed and unpacked

Spray towers serve to pre-clean, cool and humidify the dust-laden gas, usually ahead of Venturi scrubbers or wet electrostatic precipitators. They are seldom used on their own, as the sole gas cleaning equipment, because the collecting efficiencies they offer are inadequate for most present-day applications. The towers are mostly round, up to several metres in diameter, and generally two to four diameters in height. The gas as a rule enters through radial or tangential inlets at the bottom of the tower, and rises against the shower of water droplets discharged by sprays at the top of the unit. These towers are fitted with anything from one to three grid-like arrays of spray nozzles, sometimes even more. The object is to distribute the droplets as evenly as possible throughout the entire cross section of the tower, and to prevent the spray cones emitted by the individual nozzles from overlapping each other, as that might cause the droplets to coalesce into drops. In some cases, however, it is more advantageous to locate the gas inlets at the top of the unit, and the radial outlets at the bottom.

Irrespective of these differences in the basic layout, there are several requirements which are common to all these towers. The nozzles must be easily accessible and replaceable, and their cleaning should preferably not necessitate any stoppage of the process. The feed water must always be filtered, preferably by two filters in parallel, to prevent dirt blocking the nozzles. The nozzle apertures must not be so small as to foul up easily. The tendency nowadays is to use nozzles with orifices big enough to dispense large amounts of water per unit of time without any risk of clogging; to design them for a fully conical spray pattern; and to mount them in the lid of the tower, where they are easily accessible and can be replaced without any need to shut the unit down.

The gas velocity must be kept down to some 1 to 1.5 m s^{-1}, as otherwise the gas would be liable to entrain the droplets. Where a higher flow velocity is unavoidable, the gas leaving the tower should be passed through a droplet separator; but as this gas still contains a lot of solid particles, the baffles or louvres of the droplet separator will be apt to clog up fairly rapidly. The only case when an unpacked spray tower can be left to handle high flow velocities, without a droplet separator installed beyond it, is when the unit serves primarily as a cooler and delivers the gas to a wet scrubber set up in its close vicinity. In that case there is no objection to even a massive entrainment of the spray droplets. The pressure drops across unpacked spray towers are small, in the region of 100 to 200 Pa.

In an unpacked tower, solid particles will be trapped only if they meet water droplets and are wetted by them. That makes the particles heavy enough to drop to the bottom of the tower, from which the sludge is then evacuated. The likelihood of each particle meeting a droplet grows with the amount of water sprayed into the unit, with the uniformity of the droplet and gas-flow distributions over the cross section of the tower, and with the degree of atomization of the water, which in turn affects the droplet velocity too.

An altogether higher collecting efficiency is obtained in packed towers, which at one time used to be a very common item of plant equipment. Their internal fittings ranged from sheet-metal cones to rows of mostly wooden parallel walls. Perhaps the most widely used type had a number of vertically superimposed grids, generally made of timber planks some 10 to 15 mm thick and 100 to 150 mm wide, which were installed with their narrow edges facing up and down. The individual grids were usually either staggered, or else revolved around the centre line of the tower, so that the planks of one grid were set at an angle of 45 or 90 degrees to those at the next higher or lower level. Given a normal gas loading and a normal degree of wetting by the spray water, the pressure drops across these packed towers rarely exceeded some 300 to 500 Pa, unless the grids were fouled up with dust deposits. The main attraction of this design was that the grids were cheap and easy to build. They could be made of the cheapest of long-fibred timber, which withstands drenching in water far better than the more expensive hardwoods. The main drawback was the high water consumption rate, needed to keep the internal fittings uniformly wetted and to prevent the adhesion of dust on them; this could run as high as 1 to 2 litres per m^3 of gas. The packed towers were also bulky, susceptible to the formation of dust deposits, and incapable of trapping fines with a reasonable degree of efficiency. In view of all these shortcomings, they are now becoming increasingly rare, and in Czechoslovakia are no longer being built.

3.2.2 Wet vortex separators

Cyclones with wetted walls

The simplest devices in this category are ordinary cyclones in which the walls are wetted with water. This can significantly enhance their trapping efficiency, and in view of the small size of most cyclones costs only about 0.1 litres of water per m^3 of gas, but has often been found to impair the reliability of the unit. What all too frequently happens is that non-uniform wetting disrupts the continuity of the water film, and thus permits dust to settle and adhere on those wall surfaces which remain dry. As the few units of this type which have been commissioned have persistently suffered from the formation of deposits, no equipment of this kind has been introduced into routine production.

Type VTI centrifugal scrubbers

Unlike the ordinary wetted cyclones, this type has been specially adapted to operate with wetted walls. The dust-laden gas enters the upright cylinder tangentially, through a rectangular-section inlet stub set some way above the conical bottom of the unit, and ascends along a helical path, which brings the solid particles up against the walls. Water is injected through nozzles at the top of the cylinder, and flows down the walls in a continuous film. The clean gas emerges through an outlet pipe which is set into the upper part of the cylinder and passes through its lid. The water and the dust entrained by it are discharged at the bottom. The specific water consumption of these units obviously depends on their diameter, but in a scrubber of 1 m diameter is typically around 0.2 litres per m^3 of gas. Much depends on the water sprays, and especially on the water feed pressure, as the unit will function dependably only if the water film on its walls is reasonably uniform.

These scrubbers resemble dry cyclones in that their collecting efficiency is the greater, the smaller the diameter of the unit and the higher the gas inlet velocity. In practice, however, both these variables are subject to certain constraints. The general layout of these scrubbers, and the way they are usually grouped into batteries, practically preclude their construction in diameters of less than about 0.5 m. The gas inlet velocity must be confined to about 20 m s^{-1} to prevent the gas stream from disrupting the water film on the casing walls, because that would result in the formation of dust deposits as well as in excessive entrainment of the water droplets. At an inlet velocity of 20 m s^{-1}, the gas in the interior of the scrubber moves at something like 5 m s^{-1}.

Several power stations in the Soviet Union have installed units of this type, lined with acid-resistant tiles, for cleaning the flue gases from medium-sized steam generators fired with coal fines. These scrubbers, of 1.1 m diameter, are mounted side by side in banks, and attain collecting efficiencies around 90 per cent.

A further development of this basic design is the MP VTI centrifugal scrubber. In this unit the inlet duct slants downwards, and houses staggered rows of horizontal bars which are wetted by water sprays. This type again exploits the inertia of the solid particles for the separating process. Its collecting efficiency is higher than that of the basic VTI models, but so is its resistance, i.e. the pressure drop across the unit.

3.2.2.1 Vortex scrubbers

The first scrubber of this kind evolved in Czechoslovakia, and produced by the ČSVZ works, was the MVA type shown in Fig. 67. This is essentially a battery of wet centrifugal separators. Its design exploits the fact that the efficiency of these units increases as their diameter diminishes, and is intended to make the best possible use of the given amount of space without sacrificing any functional advantages in the layout.

The particles are trapped in an array of vertical tubes, which in the MVA type are of 150 mm diameter and 1200 mm long. In the later MVB design, the tubes of 150 mm diameter are shortened to 450 mm and provided with conical lower sections. The upper (inlet) ends of the tubes house deflectors which induce a swirling motion in the incoming gas. The plenum chamber above this assembly is provided with water sprays, but these are not intended to separate the solids from the gas stream: their sole function is to ensure a uniform distribution of the water among all the tubes. The deflectors divert the droplets, centrifugally, onto the inner surfaces of the tubes, which are thus covered with continuous sheets of flowing water. The gas flow pattern created by the deflectors tends to throw the solid particles against the tube walls, where they are entrained by the film of moving water. The sludge dripping off the tube ends is separated from the gas stream in a droplet separator, in effect an axial cyclone with a straight-through gas flow, which is set beneath the tube array. The trapped sludge then flows down the inclined bottom of the unit to the discharge port, while the clean gas is led off either down a vertical tube or through an elbow at an angle of 45 degrees off the horizontal.

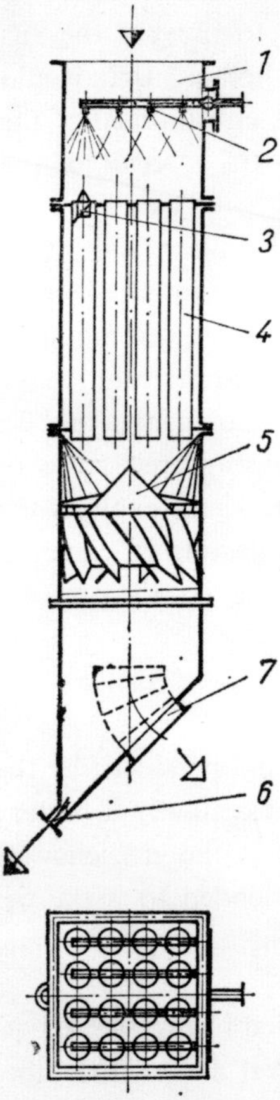

Fig. 67 Scheme of the MVA vortex scrubber:
1 — gas inlet; *2* — nozzles; *3* — deflector vanes; *4* — tubes of 150 mm dia.; *5* — droplet separator; *6* — sludge discharge stub; *7* — clean gas outlet.

The more recent MVB type differs from the original MVA unit by its shorter tubes, which also reduce its overall height; by having only one spray nozzle for every four tubes, as against one nozzle per tube in the MVA; and by employing nozzles with substantially larger orifices.

In both types, the spray water is supplied via a filter and a pressure gauge, and is injected at a gauge pressure of 0.15 to 0.25 MPa through nozzles of two different sizes. The specific water consumption ranges from 0.2 to 0.4 litres per m^3 of gas, in dependence on the temperature and dust concentration of the incoming gas, but this figure can of course be radically altered by the adoption of a water recirculation system.

The scrubbers of this type now available commercially are rated for throughflows of 5400, 9600, 15,000, 21,600, 60,000 and 86,400 $m^3\ h^{-1}$ respectively. These are the flow rates which, with a gas of $\varrho = 1\ kg\ m^{-3}$, produce a pressure drop of 950 Pa across the unit. The fractional collecting efficiency of all these types, regardless of their rating, is represented by one and the same curve. For a dust with a specific gravity of $2\ g\ cm^{-3}$, and the pressure drop corresponding to the nominal throughflow of the unit, this curve is roughly defined by the following point values: at a particle size of 8 microns, the efficiency is 100 %; at 3.5 microns, 95 %; at 2 microns, 75 %; and at 1.5 microns, 55 %.

These scrubbers have been utilized in the ceramics industry, in coal dressing plants, foundries, in the iron and steel industry, and for cleaning the flue gases from boilers fired with coal fines and sulphite extracts. When incorporated in the dust and fume exhaustor systems of ceramics factories, they have turned inlet dust concentrations of 2.2 to 3.8 $g\ m_n^{-3}$ into outlet concentrations as low as 0.02 to 0.04 $g\ m_n^{-3}$, which represents a 99 per cent efficiency. The MVA units which handle the combustion products from 50 tons per hour steam generators, and are fabricated in stainless steel, display collecting efficiencies around 98 per cent.

3.2.3 Bubble washers

Bubble washers have a closely defined field of application, especially in the chemical industry. They are used in cases when gas, with only a relatively low content of solids in the minus 5 microns bracket, is required to undergo some chemical reaction, or to be cooled down to its dew point, in the course of its cleaning. The units of this type manufactured in Czechoslovakia, by the Chepos corporation, are mostly rectangular in cross section, and roughly conform to the scheme in Fig. 68.

In these units, the dust-laden gas is bubbled through one or more layers of water resting above perforated or slotted plates. The bubbles that pass through these openings turn the water above the partitions into an unstable foam. When the openings are circular, which is the design usually preferred abroad, their diameters normally diminish from the bottom of the unit upwards; for instance, the lowermost plate may have holes of 7 mm diameter, the middle one 6 mm and the uppermost plate 5 mm holes. Slotted plates usually have slits some 2 to 2.5 mm wide. They score over the perforated plates by producing a somewhat lower pressure drop, but need an even more closely uniform distribution of the gas over their entire face areas to function properly. In practice, that means they make it rather more difficult to maintain the desired properties of the foam layer. Consequently, they tend to be reserved for the lowermost locations in washers which have to deal with high inlet dust concentrations. No matter whether the plates are perforated or slotted, the aggregate area of the openings in each plate is normally around 20 per cent of the cross-sectional area of the unit at that vertical level.

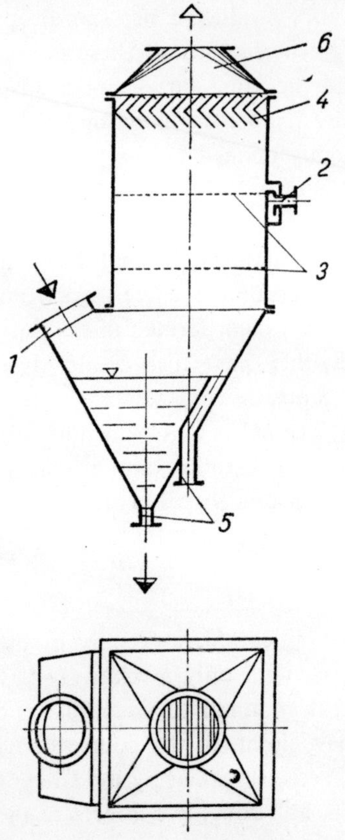

Fig. 68 Scheme of a bubble washer:
1 — gas inlet; *2* — water inlet; *3* — perforated plates; *4* — droplet separator; *5* — sludge discharge port; *6* — clean gas outlet.

One limitation of these washers is the fact that, to keep them working in the intended way, the gas velocity (averaged over their cross section) must be restricted to about 2.3 m s^{-1}, and maintained within fairly close limits.

The pressure drop across these units will obviously depend on the number of plates installed in tandem. These washers normally have one to three vertically superimposed partitions, each of which produces a pressure drop typically in the vicinity of 300 Pa. If we add the pressure losses incurred in the inlet plenum chamber and in the droplet separator, the overall pressure drop comes to something like 800 Pa for a twin-plate and 1100 Pa for a three-plate unit.

The water consumption of these washers ranges between 0.2 and 0.4 litres per m^3 of gas, in dependence on the gas inlet temperature and dust concentration. As a rule, the water is left to flow onto the plates by gravity alone, and as it need not be particularly clean, there are no problems involved in recirculating it.

The gas usually enters the washer obliquely, in a downward direction, beneath the lowest of the bubble plates. The tapering bottom of the unit can be made to hold enough as yet undischarged sludge for the incoming gas to impinge on the bath surface, depositing at least some of the coarsest particles there as the stream is deflected upwards. This arrangement also makes for a more uniform distribution of the gas over the bottom of the lowermost bubble plate, which improves the functioning and enhances the efficiency of the washer.

The droplets entrained at the uppermost bubble plate are generally arrested in a slotted-plate or baffle-flight droplet separator, which may be incorporated in the washer or set up next to it. A vital requirement on these droplet separators is that their internal fittings should be easy to clean out or replace.

A drawback of all these washers is their sensitive response to every change in the flow rate of the gas. At less than the rated throughflow, the water will tend to stream down through the openings, without ever forming the foam layer that is crucial to the cleaning action. At excessive gas flow rates, far too much water will be abstracted from the unit by the gas stream. Another shortcoming of these washers is that they cannot trap particles less than 5 microns in size with the sort of efficiency necessitated by present-day public health regulations. The only way to do so would be to fit the unit with a large number of bubble plates in succession, a solution which is only rarely acceptable.

The bubble washers now in production in Czechoslovakia are two- and three-plate units rated for gas throughflows of 1600, 2500, 4000, 6000, 8000, 10,000, 13,000, 16,000, 20,000 and 26,000 m^3 per hour respectively. Larger unit throughflows are not advisable, because at higher flow rates it becomes progressively more difficult to spread the flow evenly over the whole cross section of the washer.

3.2.4 Bath-type washers

These washers are better known by the trade names of the various proprietary types, such as Roto Clone N or Tilghman. They are largely self-contained, and are therefore at their best in plants where gas has to be cleaned at several widely separated points, i.e. where the installation of any other type of wet scrubber would entail long runs of water recirculation and sludge disposal pipings. The various designs mostly differ only in the configuration of the space where the dust-laden gas contacts the water. As evident in Figs. 69 and 70, in the Roto Clone N the gas passes through an S-shaped slot, while in the Tilghman unit the flow channel is roughly C-shaped and divergent, widening gradually along the flow direction. In both cases, the gas picks up water from a constant-level bath and carries it through the washing channel. The solid admixtures are separated by a combination of three distinct factors. Firstly, the curved flow channel, where the walls are wetted by the water that the gas has induced, subjects the particles to centrifugal forces. Secondly, the gas has to force its way through a sheet or shower of water. And finally, there is the intensive turbulence in the mixture of gas and water spray as it passes through the slot or channel.

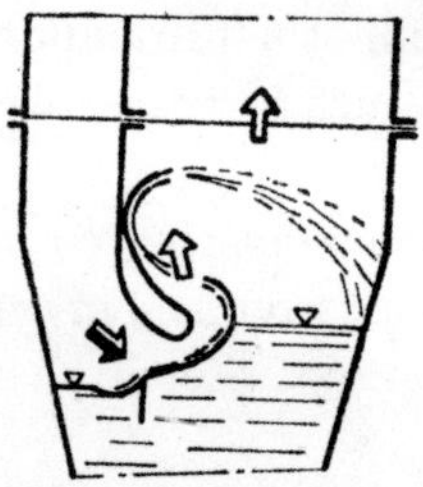

Fig. 69 Sketch of the working space of a Roto Clone N.

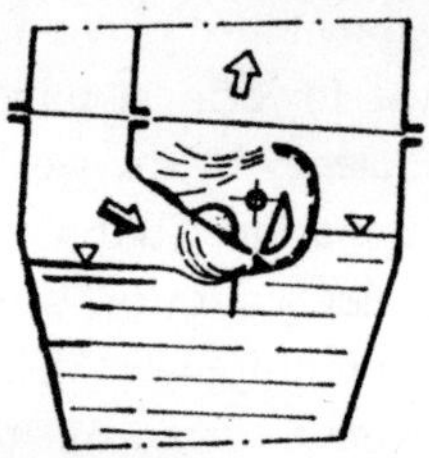

Fig. 70 Sketch of the working space of a Tilghman unit.

The water entrained by the gas remains in the outlet compartment, and flows back into the inlet compartment where, in view of the pressure drop across the washing channel, the water level is lower. This spontaneous recirculation of the process water is one of the main advantages of these designs. Fresh water is added only to make up losses by evaporation, drainage through the overflow pipe which keeps the bath level constant, and water lost in the sludge which collects at the bottom of the unit. Altogether, these losses normally total around 0.03 litres per m^3 of gas. These washers generally need two ancillaries: a continuous sludge removal system for the dense sediment that accumulates at the bottom, and. a droplet separator at the top of the unit, in the clean gas outlet area, to keep down the water losses by entrainment. Moreover, a fan is usually fitted on top of the unit casing to move the gas through the washer and/or to deliver it further downstream. That yields a compact, integrated piece of equipment comprising the washer itself, a gas conveying fan, and the sludge handling system – a fact which largely accounts for the popularity of these devices.

Apart from all their other attractions, these washers also afford a high degree of efficiency in the trapping of even very fine dusts. That explains the rapid development, over the past few years, of the Czechoslovak MHA units, the volume of their production at the ČSVZ works, and their widespread adoption especially in foundries. These units, shown in Fig. 71, have flow channels similar to those of the Tilghman type, and are presently available in sizes rated for 5000, 10,000, 15,000, 20,000, 25,000 and 30,000 m^3 per hour. The three smaller types of the range are of the single-slot configuration sketched in Fig. 70, while the three larger sizes have two horizontally opposed slots apiece.

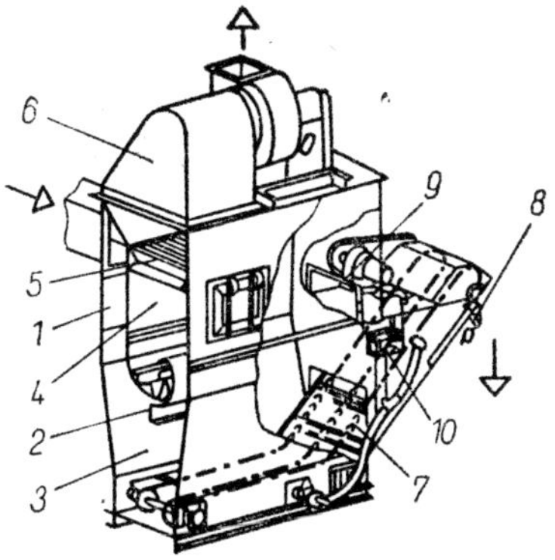

Fig. 71 Cutaway view of an MHA washer:
1 — inlet chamber; *2* — slot compartment; *3* — water tank; *4* — outlet chamber; *5* — droplet separator; *6* — collecting chamber with fan; *7* — paddle raker; *8* — wiper blade; *9* — water system fittings; *10* — overspill receiver.

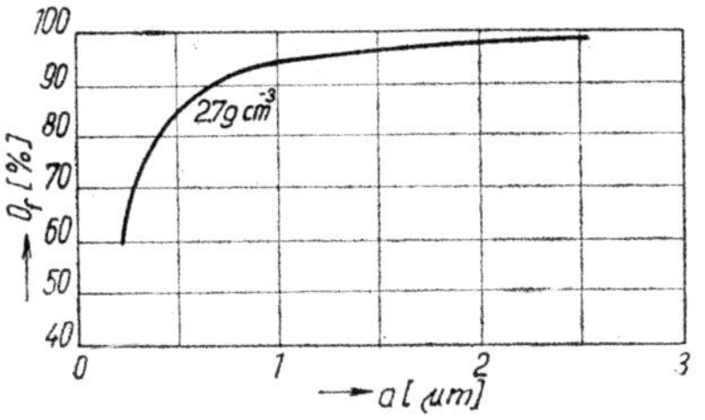

Fig. 72 Fractional collecting efficiency of an MHA washer.

Given the rated volumetric flow rate of air at room temperature, the pressure drop across an MHA washer is roughly 1300 Pa, but this value is strongly dependent on the height of the water level. The fans fitted to these units are generally selected so as to overcome the envisaged pressure drop and afford a further pressure differential, of some 500 to 2200 Pa, to cover pressure losses in the suction and delivery pipings.

When handling the most common kinds of dust, such as those that arise in foundries, these units display consistently high efficiencies which often surpass the 99 per cent level. Fig. 72 shows the fractional collecting efficiency of an MHA washer for dust with a specific gravity of 2 g cm^{-3}. Fig. 73 indicates, very roughly, how the clean gas outlet temperature in these units varies with the inlet temperature of the dust-laden gas. Fig. 74 illustrates the way the water evaporation rate depends on the gas inlet temperature.

The MHA washers have gradually been replaced on the production line by the newer MHB and MHG versions, but these share the basic design and principal technical data of the equivalent MHA types. They differ from the earlier models mainly by minor design modifications of the actual washing space, e.g. in the geometry of the slots. A more recent innovation, the MHC type, has already been supplanted by the MHF derivative. The latter incorporates the same washing unit, droplet separator and fans as the MHB and MHG types, and except for its water consumption shares their technical parameters too, but has no sludge raking device or conveyor. Sludge is evacuated from it continuously, through a pneumatically controlled closure, and consequently this type consumes water at much the same rate as the MVA wet vortex scrubbers.

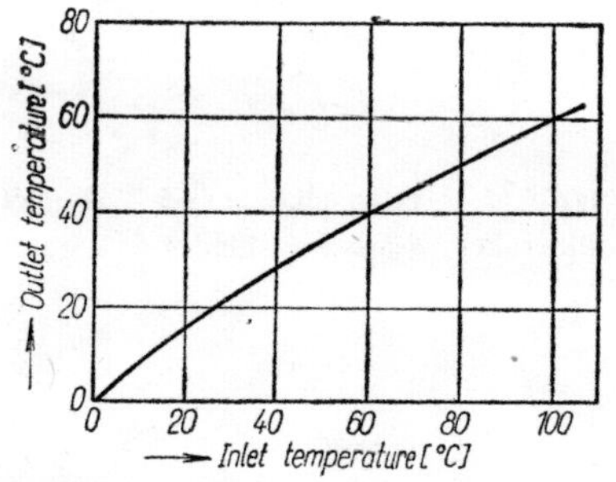

Fig. 73 Dependence of the gas outlet temperature on the gas inlet temperature in an MHA washer.

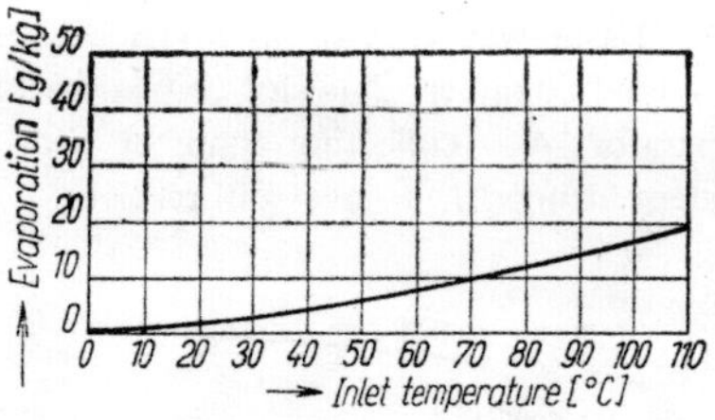

Fig. 74 Dependence of the water evaporation rate on the gas inlet temperature in an MHA washer.

The latest member of this range is the MHD type, which can handle both foam-generating and other admixtures. The MHD units are not provided with sludge raking and conveying devices, nor with fans. They are presently available in ratings of 5000, 10,000, 15,000, 25,000, 40,000 and 60,000 m^3 per hour.

3.2.5 Venturi scrubbers

None of the wet scrubbers reviewed in the preceding sections are much use when the dust contains a substantial proportion of fines in the one-micron or sub-micron size brackets. This is the almost exclusive domain of fluidic devices known as Venturi scrubbers, which work by causing two different aerosols to coagulate together. One of the aerosols is a mixture of gas and solids, i.e. the dust-laden gas to be cleaned, the other is a mixture of gas with liquid droplets which cling to the solids and thus separate them from the stream. The water droplets are atomized into the gas stream either in the convergent inlet section or in the constricted throat of the Venturi tube.

The factors which contribute to the separating action in a Venturi tube are numerous and not all properly understood, but the chief of them would seem to be the collisions between solid and liquid particles; the electrostatic charging of the particles; droplet condensation on the solids as the gas is cooled to below its dew point; and, in the case of particles less than half a micron in size, diffusion. The most important of these factors are the impacts between solid and liquid particles caused by their different velocities. In theory at least, it is immaterial whether the solids are faster or slower than the droplets, as long as the differential is sufficiently large to cause a multitude of mutual impacts.

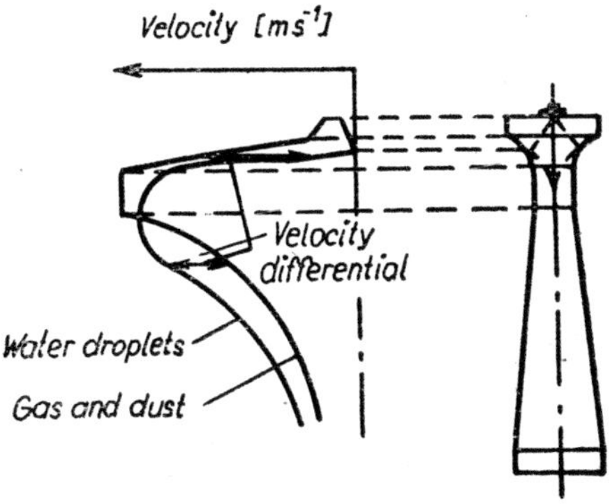

Fig. 75 Velocity profiles of solid and of liquid particles passing through a Venturi tube.

To illustrate the relationship between the solid and droplet velocities, Fig. 75 shows an example of velocity curves recorded in a Waagner Biró Venturi scrubber with the atomizing nozzle in its centre line. The peak separating effect is in this case obtained along the boundary between the convergent and cylindrical parts of the tube, where the gas velocity is highest while the droplets are only just beginning to accelerate in the gas stream. In the divergent part of the tube, the droplets actually move faster than the solid particles. The shape of these curves is of course always specific to the given type of unit, and is strongly dependent on the location and orientation of the water inlet or inlets, as well as on the water inlet velocity. In a Pease Antony Venturi scrubber, for instance, the velocity differential attains its maximum much further downstream, at the boundary between the cylindrical and divergent sections.

Apart from the velocity differential, the efficiency of Venturi scrubbers is also strongly affected by the size of the droplets. This dependence is best illustrated on an example. Let us take a fairly typical case, of oxygen steelmaking convertors emitting a virtually monodisperse dust with a particle size of one micron, and a particle weight of $2.6 \cdot 10^{-12}$ g, at a concentration level of 20 g per m^3 of gas. At this rate, every m^3 of gas will contain $7.65 \cdot 10^{12}$ particles, which ideally should be spaced an average of about 50 microns apart. If these particles are to be captured, they must be engulfed by the droplets. In other words, the surface tension of the droplets must be overcome by the kinetic energy of the particles, which is a function of the velocity differential. Extremely small droplets, no more than a few microns across, would be apt to miss the solid particles altogether; they would be far more

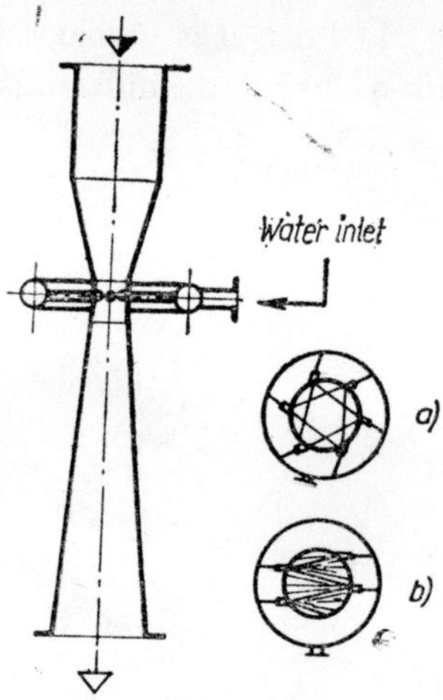

Fig. 76 Scheme of a Pease Antony Venturi scrubber.

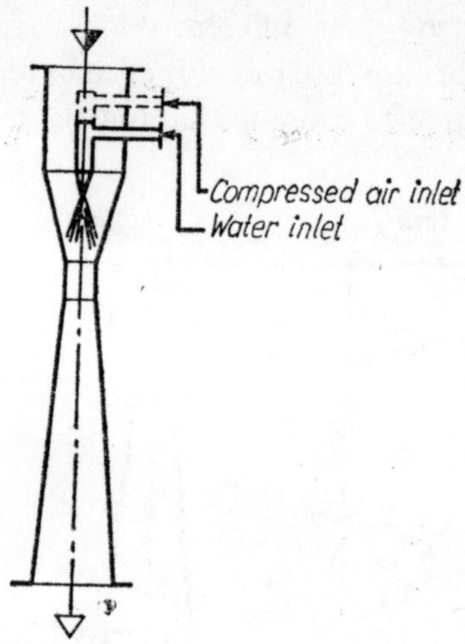

Fig. 77 Scheme of a Körting or Aerojet Venturi scrubber.

likely to traverse the 50-microns spaces between adjacent particles than to collide with one of them. Large droplets, in the 500 to 1000 microns range, would be bound to impinge on most of the dust particles, but would fail to trap them: drops of this size push a cone of accumulated dust particles ahead of them, and the small dust grains are incapable of penetrating this relatively massive barrier. For this particular duty, then, the optimum droplet size would be about 60 to 100 microns. Droplets of this size will be formed only at a certain flow velocity of the gas, which corresponds to a certain pressure drop across the unit. That explains why, in every specific application, the collecting efficiency attains a sharp peak at a certain closely defined pressure drop, and diminishes as soon as the pressure drop deviates from this value. It also explains why, in general, the absolute collecting efficiency of Venturi scrubbers drops off as the dust concentration in the incoming gas declines.

The collecting efficiency is further dependent on the amount of water atomized per unit of time. As we are out to obtain the highest feasible droplet density, we may rightly expect the efficiency to increase with the amount of water fed in – up to a certain limit, which varies from one type to another. Even this improvement, however, is achieved only at the cost of a higher pressure drop and hence a greater power consumption. Once the amount of water exceeds the optimum, the water is no longer evenly atomized throughout the volume of gas, but tends to form large drops or even continuous sheets, which naturally reduces the collecting efficiency.

Venturi scrubbers exist in a variety of basic types, which differ mainly by the ways the water is injected into them. The type in most widespread use is known

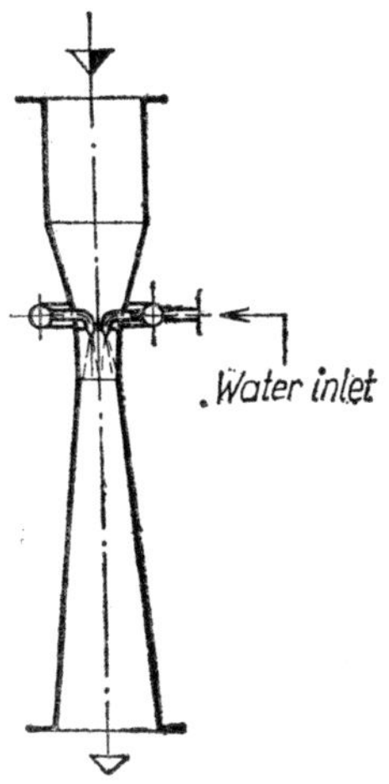

Fig. 78 Scheme of an Imatra Venturi scrubber.

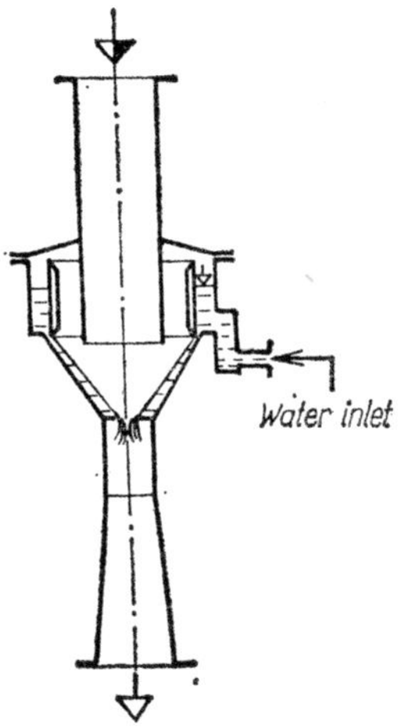

Fig. 79 A Venturi scrubber designed to handle adhesive dusts.

as the Pease Antony scrubber, and is shown schematically in Fig. 76. In this design, the water is sprayed into the throat of the Venturi tube radially, i.e. with a zero axial velocity component. High collecting efficiencies can be attained only by using a sufficient number of nozzles, and a sufficient water inlet pressure, to ensure the most uniform possible distribution of the spray throughout the cross section of the tube. Attempts to improve this water distribution have led to the evolution of various modifications. In one of them, sketched in Fig. 76a, the water is still sprayed in at right angles to the longitudinal axis of the tube, but at a small tangential angle off the truly radial direction. In another, seen in Fig. 76b, it is atomized by mutually opposed and staggered rows of nozzles. Other Venturi scrubbers, like the Körting and Aerojet types, employ twin nozzles, delivering water and compressed air respectively, to distribute the water within the converging part of the tube, as indicated in Fig. 77.

An entirely different approach is adopted in the Imatra Venturi scrubber shown in Fig. 78, where the water is driven in parallel to the gas flow. This arrangement reduces the pressure drop, but also yields a relatively low collecting efficiency. It is therefore utilized mainly in cases where the prime object is to humidify and cool the gas, rather than to clean it.

The design in Fig. 79 is intended for handling dusts which tend to bond. In other Venturi scrubbers, these dusts are liable to form deposits along the boundary between the dry and the wetted areas of the tube walls. In this unit, the problem is obviated by the elimination of that boundary: the particles leave the dry inlet tube to impinge directly on the wetted wall of the Venturi tube, where they have no chance of adhering. Water is dispensed into the throat of the Venturi tube past an annular constriction or orifice plate. This type of scrubber pays for its freedom from fouling by exhibiting a higher pressure drop than other comparable types.

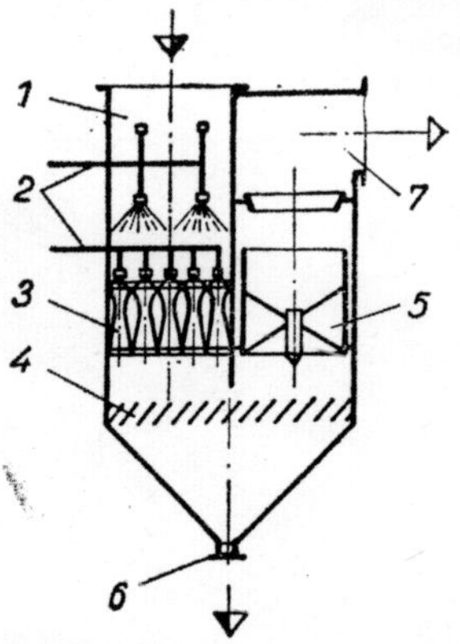

Fig. 80 The Waagner Biró multiple Venturi scrubber:
1 — gas inlet; *2* — water inlet; *3* — Venturi tubes; *4* — spray eliminator; *5* — droplet separator; *6* — sludge discharge stub; *7* — clean gas outlet.

It has already been stated that Venturi scrubbers cannot be highly efficient unless the water is atomized uniformly throughout their cross sections, and that this even water distribution is much easier to ensure in small than in large-diameter units. This is the consideration behind the popularity of multiple Venturi scrubbers, where a large number of Venturi tubes are grouped in parallel, and the individual tubes are kept down to a throat diameter typically in the region of 100 mm. Fig. 80 shows a representative design of this kind, as produced by Waagner Biró. The water, injected upstream of the bank of Venturi tubes at a pressure of several hundred thousand Pa, has to be much cleaner than is necessary in most other scrubbers.

3.2.5.1 Type MSA Venturi washers

In practice, the water available for scrubbers often contains a high concentration of solid admixtures, which the common types of Venturi scrubbers cannot handle for any length of time. Moreover, since simplicity of design is a virtue in itself, not all users are prepared to sacrifice it for the sake of a uniform water distribution, as is the case in multiple Venturi scrubbers which incorporate a large number of small-diameter tubes. In Czechoslovakia, these objections to the standard designs have led to the development and production of the MSA unit, in which the Venturi tube is set into a liquid bath.

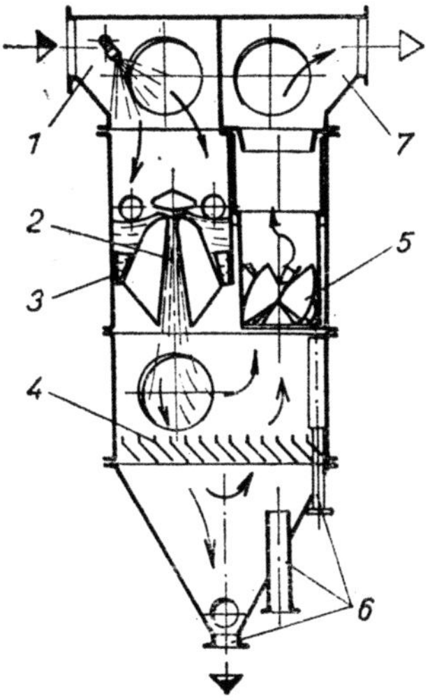

Fig. 81 Scheme of an MSA fluidic washer:
1 — gas inlet; *2* — Venturi tube; *3* — water inlet; *4* — spray eliminator; *5* — droplet separator; *6* — sludge discharge stubs; *7* — clean gas outlet.

As evident from Fig. 81, the Venturi tube of this unit is vertical, of rectangular cross section, and the lips of the inlet sleeve at its top slant downwards. The rectangular inlet opening is surmounted by a shaped cover which leaves only two narrow inlet slits, opposite each other, beneath its bottom surface. The Venturi tube passes through the bottom of a tank which is filled with water (or sludge) till the bath surface enters these narrow inlet slots. The gas passing through them then picks up some liquid, the amount depending on the exact height of the liquid level. As the gas now descends down the Venturi tube, the entrained liquid is broken up into droplets which envelop the dust particles. Finally, the gas ascends through an axial-type centrifugal droplet separator, with a straight-through gas flow pattern, and re-emerges at the top of the unit. Since the water is supplied to the tank by gravity alone, there is no objection to its containing up to 100 g of solid admixtures per litre. Consequently, the unit does not need the usual costly and complicated sludge treatment systems for water recovery. Another advantage is the rectangular plan form of the Venturi tube, which allows increased throughflows to be accommodated simply by lengthening the inlet slits; that fully preserves their small width, which ensures an even distribution of the water over the cross section of the tube.

The MSA units are now being manufactured in sizes rated for throughflows of 3750, 7500, 15,000, 30,000, 45,000, 60,000 and 90,000 m^3 per hour at a nominal pressure drop of 4200 Pa. However, they can be operated at pressure drops ranging from 2000 to 9000 Pa, which represents anything from about 75 to 150 per cent of the rated throughflows. After extensive pilot-plant trials, these units are now guaranteed to reduce the solids contents of fumes from open-hearth steelmaking furnaces, and from oxygen-blown electric arc furnaces, to less than 100 mg m^{-3}. Fumes emitted by oxygen steelmaking convertors and by twin-vessel (tandem) open-hearth furnaces can, if previously cooled and humidified, be cleaned with an average efficiency around 99.5 per cent.

The main duties for which Venturi scrubbers are currently employed are the cleaning of blast-furnace gas; of the fumes from oxygen steelmaking convertors, twin-vessel open-hearth furnaces, and oxygen-blown arc furnaces; of fumes arising in the production of non-ferrous metals and alloys; and other such applications where even fairly high running costs are acceptable, as long as the relatively large amounts of very fine particles in the gas are separated out with a high degree of efficiency. The high running costs are due to the considerable pressure drops across these units. In the cleaning of the gases encountered in non-ferrous metallurgy, for instance, these pressure drops normally range from 3000 to 5000 Pa, and in the cleaning of fumes from oxygen steelmaking convertors they have been known to exceed 7500 Pa. The water consumption of Venturi scrubbers is usually somewhere between 0.7 and 2 litres per m^3 of gas. It can be substantially reduced if the dust-laden gas is cooled, and if possible saturated with water vapour, before it enters the scrubber. That prevents the evaporation of the fine droplets, which otherwise raises the water consumption and lowers the collecting efficiency of the unit.

3.2.6 Wet rotary separators

These devices, also known as mechanical scrubbers or spin washers, can afford high collecting efficiencies even when the particles to be trapped are in the one-micron size range. For all the variety of designs in existence, these units mostly share one and the same basic principle. They comprise a spiral casing which houses an annular pattern of stator bars, and a rotor fitted with similar bars.

Water is introduced through a perforated plate at the hub of the rotor, and is removed from the gas stream by blades on the rotor circumference. The gas enters the spiral casing at its centre; is intimately mixed with and washed by the spray splashing off the system of bars; and is accelerated by rotation, so that its outlet pressure is commonly of the order of 5000 Pa. In other words, these units also act as high-pressure fans, though in view of their design their efficiency in this secondary role is inherently low.

That is about as much as need be said about these devices. For one thing, they have been more than adequately described in the literature. For another they are no longer in production, and the number of them still in service is dwindling. The type has failed to live up to the hopes once placed in it mainly because of its excessive power consumption and its complexity, which accounts for its high breakdown rate, high maintenance requirements, and an often less than satisfactory degree of efficiency under actual plant conditions. The trend nowadays is to replace these units either with wet or dry electrostatic precipitators, or with Venturi scrubbers. Only a handful of these units are still in use in what used to be their principal application, the cleaning of blast-furnace gas.

3.2.7 Droplet separators

So far we have been discussing the separation of solid particles only. Liquid droplets present quite different problems and requirements. They are mostly separated in equipment that exploits aerodynamic effects, and is as a rule run dry. In some special cases, droplets can be trapped in wet separators, and in some others they are captured in specially adapted tubular electrostatic precipitators.

The dry aerodynamic droplet separators fall into two broad categories. One consists of cyclones, the other of the diverse grids, baffle flights and vane arrays used to arrest sludge droplets in wet separating equipment.

Trapping droplets in a cyclone is in some respects easier than trapping solid particles, but in other ways a good deal more difficult. The factors which facilitate the process can be summed up as follows. Firstly, droplets are much more likely to coagulate than solid particles are to agglomerate; when two droplets touch each other, they form a larger drop which in the cyclone will be exposed to much greater centrifugal forces. Secondly, the dust impinging on a cyclone wall is apt to be re-entrained into the supposedly clean gas stream emerging from the unit, but the droplets that reach the separator wall form a continuous liquid sheet which, under any normal circumstances, will resist all attempts to disrupt or dislodge it. Thirdly, cyclones designed to separate droplets are generally lower, lighter and cheaper than those intended for trapping dust: they do not need the long cone at the bottom which leads the captured dust off, from the walls of the cylindrical section to the discharge opening, and which is often so vital to efficiency in the trapping of solid particles. Finally, in droplet separation there are no clogging hazards, so there is little risk involved in building large batteries of small-diameter high-efficiency cyclones; thus, massive throughflows of gas can be handled in batteries where, if dust were to be separated, the individual cyclones would be considered far too inaccessible.

On the other hand, the water film that prevents re-entrainment of the droplets is also one of the major obstacles in the separation process. In view of the pressure differential between the inlet and outlet areas of the cyclone, this sheet of water tends to flow along the walls towards the gas outlet opening. It must be pointed out, however, that not all the water found on the walls of the gas outlet tube comes from this source. Much of it is deposited there in the form of droplets which have evaded trapping in the cyclone, and have been thrown against the wall of the outlet tube by the swirling flow pattern inside that tube. The liquid escaping from the cyclone thus consists of two distinct components:

a) Fine droplets which have avoided contact with the cyclone walls and are carried out of the unit by the emerging gas stream, and

b) A liquid layer which flows fairly slowly down the tube walls in the direction of the gas flow; this component is known as the *layer loss*.

The magnitude of the layer loss depends to some extent on the physical properties of the liquid, mainly on its adhesive properties, viscosity, and surface tension. For example, under otherwise identical conditions it will always be greater in oil than in water. Moreover, it grows with the pressure drop across the cyclone, and increases as the distance between the inlet and outlet of the cyclone diminishes. The droplet concentration in the incoming gas affects the absolute magnitude of this loss only slightly, so that at low inlet concentrations the percentual value of this loss can reach alarming proportions.

Recent laboratory research and observations made in plant practice have clarified how various design features affect the efficiency of droplet separation in cyclones, and have led to the formulation of a set of recommendations for their design. For a start, an ordinary tangential inlet has proved superior to any helical inlet configuration, because it makes the most of the inertia of the droplets, tending to throw them against the unit walls. A spiral cyclone lid should be avoided, because it pronouncedly increases the layer losses. The conical lower section customary in dust trapping cyclones is superfluous in droplet separating units, which can therefore be shortened and kept cylindrical throughout their lengths.

The layer losses can be markedly reduced by relatively simple measures. One is to locate the tangential inlet opening at a closely specified distance, between one third and two thirds of the cyclone diameter, below the lid of the unit. Another is to suspend a splash ring from the lid; this cylindrical ring must be coaxial with the outlet tube, interposed between the latter and the cyclone walls, and its bottom edge must be roughly level with the top of the inlet tube. The liquid flowing along the lid will then collect at the lower edge of this ring, form drops, and the intensively swirling gas flow in this area will hurl these drops against the cyclone wall.

Owing to the pressure gradient between the periphery and the middle of the cyclone cross section, liquid will flow inwards, towards the centre line of the unit, not only along the lid, but along the bottom too. Consequently, a pool of liquid is apt to form in the middle of the flat bottom, an area where the gas flow is extremely turbulent. The gas vortices, which are normally incapable of detaching any liquid from the cylindrical walls, are in this region powerful enough to pick up drops of water and carry them into the outlet tube. The obvious countermeasure is to separate the cyclone bottom from the cylindrical walls by a circumferential slit, which will allow the water flowing down the walls to leave the cyclone without ever contacting its bottom. An alternative is to leave the cyclone open-ended, without any bottom at all, but in that case it will be impossible to influence the flow patterns in the lower part of the unit no matter what internal fittings we may install.

When axial cyclones are used to trap dust, a reversed gas flow (i.e. a 180 degree U-bend at the bottom of the unit) generally yields a higher collecting efficiency than a straight-through flow. In the separation of droplets, however, this rule is reversed. The much higher efficiency attained with straight-through flows is due to the lower layer losses. In a reversed-flow design, the droplets impinging on the guide vanes form into a liquid film. The pressure gradient across the cyclone cross section then impels much of this film towards the centre of the unit, i.e. towards the outlet tube, where it is easily entrained by the gas stream leaving the unit. In a straight-through design, on the other hand, the droplets that hit the vanes and tend to collect on their hub are mostly picked up by the gas flowing past the vanes, which diverts them onto the cyclone walls before they can reach the outlet tube.

The vanes serve primarily to induce an intense swirling motion in the gas stream, so as to propel the droplets centrifugally towards the cyclone walls. Still, there are also some further requirements to consider in their design. Their secondary function is to present an obstruction to the gas stream, in order to increase the droplet concentration in it and thereby to assist the coagulation of droplets into larger drops. Furthermore, they must act as targets on which the droplets will impinge by inertia, as this facilitates the subsequent removal both of the coagulated drops and of the liquid film flowing towards the cyclone walls. Experience has shown that these stipulations are best fulfilled by vanes set at an angle of 60 degrees to the cyclone centre line, and overlapping each other by about half of their pitch spacing.

Both design and functional considerations thus point to axial cyclones, with straight-through gas flows, as the preferred type of droplet separating equipment. The advantages of this type are even more marked in multiple cyclone installations. For instance, the MVA and MVB units described in Section 3.2.2, and illustrated in Fig. 67, are eminently suitable for this duty. These batteries consist of rows of small straight-through axial cyclones with no outlet tubes; the retained liquid drops off the bottom edges of the cylindrical cyclone walls, to pass into a secondary droplet separator made up of a single large axial straight-through cyclone of the conventional kind, i.e. fitted with an outlet tube. Given an adequate droplet concentration in the incoming gas, and provided that the liquid which constitutes these droplets has a sufficiently low viscosity, there is no need for any sprays at the top of the unit, above the rows of primary stage cyclones.

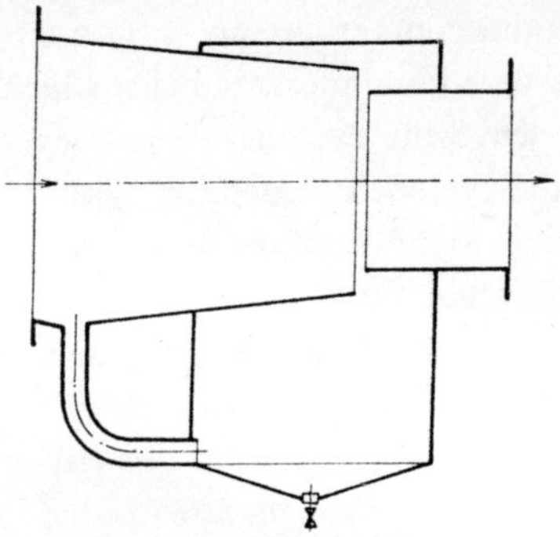

Fig. 82 A trap for capturing the layer-loss liquid film in gas outlet tubes.

Since some part of the liquid will inevitably escape along the walls of the outlet tube, it is often advisable to provide facilities downstream of the droplet separator for trapping these losses. A large proportion of the loss can be recovered by the simple device sketched in Fig. 82: the outlet tube is broken by a gap running right round its circumference, and the whole of this area is enclosed in a receptacle. Much of the liquid running down the inner walls of the tube will end up in this receptacle. Preferably, the tubing downstream of the gap should be smaller in diameter than that upstream of the trap.

A special case are the sludge droplets that emerge from some wet scrubbers. In the MVA, MVB, MSA and Waagner Biró units, these are trapped in axial cyclones with a straight-through gas flow. In other designs, such as the MHA, MHB, MHC, or various bubble washers, they are retained by grid-type or baffle flight separators. These can accept lower gas inlet velocities than the axial straight-through cyclones, but consequently also tend to take up that much more space. Furthermore, they are sensitive to excessive gas inlet velocities, at which they are liable to emit substantial amounts of entrained droplets. Besides, they clog up rather easily when handling adhesive dusts. On the other hand, unless they are clogged up, the pressure drops across them are significantly lower than those in axial straight-through cyclones.

These baffle- or grid-type droplet separators vary a good deal in design. Most of them are fitted with successive rows or grids of angle or channel sections, or similarly shaped sheet-metal baffles, which serve to deflect the gas flow in a series of abrupt turns. At each of these consecutive turns, inertia causes some of the droplets to alight on the baffle ahead of them. That implies that the liquid must be effectively drained off from each of those sections or baffles individually, as otherwise much of the arrested liquid would subsequently be re-entrained in the gas stream.

3.3 Electrostatic precipitators

3.3.1 Collecting efficiency of precipitators

The collecting efficiency of an electrostatic precipitator is still generally calculated by means of the formula evolved by Deutsch, which is based on the following assumptions:

a) The polydisperse dust enters the precipitator at a uniform concentration level.

b) All the particles within the active space of the precipitator move at one and the same velocity, equal to the velocity of the gas flow within that space.

c) The paths of the dust particles are perpendicular to the collecting electrodes; the velocity w of these particles at the surfaces of those electrodes is taken to be the mean (effective) separating brisk velocity.

d) In the course of the separating process, turbulence and the electrical wind within the active space of the precipitator incessantly mix the particles in a way which ensures that, although the particle concentration diminishes along the flow direction, the concentration will be uniform throughout any cross section perpendicular to that direction.

e) Any particle which once touches the surface of a collecting electrode is considered to be trapped, i.e. any re-entrainment is neglected.

For plate-type precipitators, *the Deutsch formula* reads

$$O_c = 1 - e^{-\frac{Lw}{Rv}} \qquad (79)$$

where L is the length of the active space of the precipitator (in m),
R – the gap between the high-tension and the collecting electrode (in m),
v – the gas flow velocity within the precipitator (in m s^{-1}), and
w – the mean (effective) separating brisk velocity (in m s^{-1}).

Equation (79) can be modified to yield

$$O_c = 1 - e^{-fw} \qquad (80)$$

where f (in s m^{-1}) is the ratio $S : Q$, with S (in m^2) representing the wetted area of the collecting electrode (i.e. the cross-sectional area it presents to the gas flow), and Q (in $m^3 s^{-1}$) denoting the flow rate of the gas. The $S : Q$ ratio thus defines the specific area of the electrode.

The residence time τ_s (in sec) which a particle spends in the active space of the precipitator can be defined as $L : v$. Similarly, the particle separating time τ_0 (in sec) equals $R : w$. Consequently, expression (79) can be reformulated as

$$O_c = 1 - e^{-\frac{\tau_s}{\tau_0}} \tag{81}$$

For tubular precipitators, the analogical Deutsch formulae are

$$O_c = 1 - e^{-\frac{2Lw}{Rv}} \tag{82}$$

$$O_c = 1 - e^{-2fw} \tag{83}$$

$$O_c = 1 - e^{-\frac{2\tau_s}{\tau_0}} \tag{84}$$

Relation (80) indicates that the collecting efficiency of an electrostatic precipitator depends both on the specific area f and on the mean (effective) separating brisk velocity w, rising with the value of the product fw. It follows that where the area f is small, a high velocity w must be attained, and vice versa. Hence, the requisite velocity w can be used as a yardstick to assess the degree of difficulty of any given separating process. The interdependences between O_c, f and w are plotted in the nomogram in Fig. 83.

Before we can set out to design a precipitator, we need a sufficiently accurate estimate of the mean (effective) separating brisk velocity that will be required for the given job under the given circumstances. This raises the question of which factors influence that velocity, and how.

Relationship between the specific area and the mean separating brisk velocity

Curve B in Fig. 40 demonstrates that the fractional separating brisk velocity is dependent on the particle size. Up to a certain particle size, the curve rises steadily. This is the domain in which the precipitator functions selectively, i.e. in which the mean separating brisk velocity depends on the size of the particles: the higher the proportion of fines, the lower is the corresponding partial range separating brisk velocity. This part of the curve is followed by another region where the separating brisk velocity is practically no longer dependent on the particle size. There, the partial range mean separating brisk velocity remains constant regardless of the amount of particles handled and of their grain size distribution.

If we want to enhance the collecting efficiency of a precipitator handling dust of a given grain size distribution, the obvious expedient, as indicated by relation (80), is to increase the specific area f. In practical terms, that means building a larger precipitator, which will trap higher proportions of both the fine and the coarse particles. This measure will also lower the partial range mean separating brisk velocity of the fines, without affecting that of the coarser particles, so that the overall mean separating brisk velocity will drop as the specific area is enlarged.

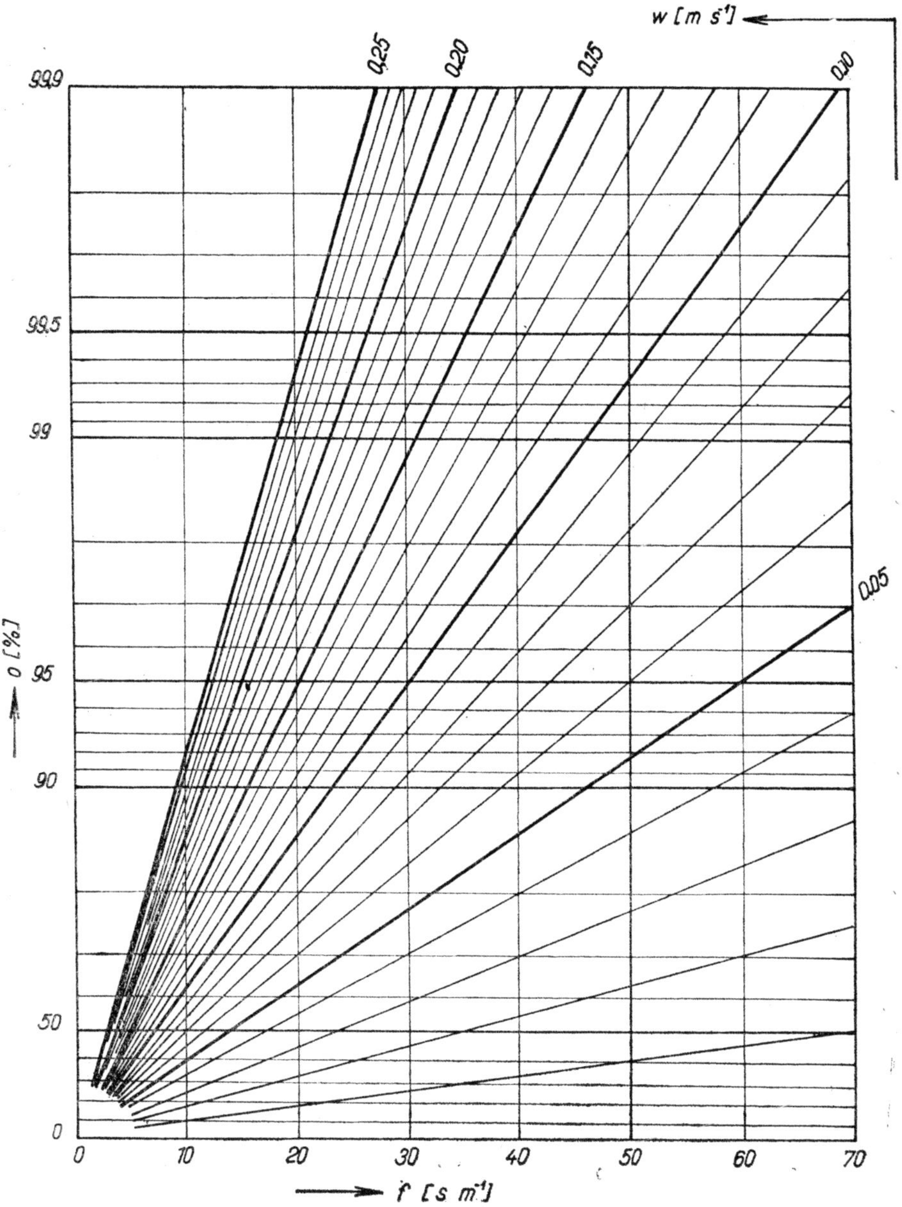

Fig. 83 Relationships between the overall collecting efficiency O_c, the mean separating brisk velocity w, and the specific collecting area f in an electrostatic precipitator.

Fig. 84 shows how, in the handling of a certain dust mixture, the mean separating brisk velocity varies with the size of the specific area. The lines for a constant product fw which are plotted in this chart indicate the conditions under which the collecting efficiency remains constant. The diagram shows that, to raise the collecting efficiency, the specific area must be increased along curve a (points A and B), while at a constant separating brisk velocity the same collecting efficiencies would be attained at points A' and B' on curve a'. The gap between these two curves is the wider, the finer the dust in question, i.e. the higher a proportion of fines falls within the region in which the precipitator functions selectively.

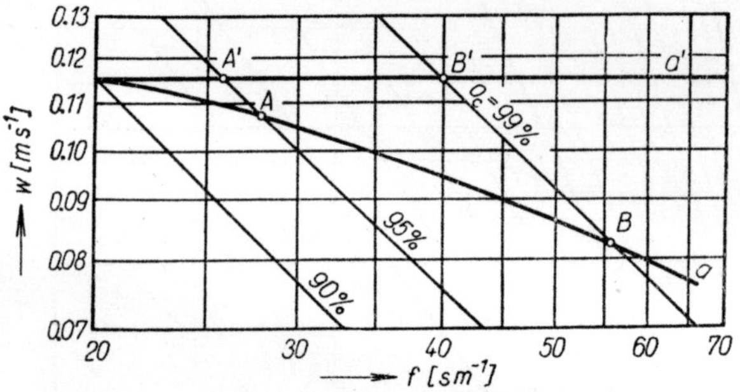

Fig. 84 An example indicating the dependence of the mean separating brisk velocity w upon the specific collecting area f.

Effect of the gas inlet parameters on the fractional separating brisk velocity

The other factors of interest, which influence mainly the shape of the fractional separating brisk velocity curve, fall into two categories. As evident from equation (71), there is one set of factors to which that velocity is directly proportional, and another set to which it is inversely proportional. The former category includes the damping coefficient β, the charging coefficient $\varkappa$, and the electric field intensity F; any rise in the magnitude of these quantities will increase the separating brisk velocity for any given particle size. That means that a large space charge of the gas ions, i.e. a high current density in the precipitator, will improve the separating brisk velocity. One way of achieving this end is to use point discharge electrodes in the primary section of the precipitator. The attainable charging coefficient varies with the permittivity of the treated dust, so a dust made up of high-permittivity particles will be easier to trap. The simplest way of raising the permittivity level is to humidify the particles, as water has a relatively high permittivity (of around 80). The electric field intensity is dependent on the applied voltage and current levels, which is why the operating voltage should be as high as is feasible. And since a safe margin must be maintained between the operating voltage and the arc-over voltage, there are obvious advantages to be derived from raising the latter voltage artificially, for instance by reducing the mobility of the gas ions, which can be done by an admixture of some electronegative gas. (The term "electronegative" describes gases in which the ion mobility is low, because electrons tend to be bound by the gas molecules; cases in point are sulphur trioxide, water vapour, or carbon tetrachloride.) This again underlines the vital importance of moisture in electrostatic precipitating processes. Finally, the arc-over voltage is also temperature dependent, dropping off as temperature rises increase the mobility of the gas ions.

The second category includes the viscosity of the gas: the higher this dynamic viscosity, the lower is the separating brisk velocity for any given particle size. The dynamic viscosity is governed primarily by the chemical composition of the gas, but also rises with the gas temperature. That is why, in electrostatic precipitating processes, high operating temperatures are generally undesirable.

Effect of the gas flow velocity and precipitator height on the separating brisk velocity

A particle which settles on the electrode adheres either to the electrode surface itself, or to the dust layer which gradually builds up on that surface and has to be removed periodically. Once trapped, the particle is exposed both to the aerodynamic forces generated by the gas flow, and to adhesion forces which tend to prevent its reentrainment in the gas stream. The aerodynamic forces grow with the flow velocity; at a certain velocity, they can overcome the adhesion forces and draw the already settled particles back into the gas stream. This re-entrainment will be the more pronounced, the greater a proportion of the surface of the deposited layer is exposed to the aerodynamic forces. That is why, in present-day precipitators, the collecting electrodes are generally shaped in such a way as to keep as much as possible of the deposited dust layer in an aerodynamically shielded area: that minimizes the amount of trapped dust which is re-induced into the gas stream.

The re-entrainment problem becomes particularly acute when the electrodes are rapped to remove the accumulated dust deposits. The trend nowadays is to shape the collecting electrodes so as to ensure that most of the dislodged dust will drop down to the outlet hopper or trough within an aerodynamically shielded space. However, the taller the electrode, the longer is the path which the dust has to cover as it drops, and the greater a proportion of this dust is likely to scatter into the gas stream. The amount of this scatter is naturally also dependent on the rapping intensity and intervals. Part of the scattered dust will always re-settle on the electrode, but the rest will leave the precipitator untrapped, reducing both the overall collecting efficiency and the mean separating brisk velocity.

Fig. 85 shows how, in a certain type of precipitator handling one particular kind of dust, the correcting coefficient of the mean separating brisk velocity varies with the flow velocity v (in m s^{-1}) and with the electrode height H (in m). One fact which emerges clearly is that, from a certain minimum velocity (called the threshold velocity) onwards, the value of this coefficient drops off rapidly.

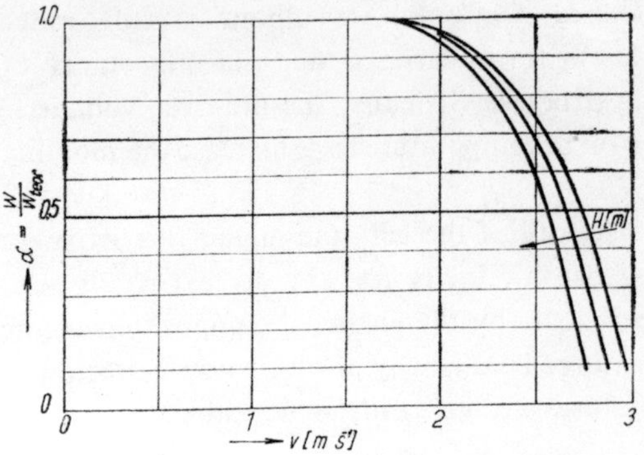

Fig. 85 Dependence of the correction coefficient α (for compensating the mean separating brisk velocity figure) upon the flow velocity v and electrode height H.

Given one particular electrostatic precipitator in which the only variable is the flow velocity, we can state that

$$f = \frac{\text{constant}}{v} \tag{85}$$

In other words, a growing flow velocity will reduce the requisite area f, and, as demonstrated in the preceding, will increase the mean separating brisk velocity for any given particle size distribution of the dust. This general dependence is plotted in Fig. 86, where the broken line applies to a theoretical state in which none of the once settled particles are re-entrained. The full line in this diagram was charted with the re-entrainment taken into account; it proves that the effect of the aerodynamic forces is to lower the mean separating brisk velocity as the

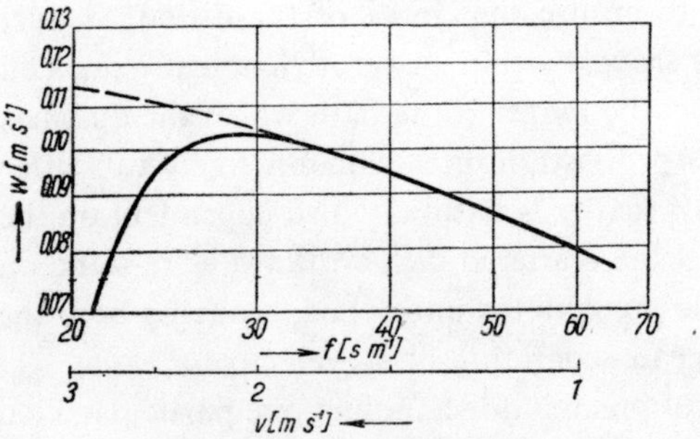

Fig. 86 Dependence of the mean separating brisk velocity w on the specific collecting area f and flow velocity v.

gas flow velocity increases, i.e. as the specific area grows smaller. Under these conditions there may be a certain optimum flow velocity, for every set of operating conditions, at which the mean separating brisk velocity will attain its maximum.

Effects of dust layers on the collecting electrodes

When a charged particle comes into contact with the collecting electrode, it is discharged. If the specific resistance of the particle material is low, say less than 10^2 ohmmetres, this process takes place so rapidly that the polarity of the particle may be switched, and the particle actually repelled from the electrode. If the repulsion force is sufficient to overcome the adhesion forces, the particle will re-enter the gas stream, pick up a new charge, and the process of its separation will start all over again. That is why highly conductive particles are so difficult to trap.

Particles with a high specific resistance are discharged more slowly, and tend to stay in place on the electrode surface, so that they gradually build up a dust layer on it. The overall specific resistance ϱ (in ohmmetres) of this layer is determined by the specific resistance of the particle material, the surface resistance of the particles, and their contact (transfer) resistance. The outer surface of this layer is continually charged by the ionized gas particles which transfer their charges to it. That sets up a potential gradient ΔU_v (in V) across the layer. Given a layer thickness of s (in m), this gradient is defined as

$$\Delta U_v = \varrho s i \tag{86}$$

The potential gradient thus depends both on the current density i (in A m^{-2}) across the layer, and on the product ϱs. The voltage drop within the layer, and the voltage drop in the gas between the discharge electrode and the layer surface, together equal the whole of the electrode feed voltage. As evident from Fig. 87, the effect of the deposited layer is to tilt the line which defines the voltampere characteristic to the right.

If a certain area on the collecting electrode is free of any dust deposit whatever, the working point shifts along the $\varrho s = 0$ characteristic. At the working point marked A in Fig. 87, the potential gradient in the gas attains a critical magnitude, and an arc-over between the electrodes ensues. Point A thus in effect limits the permissible voltage at a zero thickness of the dust deposit on the collecting electrode.

If another point of this electrode is covered by a dust layer of a certain thickness, so that ϱs is larger than zero, the working point will move along the ϱs = constant curve. At point B, however, arcing will occur through the gas layer at the place where there is no dust deposit. Consequently, point B restricts the maximum electrode-to-electrode voltage, but at this point both the current intensity and the potential gradient across the gas layer are smaller than at point A. If the dust layer had a uniform thickness and the same specific resistance all over the electrode surface, the voltage could be raised further up to point C, where the potential gradient in the gas attains its maximum. The actual electrode-to-electrode voltage is therefore governed by the point where ϱs equals zero or reaches its minimum value.

Let us now assume that the ϱs values of the dust layers at all points of the electrode surface all lie between the $\varrho s = 0$ and the ϱs = constant curves. In that case, the real working point of the precipitator will lie on the perpendicular linking points A and B, and will depend on the average value, $(\varrho s)_{\mathrm{mean}}$. The higher this mean value, the closer will the working point be to point B; the lower the mean value, the more will the working point approach point A. The maximum voltage, however, will always be constrained by the $\varrho s = 0$ characteristic, i.e. by point A on the diagram.

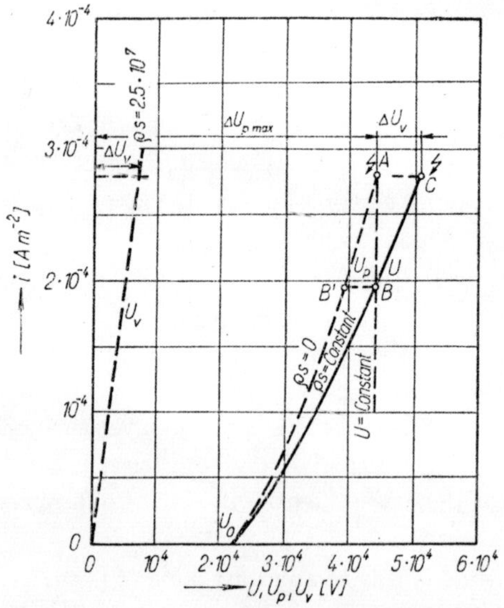

Fig. 87 How a dust layer on the electrodes affects the voltampére characteristic of a separator.

There are limits to the extent to which the potential gradient across the dust layer can be increased. At a certain specific value of this gradient, $\Delta U_v : s$, the dust layer will be punctured simply because the higher electrical stressing of the gas in the interstices between the dust particles will ionize that gas. The part of the dust layer where this abrupt ionization has occurred will then emit ions of a reversed polarity into the precipitator space. These ions will tend to neutralize the charged particles in that space, or the space charge of ions arising at the discharge electrode, and will thus interfere with the separating process. Moreover, the apparent specific resistance at the punctured point of the dust layer will diminish substantially when the puncture occurs; in our further considerations, we shall assume that it will drop to zero. The puncture of the dust layer will also reduce the breakdown voltage of the gas layer, and cause the voltampere characteristic to become steeper. This phenomenon, known as a back corona, is highly undesirable, because it markedly curtails the effective separating brisk velocity and disrupts the stability of the precipitator operation.

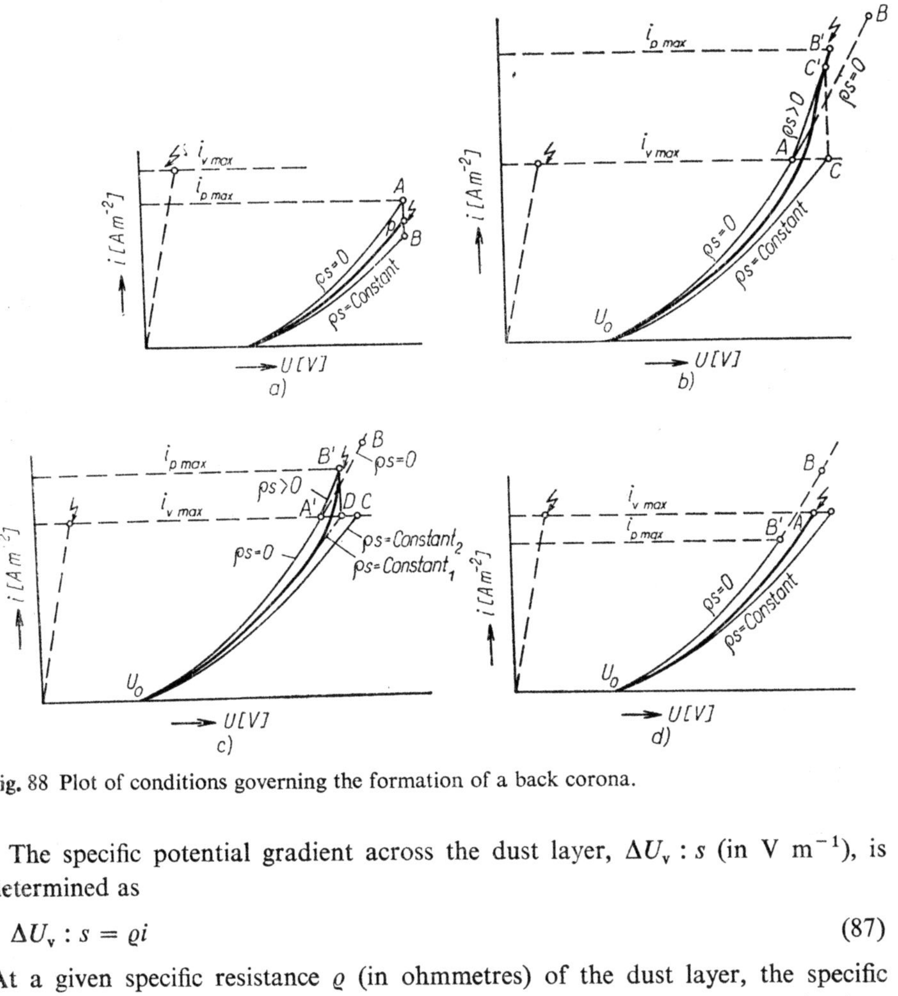

Fig. 88 Plot of conditions governing the formation of a back corona.

The specific potential gradient across the dust layer, $\Delta U_v : s$ (in V m^{-1}), is determined as

$$\Delta U_v : s = \varrho i \tag{87}$$

At a given specific resistance ϱ (in ohmmetres) of the dust layer, the specific potential gradient within that layer will therefore depend on the current density i (in A m^{-2}) of the current passing through the layer. In other words, this current density must be kept within certain limits, else the dust layer will break down.

In practice, we may encounter any of several sets of circumstances which will affect the response of the precipitator when a back corona arises. The decisive factors in this respect are two critical current densities – the density $i_{p\,max}$ at which the gas layer breaks down, and the density $i_{v\,max}$ at which the dust layer is punctured. Let us now examine the various cases one by one.

In the first case, illustrated in Fig. 88a, $i_{p\,max}$ is lower than $i_{v\,max}$, and no back corona is formed. This state represents the normal mode of functioning of an electrostatic precipitator when the dust layer has a favourable specific resistance. At a certain value of ϱs, the total voltage applied to the electrodes equals the sum of the potential gradients across the deposited dust and across the gas layer between this dust and the discharge electrode. When $i_{p\,max}$ is attained, there is an arc-over within the gas-filled precipitator space, but no puncturing of the dust layer. The current density in that layer never builds up to the $i_{p\,max}$ level, and so the specific voltage gradient across that layer is always less than $\Delta U_{v\,max}$.

In the second case, shown in Fig. 88b, $i_{p\,max}$ is larger than $i_{v\,max}$, and the breakdown voltage of the gas is higher than the voltage corresponding to point C on the diagram. At a zero thickness of the dust layer, when $\varrho s = 0$, the working point moves along the $\varrho s = 0$ curve, and can reach the maximum voltage level defined by point B. But if any part of the electrode is covered by a dust layer where ϱs is constant, the working point will shift along the $\varrho s =$ constant curve up to point C, whereupon the dust layer will be punctured. At this instant, the resistance of the dust layer will drop practically down to zero; the voltampere characteristic will assume a steeper slope; and the breakdown voltage of the gas will diminish. The working point will then jump from C to C', the dust layer will be punctured, and a back corona will form. This will be signalled by a sudden current surge at a virtually unchanged voltage level. Any further voltage rise will then displace the working point along the $\varrho s > 0$ curve from point C' to point B', at which point the gas layer will break down. Since the dust deposit on the electrode varies steplessly, between $\varrho s = 0$ and $\varrho s =$ constant, the voltampere characteristic is likewise stepless, as indicated by curve p in Fig. 88b. The voltage rise gradually widens the back corona region from the smallest to the largest s values, until the back corona covers the entire electrode surface, without any attendant breakdown of the gas layer.

A corona of this type is stable, so the voltage could be raised further until the gas breakdown voltage was reached. However, there would be no point in any further voltage rise, because at the location of the back corona no particles are trapped anyway. On the contrary, every further voltage rise will only intensify the emission of positively charged ions from the deposit layer, and step up the rate at which these ions cancel the particle charges and neutralize the negatively charged gas ions. The maximum separating effect would be obtained by operating the precipitator at or slightly above point A: there, a voltage rise could still affect the collecting efficiency more than the loss of collecting area caused by the gradual expansion of the back corona region.

In the third case, represented in Fig. 88c, $i_{p\,max}$ is again greater than $i_{v\,max}$, but the breakdown voltage lies somewhere between points A and C. Under these circumstances, we must distinguish between two different regions, defined by the ϱs values of the dust deposits there. If the latter lie between the $\varrho s > 0$ and the $\varrho s \leqq$ constant 1 limits, we are dealing essentially with the case described in the preceding paragraph; but in cases where constant 1 $\leqq \varrho s \leqq$ constant 2, the gas layer breaks down at the very instant the current density reaches the $i_{v\,max}$ level. Here too, the ϱs values alter continuously over the whole of the electrode surface, and hence the voltampere characteristic again forms a continuous curve, as seen in Fig. 88c. The resultant back corona is only partly stable, and spreads only over a certain part of the collecting electrode surface before an arc-over occurs in the gas layer. The region over which the stable back corona will expand is the larger, the closer the voltage at point B' is to that at point C – in other words the smaller, the closer the point B' voltage is to that at point A.

The fourth case, plotted in Fig. 88d, is encountered when the voltage at point B' equals that at point A, or else when the point B' voltage is smaller than that at point A, but the point B voltage is higher than that at point A. In these conditions, an arc-over in the gas layer follows immediately upon the first puncture of the dust deposit on the electrode. That puncture has to come first, because it has to reduce the gas breakdown voltage from B to B' before the arc-over in the gas can take place. We are here faced with an unstable back corona, which is instantly followed by an arc-over. Under these conditions, then, the working point must be kept slightly below point A.

Once the dust layer is punctured, what happens is that ions begin to form within it, and start moving out towards the discharge electrodes. As the polarity of these ions is opposite to that prevailing in the working space of the unit, they will tend to cancel out the space charges (of the dust particles and of the gas ions) previously created by the corona discharge. This is what we actually refer to as a back corona. It will obviously disrupt the stability of the precipitating process, and, even worse, will substantially reduce the separating brisk velocity.

There are in principle three distinct ways of combatting back coronas. One is to reduce the specific resistance of the dust layer, the usual approach being to lower the surface resistance of the particles. This is generally accomplished by narrowing down the difference between the operating temperature and the dew point of the gas. An alternative stratagem is to reduce the current density in the dust deposit, while maintaining an adequate operating voltage, by adding some electronegative gas. That reduces the mobility of the gas ions and raises the gas breakdown voltage. This device can, at least within certain limits, maintain an acceptable separating brisk velocity. The third countermeasure again involves a reduction in the current density to which the dust layer is exposed, but no attempt is made to avoid the low operating voltages and current densities which this entails. Instead, the requisite collecting efficiency is retained by considerably enlarging the collecting area, so as to allow for the very low mean separating brisk velocities which necessarily result from this approach. In practice, only the first and third of these alternatives are employed on any scale.

In the first mode, the dew point and service temperature of the gas are commonly adjusted by spraying water mist into suitable areas of the process equipment, or by adding sulphur trioxide to the gases. Another technique, used chiefly at steam

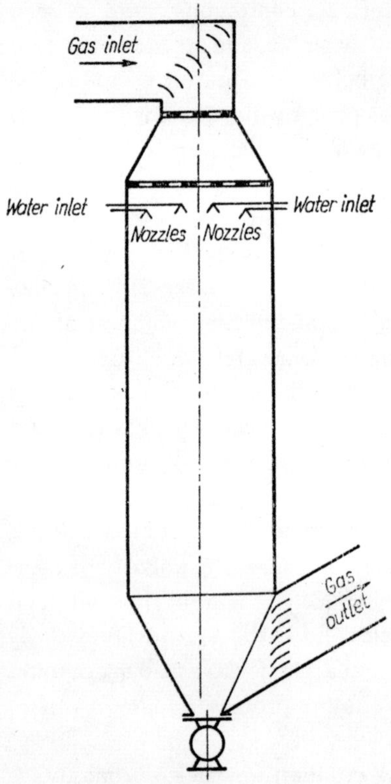

Fig. 89 Scheme of a gas conditioning unit, also known as a stabilizer.

power stations, is to control the combustion process so as to leave a certain percentage of uncombusted fines in the dust. These fuel residues reduce the otherwise high resistance of the dust deposits on the electrodes. Both these methods share the advantage of utilizing the existing process equipment for the corrective action, thus avoiding any substantial capital outlay. Where neither of them is practicable, the gas has to be treated in special-purpose conditioning chambers, one of which is schematically illustrated in Fig. 89. This is essentially a circular-section tower where the gas is held for a certain span of time while nozzles infuse water mist into it. All the moisture evaporates before the gas is allowed to leave the tower. Fig. 90 represents the resultant humidification process in terms of an $i - x$ (or enthalpy versus absolute humidity) diagram.

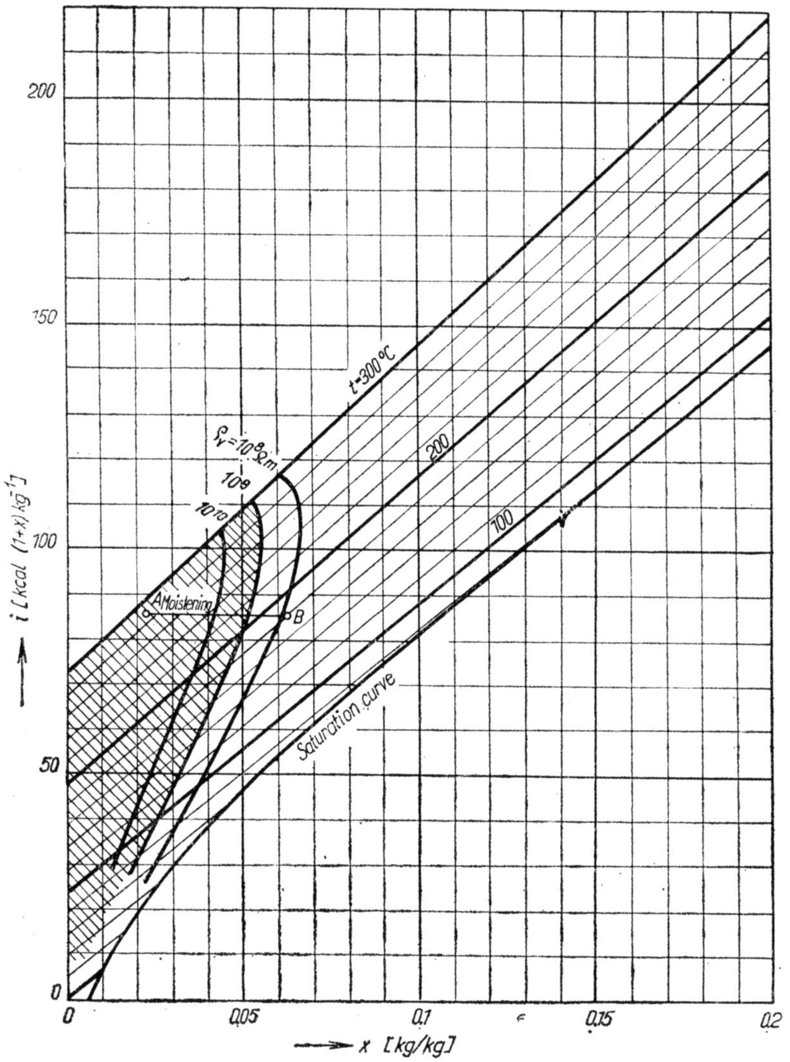

Fig. 90 An $i - x$, or enthalpy versus absolute humidity, diagram of the gas humidification process.

As evident from Fig. 90, the moistening process roughly follows an i = constant line. The curves in the diagram mark the specific resistance values of the dust at various constant values of ϱ and for various states of the gas. The hatched area denotes the domain in which the specific resistance of the dust exceeds 10^9 ohm-metres; the state of the gas must be such that operation within this hatched region can be avoided. The diagram further indicates the condensation threshold temperature; the gas inlet temperature must lie above this threshold if condensation, in the precipitator itself and downstream of it, is to be obviated. A useful point about this diagram is that it reveals both the amount of water that must be instilled into the gas stream, and the state of the gas at the outlet of the conditioning chamber.

The designers of gas conditioning equipment have to work within certain technological constraints. One of them is the minimum time which the gas has to spend in the unit. This interval will be the shorter, the higher the gas inlet temperature and the smaller the water droplets sprayed into the gas. The requisite time span is particularly sensitive to the gas inlet temperature, growing quite disproportionately as this temperature diminishes; that rules out the use of these units where the available inlet temperatures are not high enough. Moreover, to ensure that the gas will actually dwell in the unit for the specified time, the designer has to secure a uniform gas velocity throughout the unit. By the same token, he must avoid any dead spaces or pockets where the gas might cool down, and condensation might ensue. The dust that settles in the unit must be continually removed, as prolonged contact with the walls might cool it down and moisten it. This calls for special fittings at the chamber inlet and outlet ends. Special care has to be devoted to the nozzles, which govern the other really critical parameter of the unit – the droplet size. Finally, the unit could never perform consistently and efficiently without an effective automatic control system, which will maintain a closely specified state of the gas entering the electrostatic precipitator irrespective of any gas temperature fluctuations at the inlet to the conditioning unit. This latter requirement sometimes gets these gas conditioning units called stabilizers, because their prime objective is to stabilize the functioning of the precipitator which they serve.

3.3.2 High voltage sources

Industrial electrostatic precipitators are commonly run on rectified voltages of 40 to 80 kV, supplied in the form of rectified single- or three-phase currents from full-wave rectifiers. As it is vital that the operating voltage should be kept close to the arc-over voltage, and as the latter varies with the gas and dust parameters at the precipitator inlet (which in turn depend on the process conditions upstream of the precipitator), a variety of systems have been devised for automatic control of these feed voltages. Some of these systems detect and respond to the frequency and intensity of arc-overs in the precipitator; others discriminate the derivation of the voltampere characteristic; and others still simply keep the current flow through the precipitator at a constant level.

The predominant type of power source used to be rotary mechanical rectifiers, but these are no longer in general use. They required too much costly maintenance, generated high-frequency radio and television interference, emitted ozone, were noisy, and rendered relatively low unit outputs, which complicated their automatic control. Altogether, this is now considered too high a price to pay for their comparatively low first costs, so the trend throughout Europe is to resort to semiconductor sources employing selenium or silicon rectifiers. These units have an extra advantage in their small size, thanks to which even high-capacity power sources turn out relatively compact.

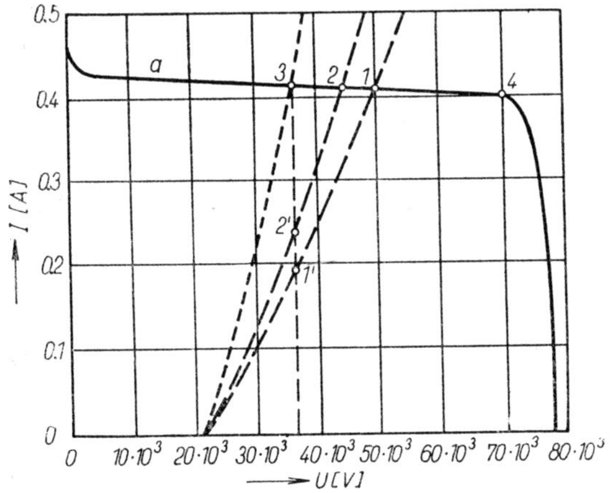

Fig. 91 Typical voltampére characteristic of a rectifier.

The relationship between the output voltage and current of the unit depends on the voltampere characteristic, which is specific to every particular rectifier type. In Fig. 91, this characteristic is represented by the area beneath curve "*a*"; the rectifier can operate at any point within the area so defined. The diagram also shows, in broken lines, the voltampere characteristics for one and for two electrostatic precipitator sections. The working point is determined by the intersection of the rectifier and precipitator characteristics. As evident from this chart, the simultaneous feeding of two precipitator sections from one and the same rectifier can cause the voltages and current intensities in both sections to drop, from the points marked *1* and *2* to points *1'* and *2'* respectively.

The rectifiers are sometimes set up at some distance from the precipitators, and linked to them by high-voltage cables, but that means inviting trouble from the transient effects induced by the capacitance of these cables. A much preferable approach, facilitated by modern semiconductor devices, is to locate the rectifiers on the precipitator roof and keep the power leads as short as possible. The controls are usually concentrated in a central control cabin. In these installations, it is advisable to cover the rectifiers with at least a light roofing, so as to protect them from direct sunlight, because rectifiers are liable to overheat when insolated or exposed to high ambient temperatures.

3.3.3 Dry electrostatic precipitators

Dry electrostatic precipitators are classified by the shapes of their collecting electrodes, and by the direction in which the gas flow traverses their working spaces. In *tubular precipitators*, the collecting electrodes are formed by a system of enclosed tubes, of circular or hexagonal cross section, which are suspended in the upright position, with the wire high-tension electrodes running down the centre lines of the individual tubes. *Plate-type precipitators* are fitted with collecting electrodes in the form of parallel plates, often variously shaped, and the parallel wire or point high-tension electrodes are located mid-way between the adjacent plates. The flow channel between two neighbouring collecting electrodes is termed a chamber. In dependence on the direction in which the gas flows through these chambers, we distinguish between *horizontal* and *vertical plate-type precipitators*. (In tubular precipitators, the flow is of course always vertical.) In all the vertical units, the gas normally flows upwards; precipitators with a downward gas flow are a rare exception.

The distinction between horizontal and vertical plate-type precipitators is by no means academic: when the gas flows horizontally, the wire or point electrodes of the high-tension system must be vertical, and vice versa. The way this affects the design of the unit will be discussed further on. A point to note is that in the plate-type units, the high-tension electrodes are always perpendicular to the gas flow, while in tubular precipitators they are parallel to it.

Precipitators are commonly subdivided into sections, each of which is powered by a separate high-voltage source. The sections may be linked together in parallel or in tandem, or in a mixed series-parallel configuration.

3.3.3.1 Tubular precipitators

Fig. 92 is a scheme of a typical dry tubular precipitator. The unit is housed in an upright casing, with a substantially larger cross section than the piping or ducting upstream and downstream of it. The raw gas enters through a vertical or inclined tube at the bottom of the casing and, as it moves upwards, decelerates from its initial 15 to 20 m s^{-1} to a mere 1 to 2 m s^{-1}. This is a key stage of the process, because a precipitator cannot function properly unless the gas velocity and particle concentration at the tube or chamber inlets are fairly uniform throughout its cross section. Consequently, the aerodynamics of the inlet plenum chamber are one of the trickiest aspects in the design of these units. The incoming gas is often distributed by means of various deflector or vane systems, but sometimes even that fails to produce the desired effect.

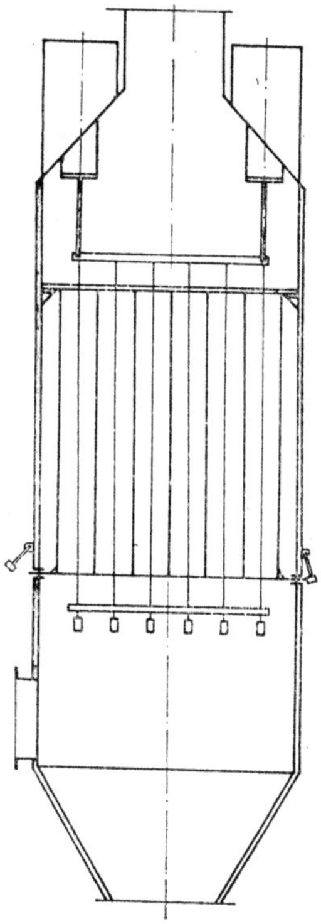

Fig. 92 Scheme of a dry tubular precipitator.

The gas next ascends into an array of round tubular electrodes, or into a honeycomb pattern of hexagonal ones; the latter layout makes the best possible use of the available space, as there are no gaps or unutilized spaces between the adjacent electrodes. The electrodes are suspended from, and fixed to the casing by, a transversal plate spanning their tops; this is sometimes supplemented by another plate across their bottoms. The carrier plate also seals off the dead spaces between the individual electrodes, so as to prevent any part of the gas from by-passing the flow channels within the tubes. The system must be complemented with a rapping mechanism, which for tubular collecting electrodes usually consists of hammers impacting against the tubes near their tops or bottoms.

The high-tension system consists of parallel wire electrodes, of circular or asteroidal sections, which are suspended from a grid at the top of the unit so as to pass down the centre lines of the individual tubes. The wires are kept taut by weights attached to their bottom ends; these weights are located by another grid, which ensures that their spacing is exactly identical to that of the wire suspension points. That, at least, is the theory. In practice, turbulence in the gas stream and other factors often caused this lower positioning grid to sway, inducing regular shorting or a succession of arc-overs in the electrode system. At one time, this used to be countered by the simple expedient of steadying the positioning grid against the unit casing by means of insulant spacers. Unfortunately, the spacers were exposed to the incoming gas stream, and were quickly covered with a dust layer on which vapour was apt to condense. That left the insulators coated with a conductive deposit, which frequently led to their shorting and/or cracking. Nowadays, therefore, these lateral supports are out of fashion, and the usual practice is to steady the lower grid by means of suspension tubes linking it to the upper suspension grid. The lower grid is designed to permit free vertical movement of the electrode weights, but to prevent their lateral displacement. That keeps the wire electrodes taut irrespective of the different rates of thermal expansion or contraction of the various components. The upper grid is suspended from the carrier beams of the casing by a system of usually four tie rods and load-bearing insulators. The latter should be carefully located and shielded so as to be inaccessible to both the gas stream and the admixture particles. Lately, designers increasingly protect these insulators against fouling with dust and moisture by enclosing them in separate housings. The high-voltage lead that feeds the wire electrode system is generally connected up at one of these suspension insulators.

Like the collecting electrodes, the high-tension system also needs rapping to keep it clear of deposits. More often than not, this is done by hammers striking the upper suspension grid from the side or from above.

The upper part of the casing narrows towards the outlet tube or duct, which presents a further set of flow control problems. The geometry of the outlet itself, and that of the tapering transition between the casing and the outlet tube, are liable to affect the flow pattern within the outlet plenum chamber. All the partitions or baffles which some makers install in these compartments cannot altogether eliminate these effects.

Tubular precipitators have one inherent drawback which no refinement in their design has so far overcome. Any dust dislodged from their electrode systems, by rapping or otherwise, has to drop through the lower positioning grid and down the inlet chamber against the ascending gas stream. Inevitably, much of it will be re-entrained before it can reach the hopper at the bottom of the unit, beneath the inlet opening. Moreover, the electrode surfaces are smooth, which affects both the retention of dust and the clearing of deposits on them. In view of these shortcomings, and of the unavoidable complications in the design and fabrication of these units, tubular precipitators are nowadays used only in isolated cases where some special circumstances, such as high service temperatures, justify the penalties involved.

3.3.3.2 Horizontal plate-type precipitators

Easily the most common type of electrostatic precipitator are horizontal plate-type units, an example of which is shown in Fig. 93. They are housed in casings,

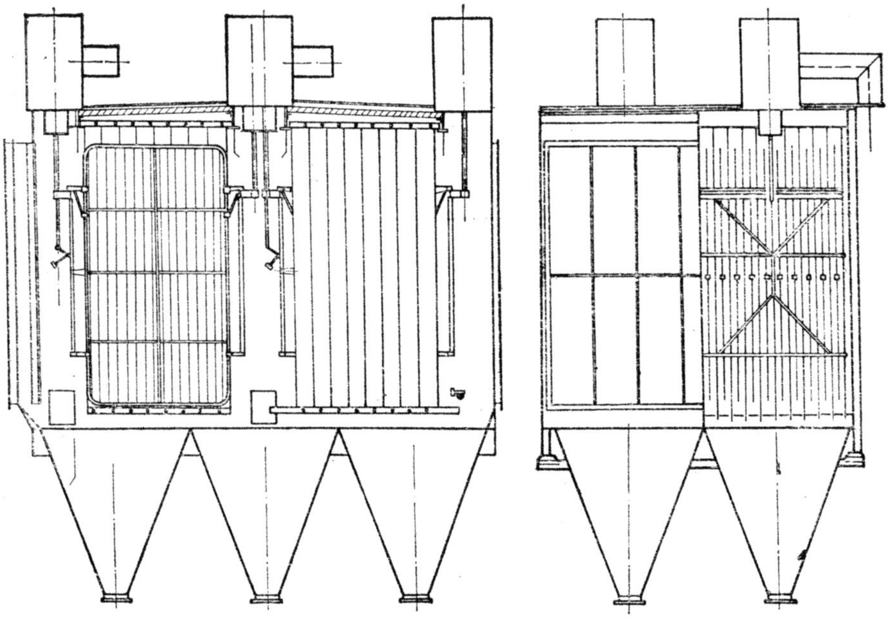

Fig. 93 Scheme of a dry horizontal plate-type precipitator.

usually of rectangular section, which are linked to the narrower tubing or ducting ahead of and beyond them by divergent and convergent transitions. The latter can either be straight, or, when the inlet or outlet tubing are set at an angle to the precipitator axis, can be combined with a bend or elbow. To secure an even flow rate around all the plate electrodes, the inlet chambers of these units, and sometimes their outlet plenum chambers too, are often provided with a variety of partitions, baffles or guide vanes.

The collecting electrodes are suspended parallel to each other in the upright position. The high-tension electrodes, located mid-way between the adjacent collecting electrodes, are formed by an array of parallel, vertically suspended and evenly spaced wire or point electrodes. The collecting electrodes consist of variously shaped or profiled strips, which are usually pivoted. Their trunnions rest on suspension beams which are supported on the main girders of the casing. The bottom ends of the electrode strips are engaged by the lugs of a rapping beam that runs down the length of the unit. This beam is free to move in the lengthwise direction, but is restrained transversally, and serves to transmit the impacts of the rapping mechanism to the electrode strips. It is important that both this beam and the upper suspension beams should be precisely located laterally, because their location governs the exact mutual spacing of the collecting electrodes, and that is crucial for the correct functioning of the precipitator.

The high-tension systems are borne by a usually tubular frame, which is subdivided by horizontal and vertical reinforcing members into discrete frame sections. The wire or point discharge electrodes are fixed to and tautened by these frame sections. The electrode frame is generally supported on transversal carrier frames, and located on them by suspension and guide arms bolted to its front and rear uprights. This arrangement keeps all the high-tension electrodes truly parallel, equally spaced, and aligned exactly half-way between the collecting electrodes that flank them. The transversal frames are as a rule rectangular structures, often stiffened by diagonal bracing, and are suspended, mostly by four tubes, from the suspension insulators. The latter may be housed in separate casings, or in enclosed compartments within the main unit casing, but must always be shielded from the main stream of dust-laden gas and adequately protected against any fouling of their internal surfaces. One of these insulators also carries the high-voltage feed to the unit.

The high-tension system described above is the one now preferred throughout Central Europe, but in principle there is no reason why these electrodes and their suspensions should not be designed along the same lines as in tubular precipitators. In that case, the wire or point discharge electrodes are suspended from a horizontal grid at the top of the unit, are tautened by weights attached to their bottom ends, and are prevented from swinging or swaying by a grid that precludes any horizontal displacement of these weights. The upper suspension and lower positioning grids are usually interconnected by spacer tubes to rule out any horizontal deflection of the system as a whole.

The collecting electrodes are generally rapped in either of two ways. The one which is now gaining ever more widespread acceptance relies on hammers which are raised by mechanical means, and then drop by gravity onto a buffer on the rapping beam. The lugs of this beam transmit the impacts to the individual electrode strips. The most common alternative is a dog or cam mechanism which shifts the whole rapping beam longitudinally, so that the lugs on this beam tilt the individual electrodes around the trunnions or pivots at their tops. Once the dog or cam disengages, a return spring drives the beam back against the similar beam in the next adjacent section of the precipitator, and this impact is again transmitted to the collecting electrodes.

The high-tension electrodes are normally rapped by hammers which strike anvils on the uprights of the tubular frame, at either one or two points, in dependence on the size of the frame. The impacts are usually delivered from the side, in the longitudinal direction. The hammers are mostly pivoted on the frames, and are raised by any of a variety of mechanisms; the latter must always be electrically insulated from their drive motors.

The hoppers at the bottoms of these precipitators are often provided with various internal fittings to keep out the gas flow and thus prevent any re-entrainment of the dust that has already settled there. It is now standard practice to mount the whole unit casing, including the hoppers, on special antifriction bearings which permit its unimpeded thermal expansion and contraction.

Horizontal plate-type precipitators as a rule consist of two sections in tandem, but for more exacting applications are sometimes split into three consecutive sections. A section, of course, is that part of the precipitator which is fed from one and the same high-voltage source. The reason why the units are split into sections is that this makes it easier to run them at the optimum operating voltages. Each section employs a differently rated power source, and a different operating voltage, which is kept as close as possible to the arc-over voltage in that section. Furthermore, a breakdown in one section, or in its power source, still leaves the rest of the unit capable of trapping at least some of the dust.

3.3.3.3 Vertical plate-type precipitators

As evident in Fig. 94, the casing of a vertical plate-type precipitator forms a vertical flow channel of a larger cross-sectional area than the tubing or ducting leading into and out of it. The gas enters at the bottom of the casing, through a vertical or inclined inlet, and emerges at the top, through an outlet which is usually set into the centre line of the unit and linked to it by a tapering transition piece. The inlet plenum chamber is generally partitioned and/or provided with deflectors, so as to ensure a uniform flow rate and particle concentration at the leading edges of all the plate electrodes.

The collecting electrodes are mostly of the enclosed type, i.e. are hollow and enclose spaces which are shielded from the gas flow. Each electrode is built up of a number of horizontal strips. The upper edge of every strip is profiled so as to catch the dust that drops down from the strip above it, and to divert that dust into the shielded interior channel within the electrode. Moreover, the strips are as a rule provided with shaped slots or pockets which serve the same end. The individual strips are pivoted on two or three vertical bars, and are arranged in mutually opposed pairs with the shielded channel running down between them. This emphasis on shielding the space through which the dust is removed is easily understood, once we consider that the dust dislodged by rapping has to drop against an ascending gas stream. The hopper is built into the base of the casing in a way which allows the gas flow to pass around it, and which ideally should facilitate an even distribution of the incoming gas and dust over the entire cross section of the unit.

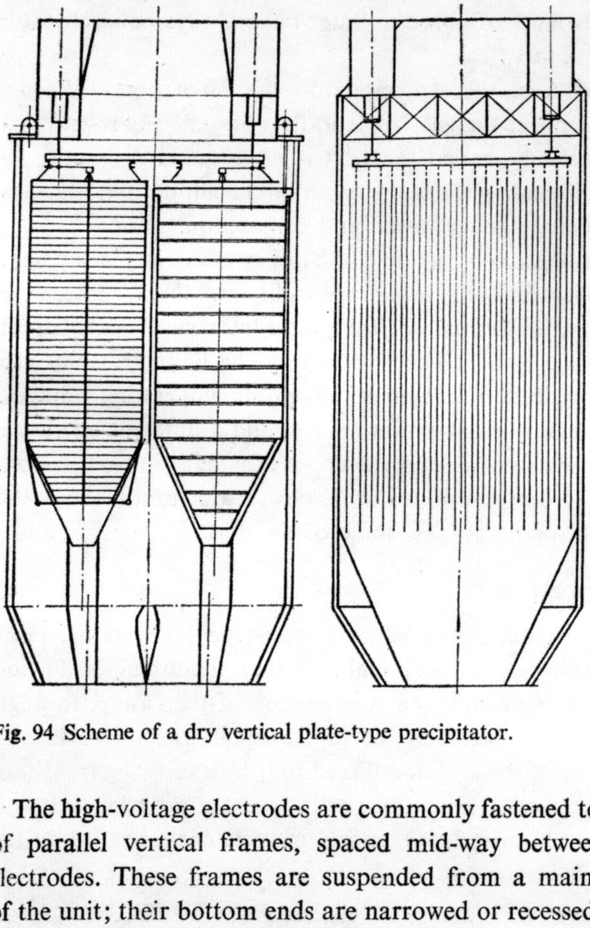

Fig. 94 Scheme of a dry vertical plate-type precipitator.

The high-voltage electrodes are commonly fastened to and tautened on a system of parallel vertical frames, spaced mid-way between the adjacent collecting electrodes. These frames are suspended from a main carrier frame at the top of the unit; their bottom ends are narrowed or recessed, and mutually interlinked by spacer bars. That in itself, however, would not prevent the whole high-voltage system from swaying with a pendulum motion. At first, the usual way of forestalling any such swinging motion was to hold the horizontal spacer bars in place by means of supporting insulators. These were generally set into housings on the flanks of the precipitator casing, but even this measure failed to prevent the accumulation of dust deposits on them. Once moistened by condensation, the deposits were prone to cause shorting of the high-voltage system. Nowadays, therefore, the prevalent practice is to provide the bottom of this system with a rigid, diagonally braced transversal frame; the resultant box-like structure is too stiff to deflect laterally.

The carrier frame which bears the entire high-tension system is mostly hung on four tubes from the suspension insulators mounted on the main girders of the casing. These insulators are commonly enclosed within sealed housings, and protected against any fouling of their interior surfaces with deposits which might impair the insulating efficiency. One of these insulators is connected up to the high-voltage source that feeds the system.

The collecting electrodes of these units are most commonly rapped by a mechanism which lifts one end of each of the carrier beams from which the array is suspended. Since the collecting electrodes form parallelograms, this action raises one end of each of the electrode strips. When the beams are subsequently released, and drop onto a buffer, the suspension rodding and pivots transmit the resultant impacts to these strips. The high-tension system is rapped by hammers which drop from above onto anvils fixed to the electrode frames.

Vertical plate-type precipitators have one salient advantage over the horizontal units: they take up very little floor space. They are therefore employed mainly where this space is at a premium. They also allow large throughflows of gas to be handled, very conveniently, by installing two or more precipitator sections in parallel, side by side, within one common casing. Their main drawback is the complicated flow pattern at their entries, which usually makes it very difficult to attain a uniform velocity distribution at the leading edges of all the collecting electrodes.

3.3.3.4 Principal components of precipitators

The principal components of any electrostatic precipitator are the housing, the collecting electrodes, the high-tension electrodes, the suspensions for both these electrode systems, and the rapping provisions for the two electrode systems.

3.3.3.4.1 Precipitator housings

Horizontal precipitators are encased in what is essentially only an enlarged part of the horizontal flue duct. The enlarged cross section is necessitated by the fact that the flow velocity in the active space of the precipitator must be kept down to only about one seventh to one tenth of that in the flue duct itself. The inlet and outlet ends of the housing must therefore be linked to the upstream and downstream ducting by conical or tapered adaptors. Depending on the amount of space available and on its configuration, the flue duct or piping may enter and leave the precipitator housing perpendicularly, coaxially, at an angle in the vertical plane, or even from the side. Given enough space ahead of the precipitator and beyond it, the simplest arrangement is to link the unit to the ducts by two straight adaptor pieces, a divergent and a convergent one, as shown in Fig. 95. These adaptors are then connected to the ducting proper by shaped transition tubes. Nowadays there is an increasing tendency to save space by combining the adaptor with a tube bend or elbow, as seen in Fig. 96.

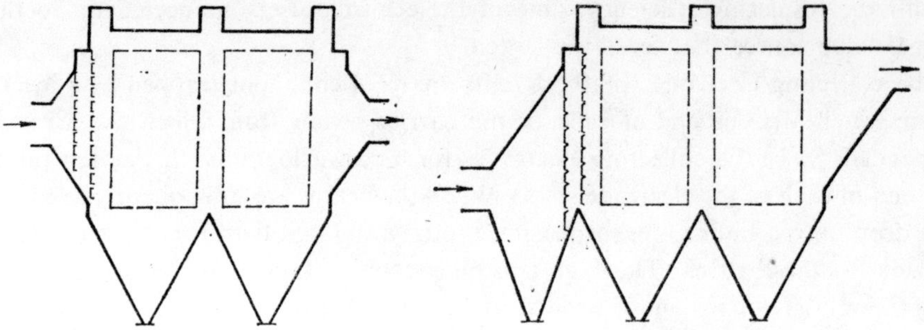

Fig. 95 Precipitators linked to their inlet and outlet ducts by straight adaptors.

Apart from reducing the flow velocity within its interior, the housing must also be designed to ensure a uniform velocity and an even distribution of the various flow components throughout the space it encloses. That usually calls for internal fittings, such as guide vanes in the shaped transition pieces and baffles in the adjacent parts of the ducting. The shapes, sizes and locations of these vanes and partitions are mostly established experimentally, on laboratory models. This is vital especially for complex systems, and where critical demands are imposed on the collecting efficiency, as the latter suffers when the flow velocity distribution is less than fully uniform. However, all the technical and economic factors involved must be taken into consideration before any decision can be reached as to whether to attain the requisite efficiency by means of a bigger precipitator, or whether to use a smaller unit and invest the money thus saved into vanes and baffles.

The hopper at the bottom of the housing must have walls sloping at a sufficient angle to ensure that the trapped dust will flow down them freely. The critical areas in this respect are the corners at the intersection of two walls, where the slope angle is smaller than on the walls themselves. These potential trouble areas are nowadays increasingly avoided by the use of trough-shaped hoppers or chutes, which afford an adequate inclination at all points. However, since there would be no point in merely shifting the bottleneck one step further down the material flow route, these troughs need highly efficient worm or other bulk conveyors to carry the discharged material clear of them (see Fig. 97). The hoppers are generally fitted with baffles to preclude any escape of the processed gas from the active space of the precipitator, as that would both impair the efficiency of the precipitation process and tend to unsettle and whirl up the dust within or approaching the hopper. When dust deposits are allowed to remain in the hoppers for any length of time, they are apt to be compacted, to cool down by contact with the walls, and consequently to pick up moisture. Once that has happened, the deposits are difficult to clear out. Moreover, if dust is left to accumulate in the hoppers, there is always a risk that the deposit will build up to the level of the high-tension system and then cause permanent shorting, thus putting the whole precipitator out of action. It is therefore of some importance for reliable operation that the hoppers should be emptied continuously rather than periodically.

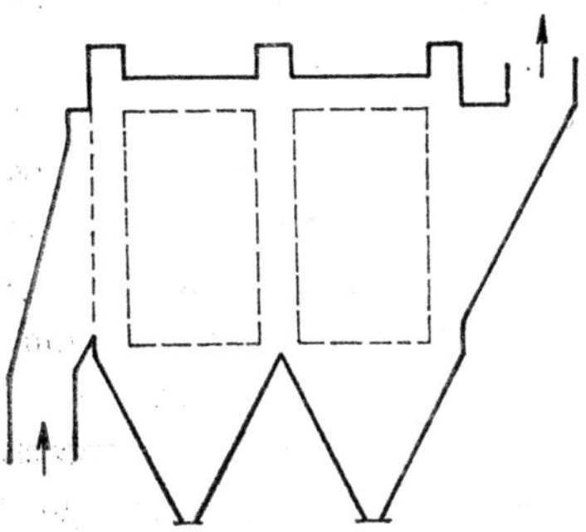

Fig. 96 A precipitator set at an angle to its inlet and outlet ducts.

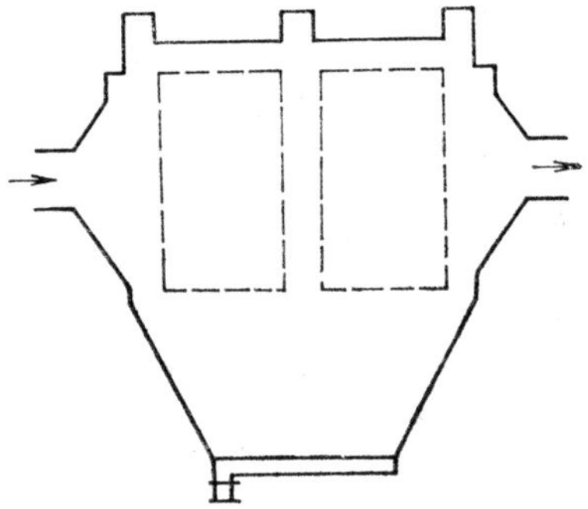

Fig. 97 A discharge hopper formed into a trough to improve the material flow.

One aspect of housing design that must not be underrated is the mechanical rigidity and load-bearing capacity of the structure. The housing has to support both of the electrode systems, and especially the collecting electrodes are apt to be extremely heavy. Both these systems are suspended from the main horizontal girders which rest on the side walls of the housing, or on pillars which form an integral part of those walls. The housings themselves are commonly borne on antifriction bearings to facilitate their thermal expansion and contraction in all directions. Furthermore, electrostatic precipitators mostly operate with a partial vacuum inside them, which exposes their walls and roofs to considerable forces (proportional to the actual pressure differential). The housings are moreover subjected to the various forces exerted by the transition pieces, adaptors or ducting. Further forces are imposed on them by the snow loading on their roofs and wind loading on their sides; the latter is usually taken up by a system of oblique struts and braces, located in the same plane as the horizontal girders of the housing. Another factor which complicates the housing design are the numerous manholes which must be provided at various points to afford access to the interior. The manhole covers must seal tightly under all foreseeable circumstances, and must as a rule be provided with locks. At each manhole a grounding lead, or a chain with a weight at its end, must be so arranged as to ground (earth) the high-tension system before anyone can actually enter the housing.

Apart from sheer mechanical strength, another design aspect that calls for special care is the tightness of the housing. In view of the internal vacuum, any leakage would result in the induction of relatively dry atmospheric air. That would both impair the stability of the precipitator operation, and lower the collecting efficiency by raising the flow velocity within the unit. Besides, the ingress of outside air could cause local cooling of the gases to less than their dew point, which always involves grave corrosion hazards. The housing roof must not only be tight, but also suitably sloped and provided with gutters, to prevent rainwater and other moisture from entering the unit or forming pools on top of it.

The top of the housing must generally be removable, to facilitate repairs and replacements of the internal components. The sealing problems are usually overcome by welding a thin sheet-metal strip or angle both to the roof and to the side walls and girders or pillars along the entire length of the housing. Before the top is dismantled, this strip is cut apart with an oxy-acetylene torch; on completion of the re-assembly work, a new strip is welded into place to seal the joint.

The interior of the housing has to be partitioned to prevent any as yet unprocessed gas from by-passing the active space of the precipitator, for instance through the gaps between the outermost collecting electrodes and the housing walls.

The high-voltage feed to the precipitator must be fully sheathed or encased, and all the suspension insulators must be enclosed within the main girder system and accessible only through lockable manholes or doors. These precautions are essential for the safety of the operating and maintenance staffs, who must be given no chance of coming into contact with live high-voltage conductors.

Precipitator housings are mostly fabricated in common structural steel grades, though special aluminium alloys or stainless-clad steels are sometimes employed when acute corrosion risks are anticipated. There were times when corrosion used to be avoided by building the housings of concrete, or, for higher service temperatures, of brick masonry, but these housings were soon found liable to spring leaks. The concrete was apt to crack, especially at the points where steel fittings, with a different coefficient of thermal expansion, were embedded in it.

Sometimes the whole housing has to be thermally insulated. This is the case when the service temperature of the unit is appreciably higher than the ambient temperature, and the nature of the operation entails a risk that the gas might be cooled down sufficiently for vapours to condense upon or within the housing. The object of the insulation is to minimize the amount of heat abstracted through the housing material, and thereby the temperature gradient along the precipitator when the latter is operating in the steady state.

The key consideration, however, is always the structural rigidity of the housing, which has to counter the strains and stresses imposed by various combinations of loading states. The foremost concern is to preserve the correct location of all the working parts, which is determined by the main girders of the housing structure. These girders naturally tend to deflect under the applied loads, so the first step is to list the principal loads that act on them. One is the deadweight of the two electrode systems; another are the forces generated by the internal vacuum, which are transmitted to the girders primarily by the housing roof; in outdoor locations, there are also snow loads to consider. Further deflections are caused by the temperature gradient within each girder. As these grow with the temperature difference between the upper and the lower chords of the girder, they can be reduced by insulating the whole girder so as to keep both its chords at much the same temperature.

The amount of girder deflection caused by the weight of the electrode systems will obviously depend on the quantity of dust that has settled on them. Similarly, the deflections induced by the internal vacuum will vary all the way from zero, when the unit is shut down, to a maximum at the lowest pressure attainable within the unit. This implies that in service, both the electrode systems will be apt to shift vertically, and that these displacements will vary from one row of collecting electrodes to the next. These factors must be taken into account throughout the design work, and particularly in devising the system that locates and guides the collecting electrodes between the rapping bars.

The side walls of the housing are subjected to buckling stresses imposed by the girders and roof, and also tend to deflect inwards under the effects of the internal vacuum. That calls for a gap between the inner surfaces of these walls and the outermost collecting electrodes, to prevent the buckling from affecting the electrode locations and suspensions. The gap must of course be sealed by partitions, to preclude any of the incoming gas by-passing the active space of the unit.

In vertical electrostatic precipitators, the housing is again in effect merely an extended section of the vertical ducting, so the requirements on it and the underlying design considerations are basically similar to these which apply to horizontal units. In the vertical precipitators, the gas inlet area must be provided with a chute or trough to receive the dust that drops down the spaces enclosed by the collecting electrodes, and to deliver it to the discharge hopper at the bottom of the housing.

3.3.3.4.2 Collecting electrodes

The collecting electrodes are among the most important working parts of any electrostatic precipitator. That has always singled them out for the special attention of the designers, which in turn has resulted in the evolution of a huge variety of diverse electrode types. The collecting electrodes of tubular precipitators differ so widely from those of plate-type units that the two categories are best examined separately.

Collecting electrodes of tubular precipitators

In tubular precipitators, the collecting electrodes always form a system of parallel flow channels, which vary widely in their shapes, and which are grouped into blocks. In the most common design, round tubular electrodes are suspended in rows or in staggered patterns, as illustrated in Fig. 98. The staggered layout is preferable when the precipitator housing is cylindrical. Other currently common configurations are honeycomb patterns formed of hexagonal-section electrodes

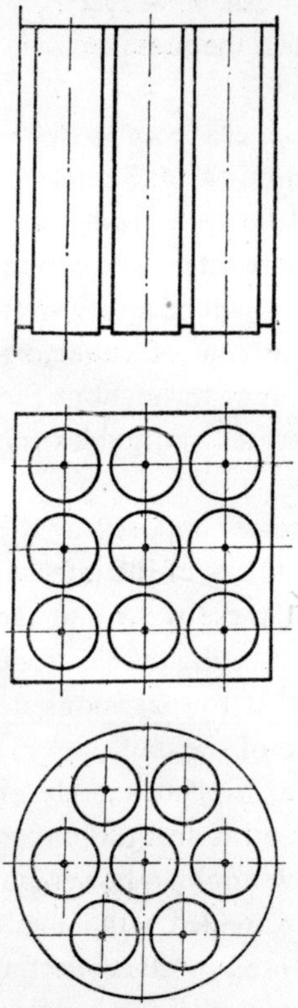

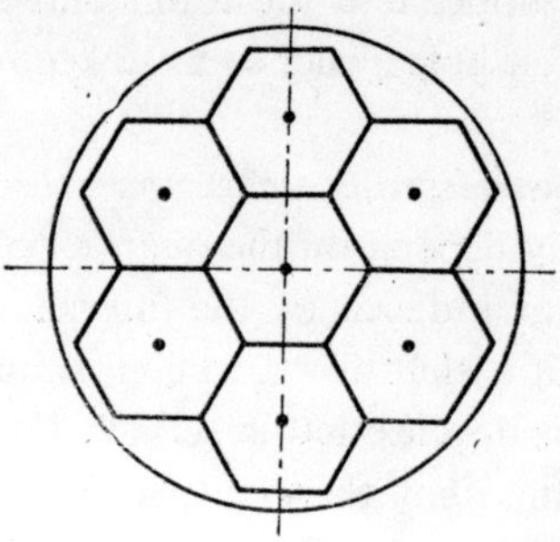

Fig. 99 Hexagonal-section collecting electrodes in a honeycomb array in a tubular precipitator.

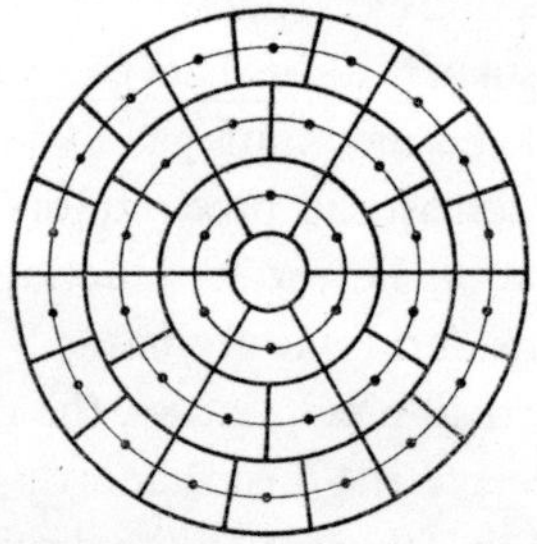

Fig. 98 Patterns in which tubular collecting electrodes are installed in tubular precipitators.

Fig. 100 Concentric cylindrical collecting electrodes forming radially partitioned annular flow channels in a tubular precipitator.

(Fig. 99), and systems of coaxial cylinders where the annular flow channels are subdivided by radial partitions (Fig. 100). This last system and the honeycomb layout are used only in cylindrical precipitator housings, but have the undeniable advantage of making the fullest possible use of the space available within these housings.

The principal requirements on the collecting electrodes of tubular precipitators are that the individual flow channels should conform exactly to the specified dimensions, and that the whole electrode array should always remain straight and precisely aligned.

Collecting electrodes of plate-type precipitators

In plate-type precipitators, the collecting electrodes are made up of discrete strips hung in a row so as to form an essentialy planar electrode surface. So many different types of these electrodes have been evolved over the years that it woud be beyond the scope of this book to review more than a few of the most important designs.

The chief requirements on these electrodes apply to their shape, which must secure the desired electrical properties; to their mechanical properties, which must ensure an adequate stiffness and prevent vibration; to their weight, which must be low; and to their aerodynamics in the trapping and disposal of dust particles. The electrical properties are generally assessed by comparing the electrode with an ideal model, consisting of a smooth and truly planer plate serving in conjunction with a system of parallel wire or point electrodes. This criterion is used mainly to judge the arc-over voltage and the distribution of the electric field intensity in various rival electrode designs.

The degree of stiffness attained in these electrodes is important because it affects the vibration of the strips, especially when the electrode is being rapped. Efficient rapping depends primarily on the most uniform possible distribution, over the whole electrode surface, of the maximum transversal acceleration (resulting from the lateral oscillation of the electrode under the rapping impacts). This aspect is now being increasingly stressed, and with good reason: in earlier times the accelerations imposed by rapping never exceeded some 100 g at the most, but it has since been discovered that really dependable removal of the trapped dust calls for some 500 g or even more.

Lightness is an obvious requirement, because the weight of the electrode system will reflect in the weights of the housing and of all the load-bearing structures associated with it. Any weight increment in the electrode system is apt to multiply as its consequences work their way through the whole unit design.

The aerodynamic properties of the electrode are expected to meet two often conflicting requirements. One is that they should allow the secondary flow generated by the electrical wind, and as far as possible the primary flow around the electrode too, to be exploited for enhancing the collecting efficiency. In other words, they must permit the existing mechanical flow forces to be utilized for improving the functioning of the unit, which is much easier said than done. The other requirement is that, in the course of the rapping operation, they must help to prevent any re-entrainment of the dislodged dust into the gas stream. As much as possible of that dust must settle and subsequently drop down within dead spaces shielded by the electrode, where it is less likely to be dispersed back into the active space of the precipitator.

The evolution of these collecting electrodes started out with smooth planar types. These were abandoned because they severely limited the separating brisk velocity, and made it too dependent on the specific collecting area (i.e. on the flow velocity in the precipitator). The root of the trouble was mainly the high proportion of dust which, when dislodged by rapping, re-entered the gas stream and the active space of the unit.

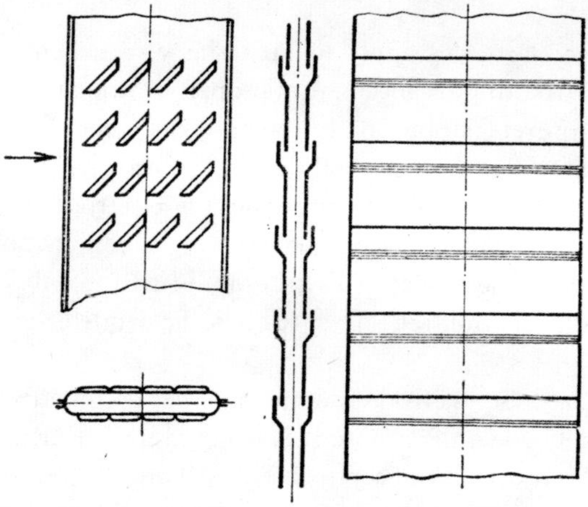

Fig. 101 Enclosed collecting electrodes for plate-type precipitators.

Development work intended to obviate this problem led to the design of various enclosed electrodes, an example of which is shown in Fig. 101. The two strips that comprised such an electrode enclosed a space between them which was open at the bottom; they were provided with pockets, slots or grooves of various shapes and sizes, orientated in different directions. On rapping, the accumulated dust dropped down the electrode surface to be caught in the pockets or slots, which diverted it into the enclosed space. Once there, it was sheltered from the gas flow, and continued its drop undisturbed to the bottom of the unit.

Unfortunately, these electrodes were heavy, laborious to produce, and the protrusions on their surfaces reduced the arc-over voltage, so that the electrical properties were inferior to those of smooth planar electrodes. True, these types afforded higher separating brisk velocities, especially in the domain of higher flow velocities (around 2 to 2.5 m s^{-1}), but they were not always an unqualified success. They proved good at handling loose, freely-flowing dusts, for instance dust with a significant sand content, but more adhesive fines soon clogged the pockets. Once that happened, only a small proportion of the dust dislodged by rapping actually passed through the hollow of the electrode, and the expected advantages of the type were practically lost. Moreover, the surfaces broken by the pockets proved difficult to clear of deposits by normal rapping. The grooved or slotted types did not clog as easily as those with pockets, so a higher proportion of the dislodged dust took its intended route through their interiors. Still, they shared another major drawback of these types – the dust was apt to fill the hollows within the electrodes, and was not readily removed from them.

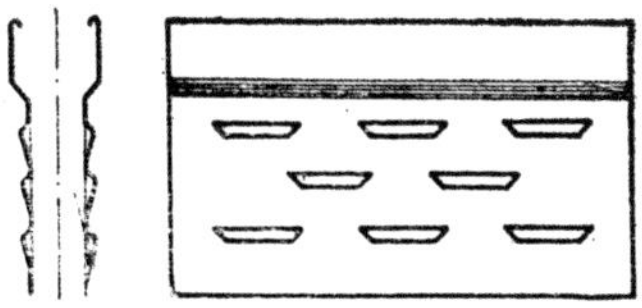

Fig. 102 A grooved plate electrode with collecting pockets.

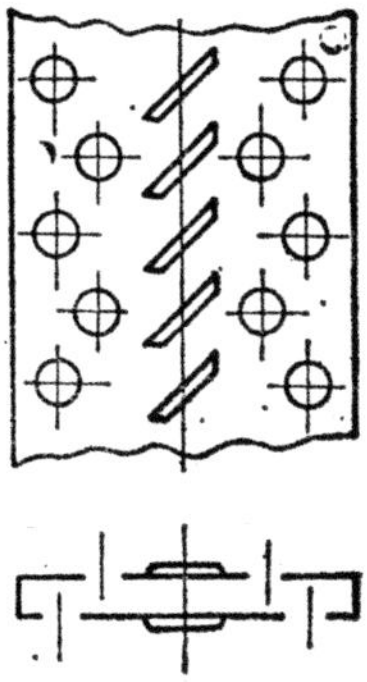

Fig. 103 A perforated plate electrode fitted with pockets.

Nowadays, grooved or slotted electrodes – sometimes with pockets in the pattern as well, – like the one in Fig. 102 are used only in vertical precipitators where the gas flows upwards. In these units, it is vital that the dust loosened by rapping should be effectively shielded from the opposed gas flow as it drops towards the hopper, since otherwise it would almost certainly be re-entrained. To improve the electrical properties in these applications, the wire electrodes are installed horizontally and mid-way between the groove locations, so as to retain the highest possible arc-over voltage between neighbouring electrodes.

The next stage in the evolution of these electrodes were perforated enclosed types, which made the most of the inertia of any particles that penetrated into their interiors. Sometimes the perforations were combined with pockets, located at points where these protrusions could not affect the arc-over voltage; one such design is shown in Fig. 103. For all their proven advantages, none of these types were ever adopted on any scale, because all of them shared two serious shortcomings. Firstly, they were so heavy and expensive as to influence the weight and cost price of the whole precipitator, without offering any really commensurate benefits. And secondly, they rendered only a poor performance when very fine fractions had to be captured with a high level of collecting efficiency.

The steadily rising demands on the performance of collecting electrodes next led to the development of profiled electrodes, sometimes known as the semi-enclosed type. This type stood out by its low weight, good electrical properties, and by its effective utilization of the electrical wind for improving the particle trapping efficiency. Also, the great majority of the trapped particles was led off through the dead spaces formed by the electrode, without any risk of its entrainment by the gas flow.

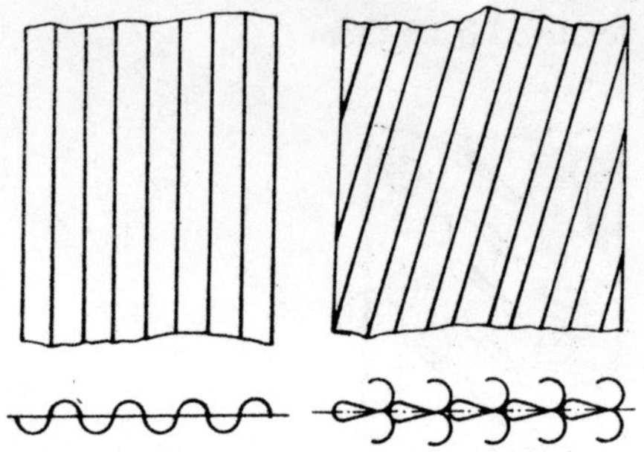

Fig. 104 Corrugated and W-section collecting electrodes.

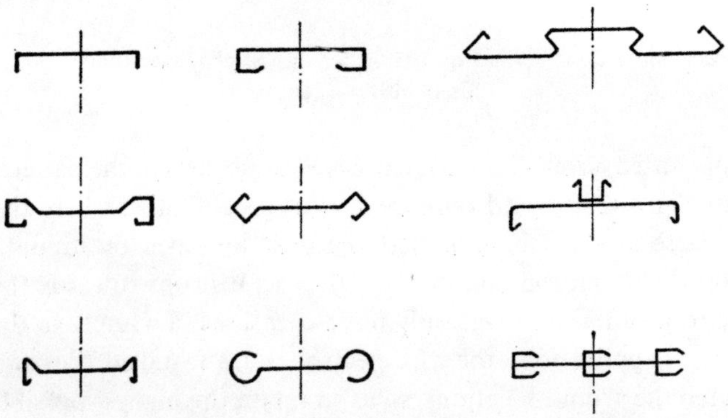

Fig. 105 A selection of common profiled electrode sections.

Some of the earliest profiled electrodes, which still fell short of fulfilling all these requirements, were the corrugated and W-shaped types seen in Fig. 104. They represented an advance mainly because the dust that settled in the dead spaces could now reach the discharge chutes or hoppers with a minimum of interference from the gas stream. Over the past decade, the development of profiled electrodes has made a good deal of progress, and has thrown up a spate of various proprietary designs. Fig. 105 presents some of these more recent innovations, all of which more or less fulfill the stipulations set out above. On the whole, the service behaviour of these types is distinctly superior to that of their predecessors. In particular, all these types lend themselves to higher separating brisk velocities, especially at higher flow velocities. They are also substantially lighter, and as most of the strip profiles can be produced by rolling, their manufacture is not nearly as laborious and costly as that of the earlier fully enclosed types. Tests of these newer profiled electrodes have demonstrated that the transversal acceleration produced by rapping is fairly evenly distributed over their surfaces, even at accelerations in the region of several hundred g. This means that even highly adhesive and fine-grained components can be dislodged from them, along with the rest of the dust deposits, without any special problems.

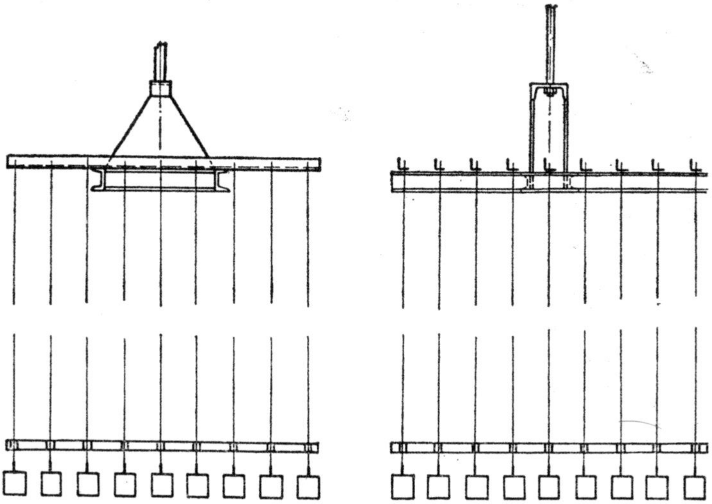

Fig. 106 The high-tension electrode system of a typical tubular precipitator.

3.3.3.4.3 High-tension electrodes

Another key working part of electrostatic precipitators are their high-tension electrodes. In tubular precipitators, these take the shape of wire discharge electrodes which run down the centre lines of the flow channels within the collecting electrodes, as evident in Fig. 106. The wires are suspended from a horizontal frame at the top of the unit, and tensioned by weights attached to the bottom end of each wire. The suspension arrangements must permit the precise alignment of every wire individually into the axis of the relevant flow channel. The weights must be heavy enough to keep the wires taut and straight under any foreseeable circumstances. To prevent the wires swinging in pendulum fashion while the unit is operating, the lower parts of the weights are laterally retained by a spacer frame.

In horizontal plate-type precipitators, most high-tension systems are made up of an array of parallel vertical wire or point electrodes spaced out in rows equidistant from the adjacent collecting electrodes. The total current flow that can be passed through the precipitator depends, among other factors, on the spacing of these electrodes along the gas flow direction: the closer this spacing, the greater the aggregate length of all the discharge electrodes, and the greater should be the current passing through the precipitator. However, too closely spaced electrodes shield each other, which reduces the current per unit of length of the wires, and therefore the maximum current value is achieved only at a certain precisely defined electrode spacing. In present-day practice, the gaps between neighbouring wire electrodes are prevalently kept to about 1.1 or 1.2 times the distance between these and the nearest collecting electrodes.

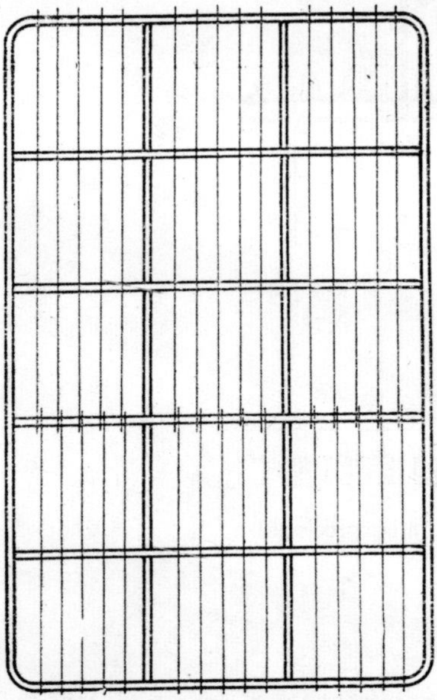

Fig. 107 Discharge electrodes mounted on an upright tubular frame.

The mountings of high-tension electrodes have evolved along two quite distinct lines. One approach is for the parallel wire or point electrodes to be fastened and stretched on vertical tubular frames, which are reinforced by a lattice of horizontal and upright tubes, as illustrated in Fig. 107. The object of this reinforcement is to render the frames rigid and planar enough to keep all the discharge electrodes correctly aligned between the collecting electrodes. The alternative technique is to suspend the wire or point electrodes from a horizontal grid at the top of the unit, as seen in Fig. 108, and to attach a separate weight to the bottom end of each of them. Another grid must be provided at the bottom of the unit to guide these weights, so as to permit their vertical movement but preclude any lateral or swinging motions. The two grids must be rigidly interconnected, by a system of tubular spacers, to prevent the bottom grid and the entire system of discharge electrodes from swinging like a pendulum.

A diversity of shapes has been devised for the wire electrodes, because their shape and layout markedly affect the key parameters of the whole precipitator – the initial critical voltage level, voltampere characteristic, and the arc-over voltage. In its simplest form, the discharge electrode is simply a round wire. The smaller its diameter, the lower is the initial critical voltage, the gentler the slope of the voltampere characteristic, and the higher the arc-over voltage. These electrical advantages of a thin wire, however, are largely offset by its inferior mechanical

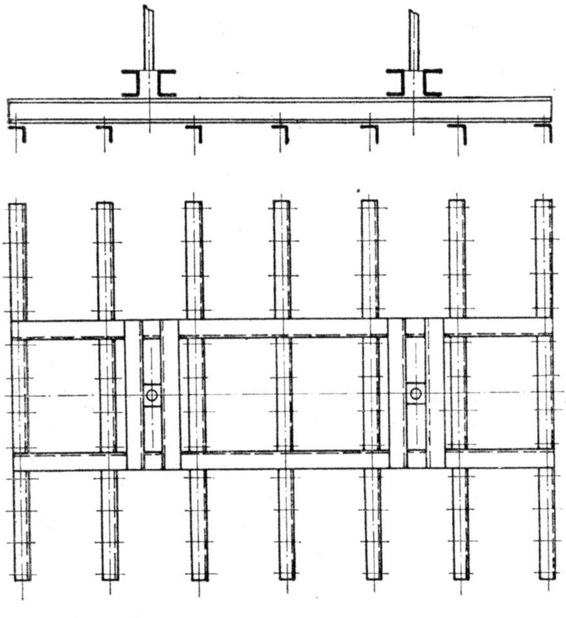

Fig. 108 Discharge electrodes suspended from a horizontal carrier grid.

strength: thin wires are prone to crack under the effects of the rapping impacts, and are also easily damaged by corrosion. They must therefore be made of some high-strength material with an excellent resistance to both corrosion and electrical discharges, which in practice means using high-alloy stainless materials such as kanthal alloys.

In an effort to avoid the mechanical shortcomings of thin wire electrodes while retaining their electrical properties, designers have increasingly turned to wires of asteroidal or other non-circular cross sections, exemplified in Fig. 109. A star-shaped section increases the net cross-sectional area of the electrode, while the sharp edges afford much the same electrical properties as a very much thinner wire. A special category are the spirally wound wire electrodes shown in Fig. 110. These are in effect springs, and are consequently very easy to mount on hooks in the suspension frame.

When it gradually transpired that high concentrations of extremely fine dust call for a radically increased charge of the dust particles, further development work was concentrated on point electrodes, where current intensities can be much higher and the initial critical voltage level much lower than with wire electrodes.

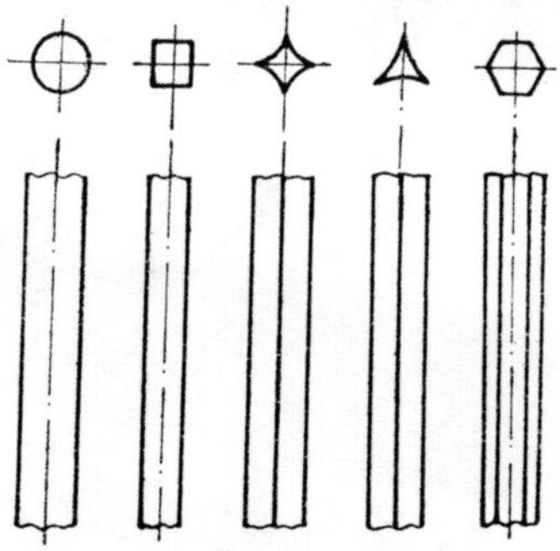

Fig. 109 Typical cross sections of wire discharge electrodes.

Fig. 110 A spirally wound wire electrode.

Fig. 111 A point electrode made of barbed wire.

In their simplest guise, point electrodes consist merely of barbed wire, like the specimen in Fig. 111. This type, however, was soon abandoned, both because of the more or less random orientation of its individual points, and because of the difficulties experienced in its installation. It was replaced by a variety of strip, angle-section and other profiled point electrodes, a selection of which is illustrated in Fig. 112. The chief requirement on any such electrode is that the length and spacing of its points should closely conform to specification. Another is that the surface should be smooth and unbroken except for the points, as that retards the build-up of deposits on it. It has also been found that points aligned parallel to the gas flow do not wear as rapidly as points set perpendicular to it.

The best of electrodes would be little use if it were not correctly mounted on its carrier frame or grid. Its fastening to the structural tubing must be sufficiently strong, and must ensure the closest possible contact between the two surfaces, which must both be metallically clean. Any infringement of these rules results in a high contact resistance, and consequently in burnt contact surfaces. The strength requirements are dictated by the rapping forces which the connection must withstand, and by the stresses arising from the constant tensioning of the electrode. Some of the more common mounting modes are depicted in Fig. 113.

A further vital consideration is the arc-over voltage of the high-tension electrodes. In horizontal plate-type precipitators, this voltage is determined not only by the discharge electrodes themselves, but equally by the frame tubes interposed between

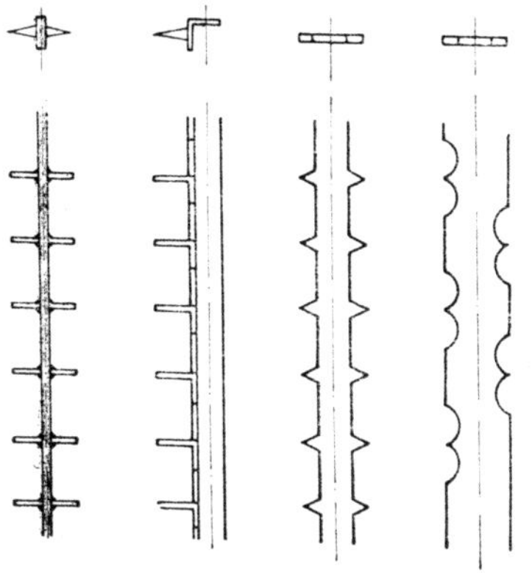

Fig. 112 Various types of point discharge electrodes.

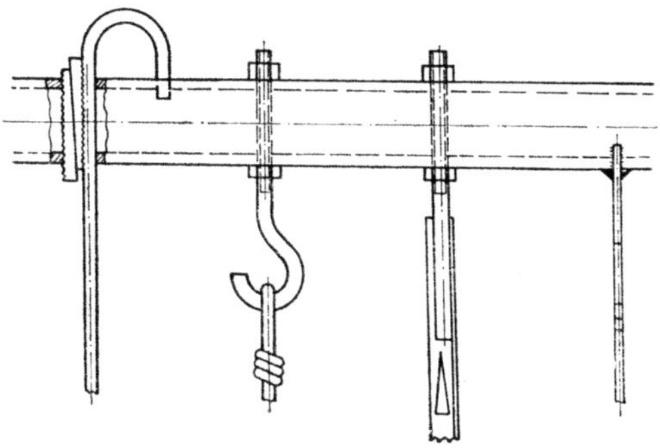

Fig. 113 Alternative suspension attachments for discharge electrodes.

the collecting electrodes. Fig. 114 indicates how the initial critical voltage and the arc-over voltage respond to changes in the radius of curvature (i.e. diameter) of the wire electrode, or of the frame tube between two collecting electrodes. Diagrams of this kind show the diameter needed to forestall corona discharges on the tubes, and also show whether, under any given circumstances, the discharge will occur on the tube or on the electrode wire. Things are much simpler when the high-tension electrodes are suspended from an upper and steadied by a lower grid. In that case, the wire electrodes and the structural members can always be keps sufficiently far away from the grounded (earthed) parts of the precipitator for the arc-over voltage there to be higher than between the discharge and the collecting electrodes.

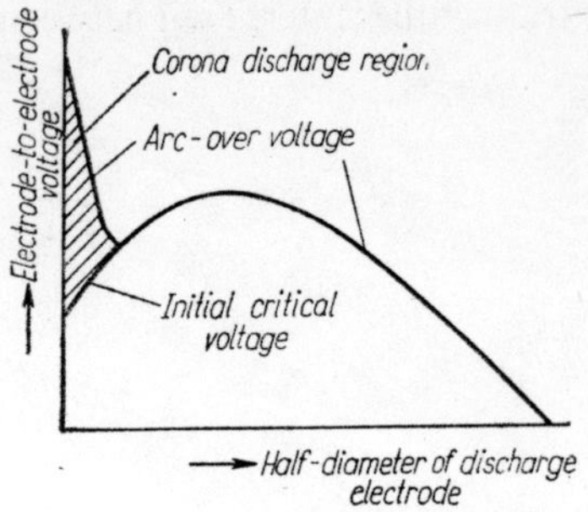

Fig. 114 Dependence of the initial critical voltage and arc-over voltage on the half-diameter of the discharge electrode.

3.3.3.4.4 Electrode suspensions

Collecting electrode suspensions

In tubular vertical precipitators, the clusters of tubular collecting electrodes are normally suspended from a suitably braced horizontal plate which is generally fixed between the flanges of the unit housing. At the bottom of the unit, the electrodes are retained and guided by another horizontal partition, which permits free thermal expansion and contraction of the individual electrodes or of whole clusters of tubes.

In plate-type vertical precipitators, the electrodes are as a rule pivoted on beams. Usually, the suspension system comprises three vertical beams, and the design allows for thermal expansion and contraction on plain or antifriction bearings

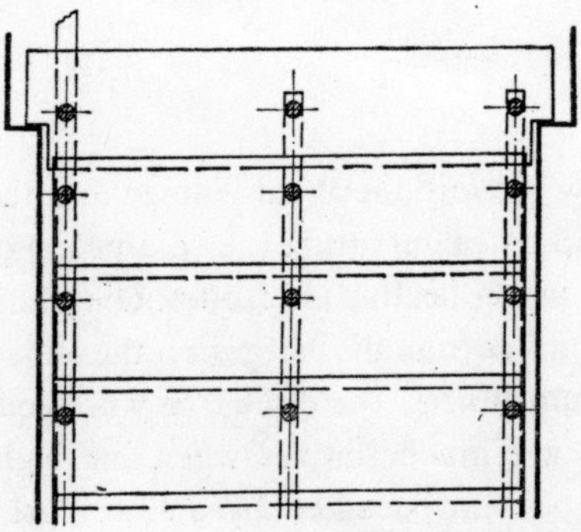

Fig. 115 Prevalent mode of mounting the collecting electrodes in vertical plate-type precipitators.

interposed between the carrier beams and the girders of the precipitator housing. Naturally, this involves providing adequate clearances for the expansion of all the components. An example of such a system is seen in Fig. 115.

Horizontal plate-type precipitators mostly utilize a similar system, as sketched in Fig. 116. The individual electrode strips are pivoted on carrier beams, with the pivot pins located so as to keep the strips in a slightly inclined equilibrium position. The bottom ends of the electrodes are retained in position by guide beams. Complications arise because the carrier beams tend to deflect under the load; because the lengths of the various electrode strips are not always fully identical; and because uneven heating or cooling will cause further differences

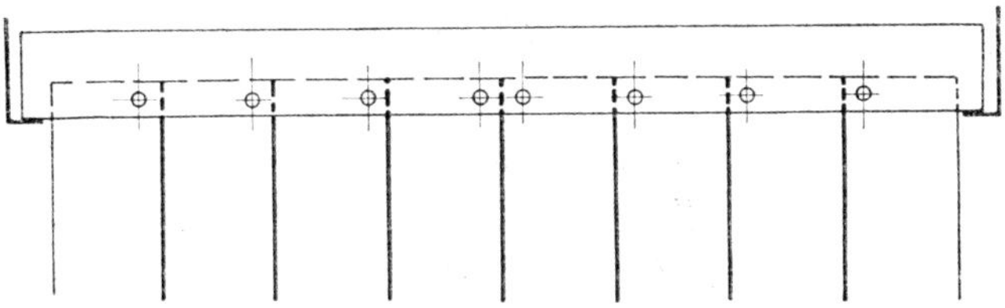

Fig. 116 A common way of mounting the collecting electrodes in horizontal plate-type precipitators.

between the actual lengths of the individual strips. To accommodate all these differences, the usual practice is to fix the guide beams only to the first and last of the electrode strips. The other strips are shaped, or recessed at their bottoms, so as to permit their free vertical movement between the guide beams. To keep the collecting electrodes spaced out at the specified pitch, the carrier beams, resting on the main structural girders, are spaced out by racks or by slotted bars with slots located at the desired electrode spacings. There are similar spacing provisions for the bottom guide beams too. These latter beams mostly also serve for the rapping of the electrodes, and are therefore sometimes called rapping beams.

This matter of correct electrode spacing is one of the overriding considerations in the design of suspension systems for collecting electrodes. The other is to prevent the deformations that can be induced by thermal expansion and by deflection of the housing girders and/or electrode carrier beams. The pivot pins, their seatings and their securing provisions also merit special care in the design stage, as pins which work loose and drop out will not make a precipitator type any more popular with its users.

High-tension electrode suspensions

The way the high-tension electrodes are suspended depends on whether the suspension system consists of an upper and a lower grid, or of a system of vertical frames. Both these alternatives can be compared in Fig. 117. In the former case, the suspension must allow for some adjustment of the wire positions, which may become necessary to align the wires (into the centre lines of the tubular electrodes, or into the plane equidistant from the adjacent plate electrodes). This adjustment is commonly effected by repositioning the longitudinal upper beams from which the wire electrodes are suspended. In the alternative system, the vertical frames must again be capable of adjustment into the plane equidistant from two collecting

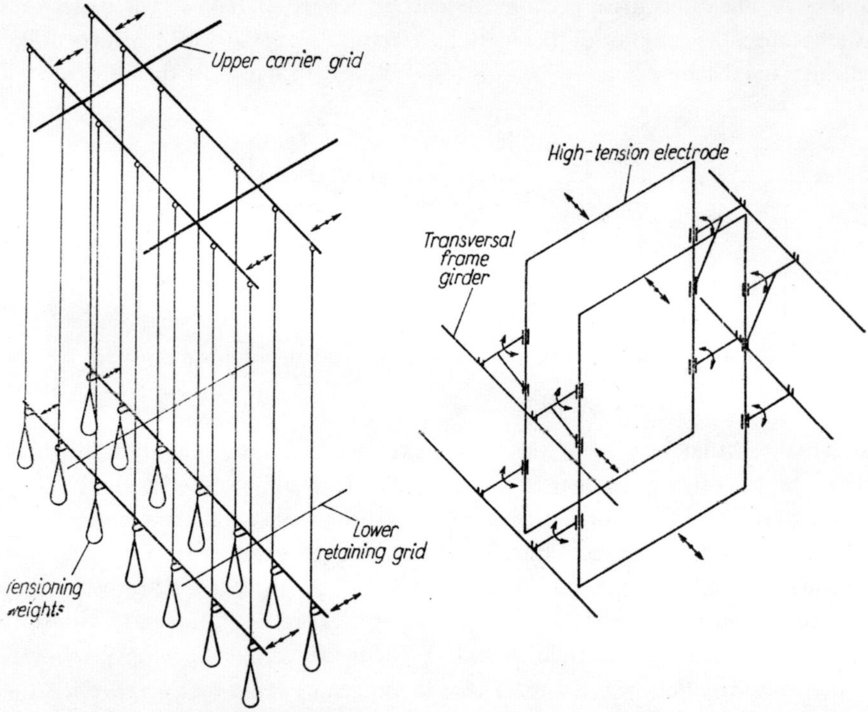

Fig. 117 Two alternative suspension systems for high-tension electrodes.

electrodes. That calls for aligning provisions on the suspension and guide arms which hold the frames on the transversal suspension members or grids. The latter in turn are mounted on tubes or bars which rest on the suspension insulators based upon the main structural girders of the housing.

The bushings through which these tubes or bars pass to the insulators are a potential trouble spot. For one thing, they could easily cause flash-overs between the tube or bar and the grounded (earthed) parts of the precipitator. For another, they might admit some as yet untreated gas to the insulators, where it would foul the internal surfaces and thereby lower the insulating capacity. To guard against the flash-overs, the bushings need a fairly large diameter and well-rounded ends, as shown in Fig. 118. To prevent dust reaching the insulators, the bar and bushing assembly is made to form a sufficiently long discharge-free zone, so that any particles which penetrate between the bar and the bushing are electrically attracted and forced to settle there. Furthermore, either the base or the cover of the insulator unit is usually provided with holes through which a very small quantity of outside air is induced by the partial vacuum inside the precipitator. This air flowing through the gap between the bar and the bushing forms an effective barrier against any escape of dust-laden gas into the vicinity of the insulator. The amount of air drawn in will naturally depend on the suction pressure within the precipitator housing, and is therefore best controlled by blanking off part of the area of the induction holes.

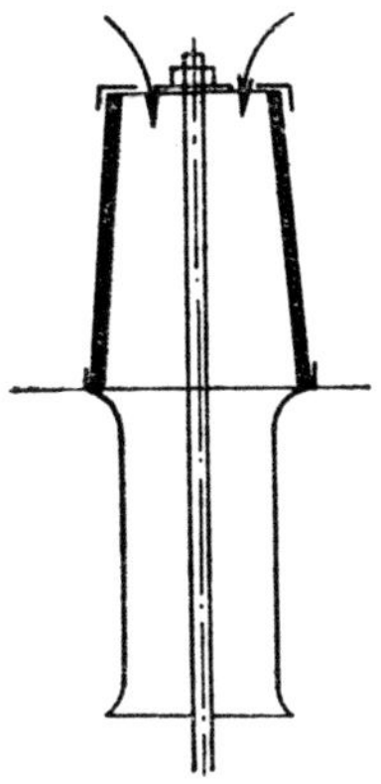

Fig. 118 Typical bushing through which an electrode-system suspension bar passes through the casing to the suspension insulator.

The insulators which support the whole assembly are highly stressed mechanically, electrically and thermally. The mechanical stressing is imposed by the deadweight of the whole high-tension system. The resultant specific compressive load on the insulator can be kept within acceptable limits only by increasing the annular plan area of the insulator, but the prospect of using large and heavy insulators is not an attractive one.

The electric strength of the insulator obviously depends upon that strength of the material of which the insulator is made. Given this limitation, however, the strength is also governed by the surface conductivity of the insulator, and that is why these insulators are generally made very much longer than would seem necessary in view of the electric strength of their material. Even so, however, the surface conductivity can be kept down only as long as the insulator surface is dry and clean: a dust layer or moisture film will instantly and very markedly reduce the electric strength of the insulator. The precautions normally taken to prevent fouling of the internal surfaces of insulators have already been described in the preceding paragraphs; as for the outer surfaces, there is nothing for it but to clean them regularly and frequently. Even the cleanest of insulators, however, has to be protected against the condensation of moisture both within and upon it. This risk is especially acute when the processed gas has a high dew point and a relatively low temperature at the precipitator inlet. In such cases, often the only effective countermeasure is to heat the whole vicinity of the insulators, and in extreme cases, the outside air drawn in through their intake holes as well.

The thermal stressing to which the insulators are exposed in service varies within wide limits. As a general rule, the insulator is heated from within and/or from below, and cooled by the induced outside air from above. Since insulators are by definition very poor heat conductors, considerable temperature gradients can build up within them, and the concurrent expansion and contraction can set up mechanical stresses sufficiently large for the insulator to crack.

Nowadays there is a distinct swing away from porcelain towards quartz insulators, as the latter afford superior mechanical, thermal and electrical properties.

3.3.3.4.5 Electrode rapping systems

Rapping, or the process of removing the dust that has settled on both the electrode systems, is a crucially important operation. Unless it is performed effectively, the precipitator will cease to function properly, and its collecting efficiency will diminish. Dust deposits on the discharge electrodes will increase the effective diameter of the wires, and in view of the potential drop across the dust layer, the effective voltage remaining available on the surface of this layer will diminish too. That will impair the ionizing action of the wire electrodes, thereby raising the initial critical voltage and curtailing the current intensity. As a result, the dust particles in the gas stream will be insufficiently charged, the electric field within the gas will be inadequate, and the effective separating brisk velocity will drop off disastrously.

Dust sediments are no more acceptable on the collecting electrodes either. They will again cause a potential drop across the deposit, with consequent losses of effective voltage within the gas flow and of current intensity at any given voltage level. If left to accumulate long enough, the dust layers on the collecting electrodes would finally reduce the current flow to the point where the whole separating process would come to a halt.

Both the discharge and the collecting electrodes therefore need rapping, at intervals short enough to rule out any serious interference of the dust deposits with the separating process. The two factors which between them decide the success and efficiency of the operation are the rapping intensity, in terms of the transversal acceleration attained at the electrode surface, and the rapping intervals, or the span of time that elapses between consecutive rapping operations.

The rapping intensity must obviously be adequate to ensure complete and dependable removal of all the dust from the electrode surfaces, but that statement does not help much towards defining the actual requirements. Coarser particles, and dust with a low electrical resistance value, are easier to dislodge, and therefore impose no great demands on the rapping intensity. Very fine particles which tend to agglomerate or adhere, as well as dusts with a relatively high electrical resistance, will cling to the electrodes and need fairly hefty rapping to shake them loose. The point is that the rapping intensity must be only just sufficient, and never excessive. Too intense rapping would break up the caked deposits as they are shaken free, and thus release clouds of discrete particles; these would mostly be re-entrained in the gas stream, and leave the unit untrapped. Some degree of re-entrainment is of course unavoidable in the rapping operation. The gas stream will pick up mainly the dust settled on those electrode surfaces which are exposed to that stream. But, given a well-selected rapping intensity, at least the dust accumulated in the shielded areas will land in the hoppers with next to no losses through re-entrainment.

The choice of a correct rapping interval naturally depends on the rate at which the deposits build up. The thicker the dust layer, the more densely will it be compacted, and the less comminution and scatter will attend the rapping operation. Too thin a layer should never be rapped, as the amount of dust scattered is apt to defeat the whole object of the exercise. On the other hand, a thick deposit on the electrodes will entail a greater voltage drop across it, and thus, at a constant feed voltage, reduce the potential gradient in the gas. That in turn will lower the current and electric field intensities, so that an increased proportion of the solids will escape trapping. Automatic voltage control can at least to some extent counter the effects of the rising voltage drop as the dust layer accumulates, i.e. can keep the potential gradient within the gas more or less constant by gradually stepping up the feed voltage. Without this automatic control, however, a very fine balance has to be struck between the increased dust emission due to scatter when the electrodes are rapped prematurely, and the increased dust emission caused by the voltage drop in the deposit when the electrodes are rapped too late.

Fig. 119 shows a typical dependence between the rapping interval and the mean rate of solid particle emission from the unit in question, and indicates how this latter rate mounts as the dust deposits on the electrodes grow thicker. Let us assume, for the sake of simplicity, that every rapping operation produces one and the same amount of dust emission caused by scatter. On this assumption, we obtain a hyperbolic dependence of the mean particle emission rate caused by rapping, with the particle emission per unit of time dropping off as the rapping interval increases. The sum of the two mean emission rates indicates the dependence of the overall mean particle emission rate upon the rapping interval. In the example in Fig. 119, there is a clearly discernible minimum of overall dust emission at a certain optimum duration of the rapping interval. Fig. 120 presents a similar plot for a unit where the feed voltage is regulated automatically. Curve *A* is in this case obviously much flatter, and the summation curve displays no clear-cut minimum. From a certain minimum rapping interval onwards, however, this system affords no further significant drops in the total emission rate, no matter how much further we increase the rapping intervals.

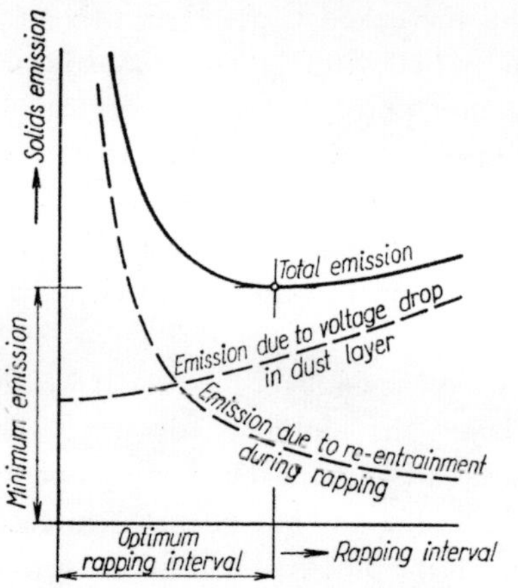

Fig. 119 Dependence of the solids emission rate on the electrode rapping interval.

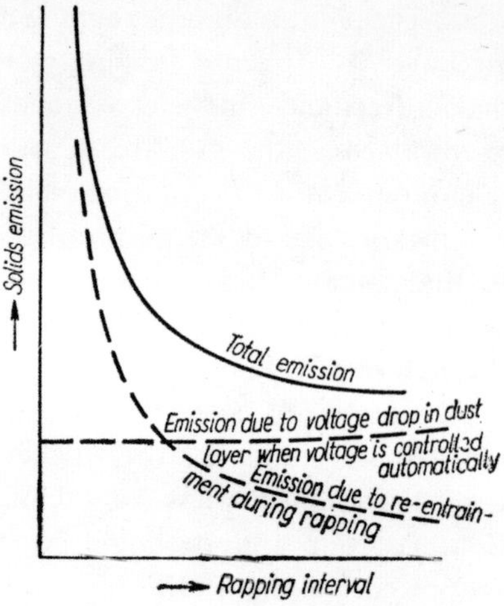

Fig. 120 Effect of automatic voltage control on the solids emission curves.

With this much said about the factors which govern the rapping operation, it is time to examine the actual rapping mechanisms. High-tension electrodes are almost invariably rapped by hammer blows directed at their vertical frames or upper suspension grids. The rapping intensity depends on the weight of these hammers and their stroke length. The hammers mostly swing up with a pendulum motion, and drop by gravity. They are raised by any of a variety of mechanisms, which are usually powered by gear motors acting through insulating couplings (because the hammers are live). The drives are generally located outside the precipitator housing itself, and linked to the hammers by transmissions which pass through some of the suspension tubes.

Fig. 121 schematically represents three of the most common hammer raising mechanisms. In the first, the crank that actually lifts the hammers is actuated, through a pull-rod, by a rocker which slips out of engagement as its retaining chain is tautened; that releases the hammers, which then drop onto an anvil plate fixed to the vertical frame that carries the electrodes. In the second mechanism, the motor drives a vertical shaft that runs through one of the suspension tubes. This shaft bears a horizontally revolving cam which first raises the hammers and then, as further rotation presents its back to the lifting mechanism, lets them drop onto the rapped frames. In the third mechanism, the drive power is applied to a dog that revolves a lever which is freely rotatable on its shaft; this lever operates the pull-rod which actually raises the hammers. Upon reaching its top dead centre position, the lever topples over to its bottom dead centre

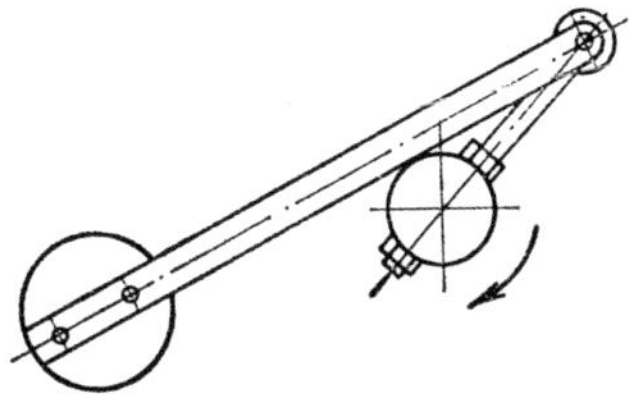

Fig. 122 A typical rapping mechanism for the collecting electrodes of plate-type precipitators.

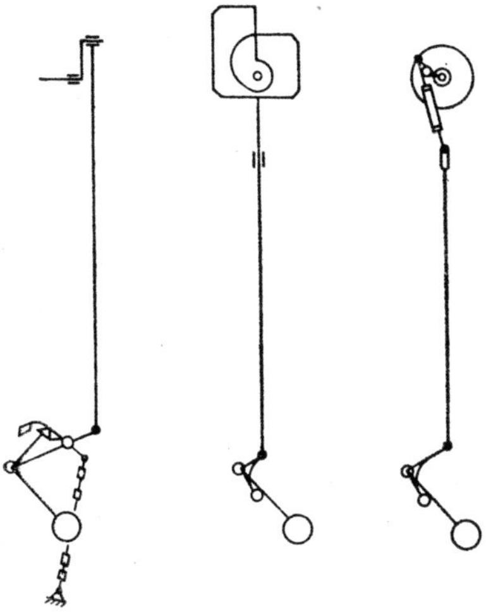

Fig. 121 Typical hammer raising mechanisms for the rapping systems serving high-tension electrodes.

position, and thereby releases the pivoted hammers. This last mechanism differs from the other two in having its crank, i.e. its key transmission component, located outside the precipitator housing. That both facilitates inspection and maintenance, and reduces the wear and tear, which in mechanisms housed within the precipitator casing can be quite a problem. True, there are rapping drives which employ a slipping rocker type power transmission placed outside the precipitator casing, but none of them has so far been adopted on any scale.

Collecting electrodes are mostly also rapped by hammer mechanisms. Fig. 122 illustrates a system typical of those used in plate-type precipitators: the hammers strike the bottom rapping beam, which transmits the impacts to the individual electrode strips. The hammers swivel on pivots seated in collars which are welded to a slowly rotating shaft. Upon reaching its uppermost position, each hammer topples forward to hit the anvil plate on the rapping beam. An alternative system, in widespread use until fairly recently, employed a power-driven cam which displaced the spring-loaded lower rapping beam, and all the electrodes with it. As soon as the apex of the cam profile moved out of engagement, presenting the flat rear side of the cam to the beam, the return spring hurled the beam back, beyond its lowermost (or equilibrium) position. On the upward part of its arc of travel, the beam struck the rapping beam of the next adjacent electrode array; the last beam down the line usually struck a fixed buffer. That generated a whole succession of impacts, which were transmitted to the electrode strips.

The desired rapping interval is more often than not set up on timer relays which switch on the drive motors of the rapping devices. The device employing toppling hammers on a slowly revolving shaft is sometimes timed by means of a quadruple crank mechanism fitted with a ratchet and pawl, where the rapping interval is adjusted by altering the effective stroke of the crank on the shaft of the drive motor.

The whole rapping mechanism, like all the other parts of the precipitator, must fulfill one condition which is vital to the efficient functioning of the unit: the gaps between the live and the grounded (earthed) components must be wide enough to permit operation at the highest attainable voltage. Gaps of this width are easy enough to provide in the design and assembly stages, but not nearly as easy to keep constant in the face of deflections caused by the vacuum within the unit or by thermal expansion and contraction. Other important but often conflicting requirements are that the rapping mechanism should be utterly reliable; should not affect the tightness of the precipitator housing, nor the electrical strength of the insulators; and that efficient rapping should be matched by equally effective provisions for the removal of the dust dislodged by it, i.e. by preferably continuous emptying of the hopper at the bottom of the unit.

3.3.4 Wet electrostatic precipitators

Wet electrostatic precipitators can be either vertical or horizontal, and can in either case be used to trap either liquid or solid admixture particles in the gas. When the admixtures are liquid, the electric field between the electrodes ionizes the gas and charges the droplets. The latter then settle on the collecting electrodes and flow down them, in a thin sheet of liquid, to discharge troughs at the bottom of the unit. Dust-laden gas, on the other hand, is generally first treated in a wet scrubber or washer, where some of the solids are retained. The gas then carries the remaining particles into the precipitator, where it is almost fully saturated with water vapour by sprays; these flush the electrodes, and also dispense mist into the spaces between them. The solids are now separated along with these droplets, and flow down the collecting electrodes as sludge films, to end up in discharge troughs. The spray systems are often fitted with several different kinds of nozzles. Some of these continuously inject a fine mist into the active space, while others intermittently rinse down the electrodes. Sometimes, the nozzles which flush the collecting electrodes further differ from those that serve the discharge electrodes. Fig. 123 shows a fairly typical example of the internal layout of a wet electrostatic precipitator.

The housings of these units are essentially similar to those of dry precipitators, but differ by having no hoppers. Instead, they are provided with only mildly sloping troughs which channel the trapped liquid or sludge to the discharge openings. As these units are mostly run with a partial vacuum inside them, their liquid discharge openings must be fitted with air-tight seals, usually in the form of siphon traps. The housings must also be tight, and their internal surfaces should be as smooth as possible, with no protrusions where deposits might settle. As long as the gas is non-corrosive, the housing and other components are generally fabricated in steel. However, to ensure an acceptable service life, the sheets are

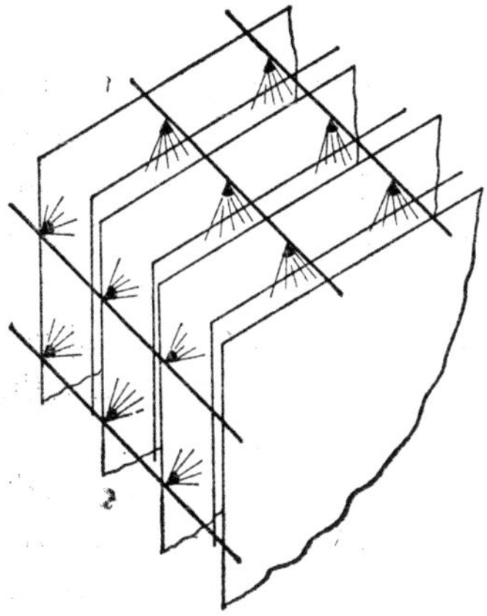

Fig. 123 Scheme of nozzle locations in wet electrostatic precipitators.

as a rule thicker than those used for the housings of dry precipitators; typically, they are some 5 to 10 mm thick. When aggressive constituents are handled, the normal practice is to build the load-bearing structure in steel, and clad it with sheets of some chemically resistant plastic.

The discharge electrodes are commonly strips or wires, attached to and stretched out on vertical or horizontal frames. The collecting electrodes of plate-type wet precipitators are mostly smooth, heavily stiffened plates fixed to the supporting girders. Vertical wet precipitators are more often than not fitted with tubular collecting electrodes. In either case, the electrodes are always rigidly fixed, because there is no need for their rapping. Instead, the sludge or liquid are washed off by sprays or flushing systems, which are installed above the electrodes and fed through one or more water distribution circuits. The feed water pressure need not be particularly high, a gauge pressure of 0.1 to 0.2 Pa being adequate for most applications. The nozzle orifices must be large enough to be reasonably safe against clogging by dirt particles. A few typical electrode rinsing systems are sketched in Fig. 124.

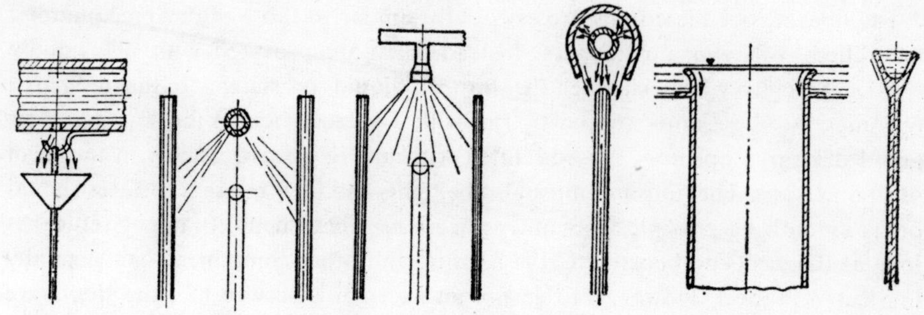

Fig. 124 Alternative electrode flushing arrangements in wet precipitators.

The insulators that support the high-voltage system are either of the oil-filled type, as shown in Fig. 125, or else conventional porcelain or silicon insulators. They must be shielded against the penetration of dust and droplets just as in dry precipitators, and must moreover be heated to prevent the condensation of moisture on their internal surfaces.

In wet precipitators, there is next to no risk that particles once captured will be re-entrained, either by the gas stream or by the effects of the electric wind in the active space. Wet particles have a much higher permittivity than dry ones, and solid particles which have adhered to or been engulfed by a droplet naturally have a larger apparent diameter than previously. All this contributes towards a more effective separating process. In fact, in these units the relation between the separating brisk velocity and the specific collecting area is virtually constant. There is no need to introduce correcting coefficients dependent on the flow velocity

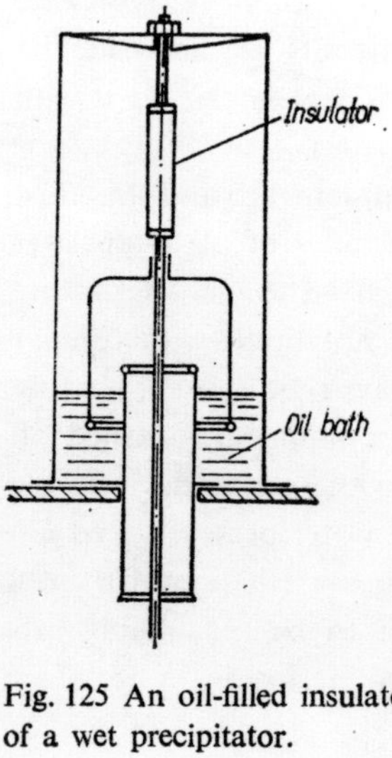

Fig. 125 An oil-filled insulator housing for the high-tension system suspension of a wet precipitator.

or on the height of the precipitator, because these coefficients would almost invariably equal unity. Furthermore, in wet precipitators there are none of the otherwise usual problems with the resistance offered by accumulated dust layers, nor is there any likelihood of a back corona disrupting the process.

These units mostly handle gases at temperatures between 20 and 40 °C, and are hardly ever used where the normal service temperature exceeds some 70 °C. They afford a high collecting efficiency, do so more dependably than most other types of equipment, and tend to be more stable in operation than their potential rivals. On the other hand, they use up considerable quantities of water, and have to be supplemented with sludge handling systems which often present intricate problems of their own.

3.4 Cloth filters

One of the oldest-established ways of trapping pollutant particles is by filtration. In this process, the carrier medium is passed through some material with pores or interstices large enough for it to penetrate, but small enough to retain the admixture particles it bears. Yet cloth filters, by far the most widely used of all gas filtering devices, have only fairly recently emerged as one of the major categories of industrial gas cleaning equipment. Over the past few years, industrial gas filtration has undergone a spectacular process of evolution, and cloth filters have invaded many spheres which had previously been the domains of other types of equipment. Part of the explanation is that the main rival types—electrostatic precipitators, mechanical separators, and scrubbers—are all nearing the limits of their development, as far as their collecting efficiencies are concerned; and yet none of them can reduce pollutant emission rates as radically as even the presently existing cloth filters. Moreover, cloth filters still have a very long way ahead of them before their development potential is anything like fully exploited. The best of the contemporary types, which utilize the latest advances in man-made fibres and in the technology of filter cloths, can keep the pollution levels at their outlets to less than 10 mg m^{-3}, but recent developments indicate that it will soon be feasible to maintain these levels consistently below 1 mg m^{-3}. Besides, new synthetic fibres, with an enhanced heat resistance and improved strength properties, look like extending the scope of cloth filters to even some of the high-temperature applications from which they have up to now been largely excluded.

3.4.1 Filter cloths

The properties of the filter cloth directly affect both the efficiency and the economics of gas filtration processes. The most important of these properties are the collecting capacity of the cloth, the pressure drop it induces in the gas flow, the ease with which it can be cleaned after use, and its wear resistance.

An analysis of the filtration process shows that there are several distinct mechanisms involved in it, each of them exploiting a different set of forces or factors. The theoretical and experimental research performed so far has shown that, invariably, filtration processes are based on one or more of the following effects:

a) *Impact phenomena*: particles impinging on an obstacle (a fibre of the cloth) are trapped there by their own inertia;

b) *Diffusion phenomena*: very small particles settle on the fibres in consequence of their Brownian motion;

c) *Sedimentation phenomena*: particles settle on the filter layer under the action of gravity;

d) *The mesh effect*: this is the direct retention of particles larger than the pores or interstices of the filter material;

e) *Electrostatic phenomena*: these are the outcome of electrostatic forces acting between the particles and the surface of the filter cloth.

Whether and to what extent each of the above factors participates in any given filtration process depends on the sum of the properties of the admixture particles, of their carrier medium, and of the filter cloth itself. The particle properties which most strongly affect the outcome are their size, shape, specific gravity, their electric charges and dielectric constants, and their surface properties. The key properties of the filtered fluid are its viscosity, specific gravity, temperature, humidity, and filtration velocity. The most important variables of the filter material are its porosity, thickness, the size and type of the fibres which constitute it, its surface properties, permittivity, and electrical charge.

This list of quantities which govern the resultant filtering action is far from complete. Even so, it shows that filtration processes are far too complex for any theoretical analysis ever to describe them completely. Even if we limit our considerations to the trapping of solids from a gas stream, the quantitative effects of each of the different mechanisms involved will still largely depend on the type of filter cloth we employ. Apart from the material of which the cloth is made, the main distinction in this respect is between the conventional woven and the more novel needled filter cloths.

Woven cloths used to be one of the most common filter materials as recently as the sixties. One of their great attractions is that, by careful selection of the material and weaving technique, a wide variety of cloths can be obtained to suit almost any intended application. Extra-smooth cloths can be woven from smooth, straight synthetic or glass fibres; or normally smooth cloths from suitably processed cotton or woollen yarns; but coarser hairy textures of the napped cotton or worsted types are equally easy to produce. The smooth and extra-smooth cloths filter out solids mainly by the mesh and impact effects. These effects predominate as long as the cloth is new and relatively clean, and are still in evidence later on, when its upstream side is covered with a dust layer. The smoother cloths are much easier to clear of dust deposits than the coarser ones, but are also much quicker to clog up: the pressure drop across them rises rather abruptly, as compared to that across the coarser types.

Needled cloths, a relatively recent development, have already caused something of a revolution in the sphere of industrial gas cleaning. They afford much higher collecting efficiencies for lower pressure drops than their woven counterparts, and are so much easier to clean that the filters which employ them can in practice be run at substantially lower pressure losses. These cloths are made by a technique in which a core of carrier fabric is covered on both sides with layers of freely deposited and evenly spread fibres, and the latter are then "needled" fast on a machine fitted with rows of triangular-section needles. The edges of those needles are formed into hook-shaped recesses which engage the fibres and draw them into the depth of the material. That forms a dense system of loops which reinforces the layers. The upstream side of the cloth is usually processed, by scorching or otherwise, so as to singe off or otherwise remove any protruding fibres. The resultant smooth surface facilitates the removal of dust deposits when the cloth is being cleaned.

This smooth surface is vital, because any fibres or hairs which protrude from it will tend to retain the dust settled in their vicinity when the cloth is being cleaned by a reverse air blast. On a napped surface, the dust deposit is never blown off in its entirety, but comes off in separate flakes, leaving patches of dust still clinging to the cloth. On the smoothed surfaces of needled filter cloths, the dust is dislodged in large coherent sheets. That not only makes for a more complete reconditioning of the cloth, but also increases the speed at which these larger and heavier cakes of dust drop towards the hopper, and thereby reduces the amount of dust which the reverse blast carries back into the other chambers of the filter.

The enhanced collecting efficiency of needled filter cloths is due to the uniform distribution of their fibres both over the surfaces and throughout the depths of the filter layers. In woven cloths, this uniform distribution is easy to achieve, but almost impossible to maintain for any length of time. When the cloth is being cleaned, the reverse air blast will obviously follow the line of least resistance. Hence, it will concentrate at the interstices where the individual fibres or strands are separated by gaps, and will gradually enlarge these openings. When the normal flow direction is restored, the dust-laden gas will tend to flow through these enlarged gaps between the weft and the warp, which after a few flow reversals will be wide enough for a substantial proportion of the dust to pass right through them. As filtration proceeds, these gaps which admit the bulk of the throughflow will quickly clog up with dust deposits, which then in effect take over as the main filtering medium. Consequently, a freshly cleaned woven cloth will usually exhibit an abrupt if temporary drop in collecting efficiency, which impairs the overall long-term efficiency of the whole filter. Needled filter cloths, on the other hand, present a much more homogeneous surface to the reversed blast, and in service are quickly coated with a fine and uniform dust layer which forms a high-efficiency filter. Cleaning by reversed blowing produces little or no deterioration of the cloth, and practically no effect on its collecting efficiency.

If the impact of needled cloths on the technical aspects of gas filtration has been considerable, their effect on the economics of the process has been even more pronounced. The savings that accrue from their longer service life, reduced maintenance requirements, improved collecting efficiencies, and lower pressure drops, are truly remarkable. Long-term experience in various industries has confirmed that these cloths usually render more than twice the service that could be obtained from woven fabrics. Thanks to their smoothened surfaces, these cloths display pressure losses which are on average 30 per cent lower than in comparable woven cloths, and that naturally shows up in the power consumption of the filters.

3.4.1.1 Aerodynamics of filter cloths

There is no hope of ever establishing the aerodynamic properties of a filter cloth by a purely theoretical approach, because there is no way of adequately representing the degree of their porosity nor the mean diameter of their fibres. However, a set of simplifying assumptions allows us to express the pressure drop across a clean filter cloth, in N m^{-2}, as

$$\Delta p_F = k_1 \cdot \eta^{\alpha} \cdot v_F^{\beta} \tag{88}$$

where k_1 is a constant of the given cloth, which must be ascertained experimentally,

η – the dynamic viscosity of the gas (in Pa s),

v_F – the gas flow velocity through the cloth (in m s^{-1}),

α, β are exponents which again must be established experimentally for every particular cloth.

As the cloth gradually clogs up with retained dust, a gas flow of the same velocity will encounter a steadily increasing resistance. The way this raises the pressure drop has been investigated by a good many authors; an example of their findings in the formula by which *Rekk* defines the pressure drop increment in dependence on the elapsed filtration time τ (in sec):

$$\Delta p_F = (C_1 k_p v_F \tau O_c + C_2)\, v_F \qquad (89)$$

where C_1 and C_2 are experimentally determined constants for the filter cloth and dust in question,
k_p is the inlet dust concentration (in g m^{-3}), and
O_c – the overall collecting efficiency of the cloth (in per cent).

3.4.1.2 Sorption properties of filter cloths

Since the gas to be filtered will very often be moist, it is important that the filter cloths for these applications should have the right combination of hydrophobic qualities. Fibres which are not easily wetted or permeated by water have superior electrical properties, and cloths made of them will not clog up with slime as easily as other cloths when the filtered gas is humid. However, the outcome is always dependent not only on the properties of the cloth itself, but equally on the sorption properties of the dust, and on the way the surface properties of the dust respond to the presence of moisture.

Highly hygroscopic cloths should be avoided, because capillary condensation on these fabrics or in their fibres can moisten the dust by its direct contact with the condensed water. That makes the dust adhesive, and difficult to remove. Even a non-hygroscopic cloth may cause trouble when the trapped dust itself is hygroscopic: the dust moistened by capillary condensation will tend to form arches within the structure of the cloth, and will then resist most attempts to dislodge it. Therefore, the sorption properties of a filter cloth must always be examined in conjunction with those of the dust which it is to handle. Where these properties are ignored, the penalties that accumulate over the years can be very heavy indeed.

3.4.1.3 Selection of filter cloths

Cloth filters cannot perform satisfactorily unless their filter cloths have been correctly selected to suit the given duty and service conditions. The factors to consider in this process of selection are as follows:

1. The physical and chemical properties of the filtered gas, especially:
– The maximum and minimum temperatures at which the gas may enter the filter;
– The presence of constituents which might attack the cloth, such as sulphur dioxide or trioxide, carbon monoxide, chlorine, etc.;

– The humidity and dew point of the gas; the gas temperature should always be at least 15 °C above the dew point;
– The acid, alkali, and solvent contents, if any;
– The explosion hazards inherent in the gas.

2. The physical and chemical properties of the particles dispersed in the gas stream, particularly:
– The particle size;
– The particle concentration at the inlet;
– The abrasive effects of the dust;
– The chemical activity;
– The adhesive properties;
– The degree of hygroscopicity;
– The tendency of the particles to agglomerate;
– The electrical properties of the particles.

All these data should of course be ascertained in the plant, and at the spot, where the filters are to be installed. Unfortunately, detailed information of this kind is all too often unavailable, and the designers then have to resort to know-how gained in previous roughly similar installations which operate under analogical service conditions. Even that approach, however, is better than simply neglecting one or more of those factors. No matter what a dearth of information we are facing, we must always bear in mind that no one can responsibly pick a filter cloth without taking into account every one of the factors set out above.

3.4.1.4 Testing of filter cloths

Filter cloths are commonly subjected to two different kinds of testing. One set of tests is run to investigate their *physical* and *mechanical properties*, another is intended to establish their actual *filtering performance*. The former tests need not be discussed here, since they are performed by the routine methods employed in the textile industry for ascertaining the tensile strength, elongation, chemical resistance, porosity, and other such characteristics of fibres and fabrics.

As for the filtration characteristics, the test methods now standard in Czechoslovakia have been evolved at the Institute for Air Engineering in Prague. The first of them, used for simple and rapid assessments of the collecting efficiencies of filter cloths, consist of the trial filtration of a standard oil mist. This aerosol is prepared by the condensation of oil vapours on condensation nuclei, which yields a mean droplet size of 0.28 to 0.33 microns. The standard inlet concentration of 2.5 g m^{-3} is adjusted with the aid of a Zeiss nephelometer, the outlet concentration beyond the filter cloth is determined optically by means of a Koll-90 instrument. The filtration velocity is kept constant at 1.67 cm s^{-1}, which means that the cloth specimen is exposed to a specific throughflow of 1 m^3 of gas per 1 m^2 of area per minute. This test serves for a rough classification of the cloths, by their overall collecting efficiencies, into four categories:

1. Inefficient cloths, with O_c values below 10 %;
2. Low efficiency cloths, with O_c ratings from 10 to 20 %;
3. Medium efficiency cloths, with O_c between 20 and 30 %;
4. High efficiency cloths, where the O_c value exceeds 30 %.

In view of the low efficiency levels involved in the tests, this technique permits a fairly fine discrimination of the relative collecting efficiencies. Its findings have been amply corroborated in practice: the cloths which have proved a success in actual plant service had all been classed in categories 3 or 4, while those which users in industry quickly abandoned as inefficient had all finished this evaluation in category 1 or 2.

Apart from this quick and convenient oil mist test, the filtering efficacy and ease of cleaning of various cloths are also investigated on an automated test line where the samples are made to trap a standard type of dust. This is a polydisperse dust known as Tripol D, which is supplied by the glass industry; it contains 85 per cent of SiO_2, and 80 per cent of its particles are in the minus 5 microns range. The cloth specimens can be tested in various bag, sleeve or other configurations, and can be cleaned, as part of the test cycle, by reversed air blasts, mechanical rapping, shaking, and various other means. The whole test sequence is run off automatically, in conformance with any of nine standard test programmes, and is controlled from a central control post.

Another aspect which calls for close examination is the way a filter cloth withstands elevated temperatures. Every type of fibre has a certain critical temperature, at which its physical and mechanical properties begin to suffer. The tensile strength, elongation, and wear resistance of cloths are all temperature dependent. Once a certain temperature has been exceeded, they deteriorate to the point where the cloth disintegrates, burns, or fuses. Particularly important is the temperature response of the elongation value, because filter bags are sometimes apt to expand or contract at high service temperatures; and such unforeseen changes in the filter dimensions are hardly conducive to efficient and dependable operation. The heat resistance of filter cloths is commonly investigated by their static exposure to high temperatures for various lengths of time, but the results of these tests convey very little about the way the cloth will behave under actual high-temperature service conditions. Service stressing often produces elongation changes which the previous static testing at the same temperature had failed to reveal. To eliminate these discrepancies between the laboratory findings and the actual service behaviour, a new test has been devised, in which the cloth sample is vertically stressed by a tensile force of 15 N cm^{-1} while hot air is blown through it. The results are consistently in close agreement with the findings made in subsequent observations of the same cloth under plant service conditions.

To illustrate the sort of parameters we can expect of good present-day filter cloths, Table 15 lists some of the basic properties ascertained in the testing of three selected needled cloths made in Czechoslovakia by MITOP of Mimoň.

TABLE 15

Technical data of three Czechoslovak needled filter cloths

Cloth designation		Finet PES-1	Finet POP-1	Finet PAN-1
Fibre type		100% polyester	100% polypropylene	100% polyacrylnitrile
Specific weight	($g\ m^{-2}$)	530	500	500
Porosity of surface	(%)	89	87	87
Tensile strength in				
direction A	(N)	750	600	450
direction B	(N)	850	1000	450
Permeability	(litres $m^{-2}\ s^{-1}$)	500	600	500
Max. service temperature	(K)	423	363	413
Dimensional change after 2 hours of stressing by 15 N/cm at the maximum service temperature, in				
direction A	(%)	+3.4	−0.3	+7.7
direction B	(%)	−4.5	−3	−4.3
Trapping efficiency in standard oil mist test at filtration velocity of 1.67 cm s^{-1}	(%)	27	37	33
Pressure drop at filtration velocity of 1.67 cm s^{-1}	(Pa)	12	12	11
Penetration by Tripol D test dust at 2 cm s^{-1}	(mg m^{-3})	1.07	0.720	0.825
Trapping efficiency on test dust at 2 cm s^{-1} and an inlet concentration of 10 g m^{-3}	(%)	99.9892	99.9928	99.9917

3.4.2 Review of cloth filter designs

The design of cloth filters is still largely in a fluid state, but the trend that has so far emerged is to build a series of repetitive filter elements, or sets of unit-construction chambers, into a mostly box-shaped casing. The filters are generally operated in cycles, in which filtration periods alternate with filter cleaning periods. Within this general concept, the detail design of the main components depends chiefly on two factors: the way the cloth is installed in the filter, and the way it is cleared of the accumulated dust deposits. By the former criterion, we distinguish between:

1. *Bag filters*, where the cloth is formed into hoses of various diameters and lengths. The length to diameter ratio is usually somewhere between 15 : 1 and 25 : 1. Especially when the raw gas is induced into these bags, rather than expelled through their walls, the bags often need reinforcing with hoops fastened within them so as to maintain their desired cross-sectional shape.

2. *"Pocket" filters*, in which the cloth is formed into flat pockets of rectangular or triangular shapes. These pockets must nearly always be stiffened with wire mesh or frames to keep their sides apart as the raw gas is drawn in through them.

As for the mode of cleaning, filters of either of these two categories can fall into any of four classes:

a) Filters cleaned by a reversed gas flow only; the one or more reverse blasts are sometimes followed by a settling interval, in which the filter is shut down while the dislodged dust drops into the hopper.

b) Filters cleaned by mechanical rapping, which is usually repeated, and which may but need not be accompanied by reversed gas blowing; in either case, the procedure is sometimes followed by a settling period when the gas flow is cut off while the dust descends to the hopper.

c) Filters cleaned by mechanical or ultrasonic vibration, which again may but need not be combined with reversed gas blowing and/or followed by a settling period.

d) Filters cleaned by blasts of compressed air, either while the unit is shut down or while it is still in operation.

Despite the bewildering variety of cloth filters now available on the market, development of these units is still continuing, mainly in two directions. One aim is to cram the specified area of cloth into the most compact space, another is to evolve cleaning systems which will inflict the least possible mechanical stressing on the cloth as they clean it. Naturally, both of these improvements must be achieved without sacrificing any of the collecting efficiency, dependability, or life expectancy of the unit. For all the recent development work, however, one of the most widely used types of cloth filters is still the earliest of them all, the bag filter cleaned in the way evident from Fig. 126.

In this unit, the filter bags (*5*) are suspended within casing (*4*) on plugs set into a frame (*10*). This frame is mounted on a rod that is free to slide vertically within the guide bushing (*11*). The lower ends of the bags are fixed to the bottom plate of the filter chamber. When the unit is filtering, the bags are stretched out; the gas flows past partition (*3*) into hopper (*2*) and up into the interiors of the bags, to be cleaned as it passes through their fabric towards the outlet. For the filter cleaning cycle, the outlet flap (*8*) is closed, and flap (*9*) is opened to admit a reversed flow of outside air. This air can either be forced through the bags by a fan, or induced by the partial vacuum within the filter, i.e. flow in spontaneously, opposite to the normal flow direction. Simultaneously, a cam mechanism repeatedly raises frame (*10*), which on being released drops onto a resilient stop. This combination of shaking and reverse blowing produces an intensive cleaning effect, but the system has its drawbacks too. The moving parts of the mechanism are a potential source of trouble; cleaning causes the bags to rub against each other;

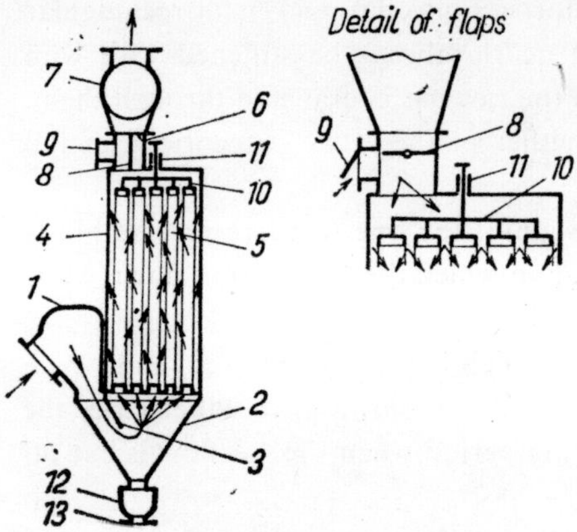

Fig. 126 Scheme of the type FTB filter:
1 — gas inlet; *2* — hopper; *3* — partition; *4* — filter casing; *5* — filter bags; *6* — gas outlet stub; *7* — delivery piping; *8* — clean gas outlet flap; *9* — air inlet flap for reverse blowing; *10* — rapping frame; *11* — rapping-rod guide bushing; *12* — worm conveyor; *13* — dust discharge port.

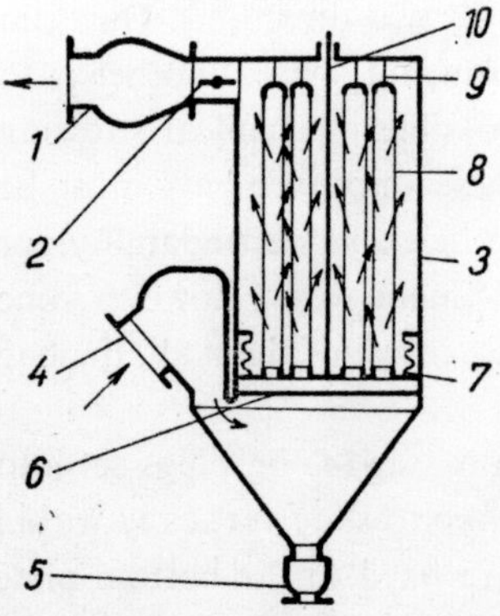

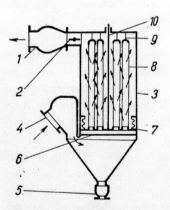

Fig. 127 Scheme of the "Hopex" filter:
1 — gas outlet piping; *2* — outlet closure flap; *3* — filter casing; *4* — gas inlet; *5* — worm conveyor; *6* — buffer; *7* — moving base plate sealed by bellows; *8* — filter bags; *9* — suspension springs; *10* — lifting rod of rapping mechanism.

not all the bags are cleaned with an equal efficiency; the filter cloth is always bent or folded over at the same places; the mechanism is sensitive to any shortening of the bags by the humidity-induced shrinkage of their cloths; and the filter can be cleaned only when it is shut down, which implies intermittent rather than continuous operation of the unit. Moreover, the flaps which admit the reverse blowing air will not afford a tight seal indefinitely. Wear will soon cause them to pass some outside air into the clean gas stream, and users report the induction of some 15 per cent of air as a typical figure.

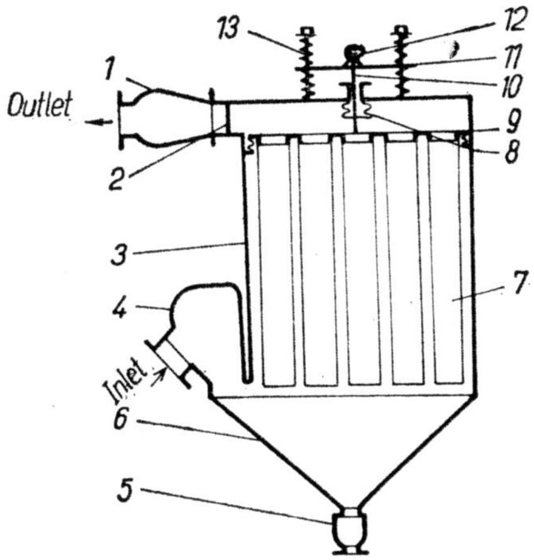

Fig. 128 Scheme of a filter with a vibratory rapping system:
1 — gas outlet piping; *2* — outlet closure flap; *3* — filter casing; *4* — gas inlet; *5* — worm conveyor; *6* — hopper; *7* — filter bags; *8* — sealing bellows; *9* — moving ceiling plate; *10* — rapping rod; *11* — vibrator carrier beam; *12* — vibrator unit; *13* — springs.

The Hopex filter, sketched in Fig. 127, avoids some of these shortcomings by employing only a mechanical rapping system, with no reversed air flow. The filter bags are suspended by plugs and springs (*9*) from the top of the casing. Their lower ends are fixed to the pipe stubs on the movable base plate (*7*), which is sealed within the casing (*3*) by a strip of stretchable fabric or by a bellows system. For rapping, this base plate is lifted by some 6 to 8 cm by pull-rod (*10*), and then released to drop onto stops (*6*); throughout this procedure, flap (*2*) is closed, but no reversed flow is admitted. The suspension springs (*9*) serve two distinct purposes: they compensate for any contraction of the filter bags, and to some extent even out the rapping intensity over the whole length of each bag. Even this system, however, is far from an ideal solution. The intensive rapping action subjects the cloths to severe stressing; the cam mechanism which actuates it is apt to jam or seize up; the suspension springs are liable to break; and the bags are inclined to burst at their lower ends, close to the bottom plate.

Another alternative is to clean the bags by vibration. In the design shown in Fig. 128, casing (*3*) houses filter bags (*7*) which are held in shape by wire reinforcing frames. The bottoms of the bags are closed, the tops are attached to the stub pipes of a movable ceiling plate (*9*) which is sealed against the unit casing by bellows (*8*). Rod (*10*) links the ceiling plate to vibrator (*12*), mounted on springs (*13*). The raw gas enters at the bottom of the unit, ascends along the bags, and is drawn into them, so that the dust deposits build up on the outside surfaces of the bags. The filter section to be cleaned is shut off by means of flap (*2*) and the deposits on its bags are shaken off by vibration, without any reversed gas or air flow. Very little is so far known about the performance of this type under actual service conditions, so it is too early yet to attempt any assessment of its relative merits and disadvantages.

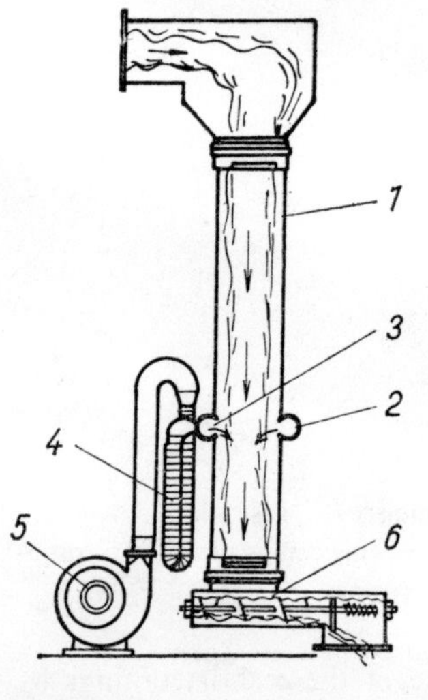

Fig. 129 Scheme of the Junkmann filter:
1 — filter bag; *2* — travelling circular-section ring; *3* — reverse blowing slit; *4* — air supply hose; *5* — high-pressure fan; *6* — worm conveyor.

If all the units described in the preceding paragraphs have to be closed down for cleaning, and therefore operate intermittently, the type illustrated in Fig. 129 is cleaned continuously while filtration is actually in progress. The bags are fastened to, and stretched out between, the top and the bottom of the chamber; the raw gas enters them at the top, and passes through their walls as it descends, thus depositing the dust on their inner surfaces. Each bag is closely encircled by an annular ring (*2*), which continually travels up and down the length of the bag. A high-pressure fan (*5*) delivers air through a hose to the interiors of these rings, from which the air is discharged through slits (*3*), some 0.6 to 1.2 mm wide, around their inner circumferences. The air jets leave those slits at usually something like 15 to 30 m s^{-1}, which is enough for them to penetrate the cloth and blow the dust off its inner surface. In practice, all the ring nozzles of each chamber are usually joined into a single rigid grid. That makes the mutual alignment of the individual rings and bags a crucial factor, because to achieve an efficient cleaning action, the inner surface of each ring must lie close up against the bag it serves. That is generally the downfall of these filters, because plant experience shows that the resultant friction causes fairly rapid wear of the filter bags. Besides, the complicated design and construction of these units, and particularly of their moving systems of ring nozzles, raises both their cost price and running costs,

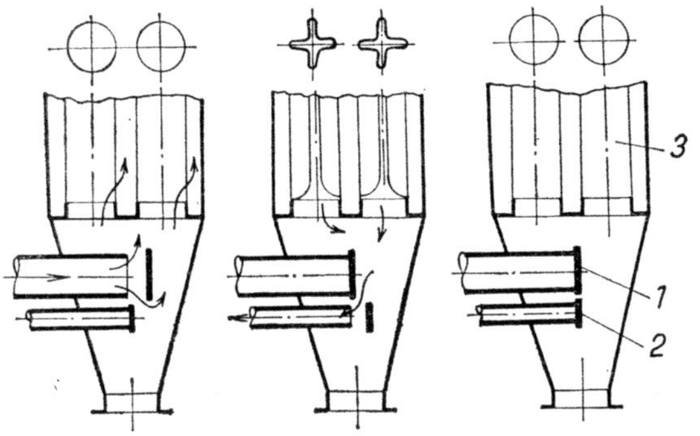

Fig. 130 Three successive stages in the operation of the Amertherm filter cleaning system: *1* — gas flow stop valve; *2* — reverse blowing valve; *3* — collapsible filter bags.

increases their maintenance requirements, and makes their long-term reliability rather dubious. The nozzles are effective only while they are moving downwards: their upward return travel does very little to clean the cloths. Still, there is no denying the highly efficient and dependable cleaning action of this system at least in the short term, and in many applications all its inherent disadvantages are accepted as quite a reasonably price to pay for the continuous, uninterrupted operation of the filter.

One of the most widely used methods of cleaning filter bags is their reversed blowing accompanied by a collapse of the bags. This is the system used on all the Amertherm cloth filters made by CEAG; its principle is schematically indicated in Fig. 130. The cloth is blown in the reverse direction by one or more short blasts of air, each of which causes the bags to cave in, from their normal round cross section into a star-shaped one. That breaks up and then shakes off the dust layers on their surfaces. The filter is next shut down long enough for the dust to drop into the hopper without any risk of its re-entrainment. The raw gas and back-blowing air inlets are generally provided with pneumatically operated disc-type stop valves. This system is employed mainly in conjunction with the glass fibre filter cloths intended to handle large throughflows at high gas temperatures. Thanks to its inherent simplicity, it has proved highly reliable. Its chief drawback is the relatively low speed at which the dust-laden gas and purging air pass through the filter cloths.

Filters where the cloths are formed into pockets are often cleaned mechanically, without any reverse blowing, by the system sketched in Fig. 131. The raw gas enters at the bottom of the unit, and rises up the narrow slits between the parallel cloth sheets (*4*). For rapping, the gas flow is stopped; rod (*1*) lifts the frame that carries the filter pockets, and then releases it to drop onto buffers (*5*). The whole block of cloth sheets is sealed against the unit casing (*3*) by bellows (*2*). The great advantage of these filters is that they pack a maximum of cloth area into a minimum of space.

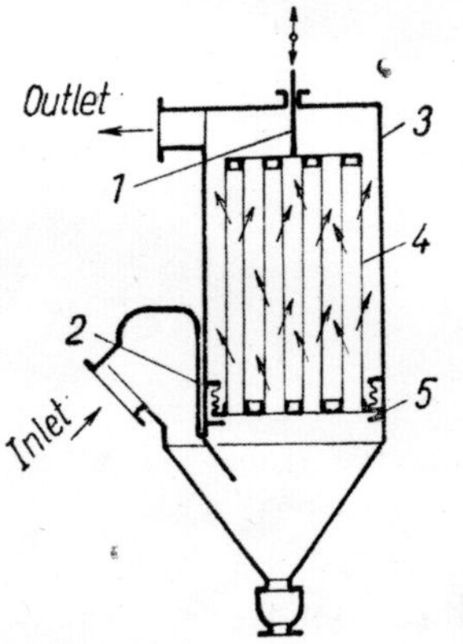

Fig. 131 Scheme of a filter where the cloths are folded into pockets:
1 — lifting rod of rapping mechanism; *2* — sealing bellows; *3* — filter casing; *4* — cloth pockets; *5* — rubber buffer.

A more recent innovation is the Mikro-Pulsair filter, shown schematically in Fig. 132. Where the bags are cleaned by short intensive blasts of compressed air, admitted opposite to the normal flow direction, while the unit is still filtering. The bags (*3*) are held in shape by rigid cylindrical frames, are closed at their bottoms, and are fastened at their tops to the stationary top plate of the chamber. Their open upper ends are provided with narrowed throat rings, essentially Venturi tubes, which are surmounted by a compressed air distributor piping (*4*). The raw gas enters the bottom of the casing, deposits its dust content on the outsides of the bags, and rises within these bags and through their constricted throat rings to the outlet opening (*6*). The compressed air pipe over each row of bags has its openings or nozzles aligned above the centres of the throat rings. When the automatically controlled solenoid valve (*5*) admits a short blast of air from a pressure reservoir into the piping, the air jets entering the Venturi tubes induce enough of the surrounding gas or air to produce a substantial pressure pimact inside each of the bags. The rows of bags are thus cleaned one by one, very effectively, in a sequence that is usually governed by a timer relay which opens the various solenoid valves in succession. The timing of the series of blasts depends on the pressure drop across the filter.

A Czechoslovak filter where the cloths are cleaned by compressed air blasts is the FTA type, schematically represented in Fig. 133. The filter cloths are borne on wedge-shaped wire mesh supporting structures (*2*). The space within each wedge is enclosed by a plate (*3*) which is provided with a number of cylindrical ejector stubs (*9*), and which also bears the compressed air supply tube (*4*). The

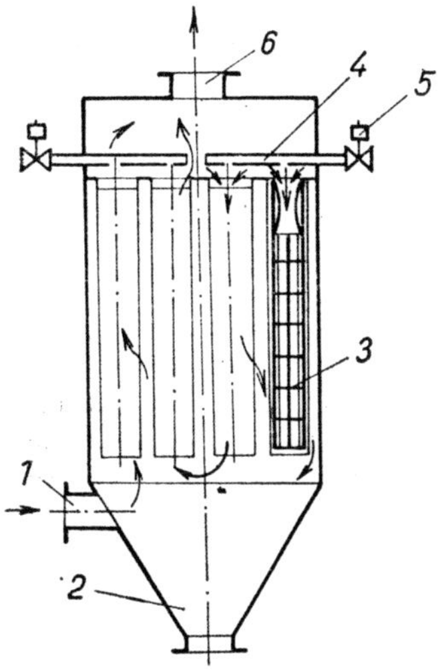

Fig. 132 Scheme of a Mikro-Pulsair filter:
1 — gas inlet; *2* — hopper; *3* — filter bags with wire frames; *4* — compressed air distributor piping; *5* — solenoid valve; *6* — gas outlet.

nozzles (*10*) of this tube lie opposite the centres of stubs (*9*). The raw gas passes through the cloth (*1*) into the interior of the pocket, and emerges through the stubs (*9*) into the outlet plenum chamber. When solenoid valve (*5*) sends a short blast of compressed air into tube (*4*), the air leaving the nozzles entrains some of the cleaned gas or air in the outlet plenum chamber. The cloth is not only blown through in the reverse direction, but also blown back against a slotted partition (*11*); the impact helps to shake off the dust deposit, which then drops down the interior of the pocket into the hopper.

Filters of this kind, where the cloth is formed into pockets, have the twin advantages of being compact and being easy to clean by reverse blasts of compressed air. Unfortunately, they tend to be less reliable than the more conventional bag filters. Their weakest links are the solenoid valves, which are the heart of the cleaning mechanism. When one of those valves goes wrong, the consequences can be so serious that the prospect has discouraged many users from choosing any equipment of this type.

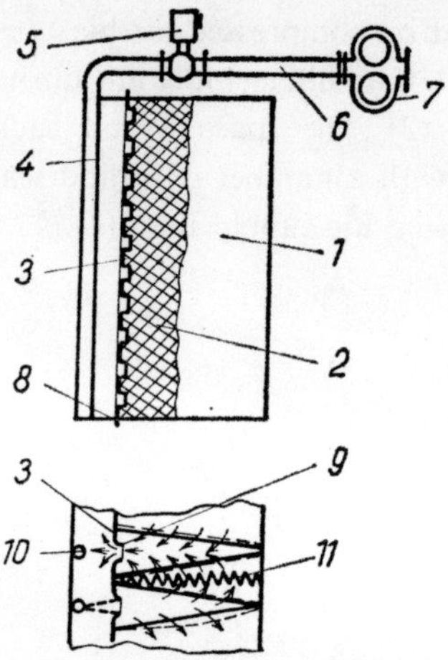

Fig. 133 Scheme of the type FTA filter:
1 — filter cloth formed into pockets; *2* — wire-mesh supporting wedges; *3* — plate with stubs; *4* — compressed air distributor pipe; *5* — solenoid valve; *6* — compressed air supply line; *7* — compressor; *8* — spacer plate; *9* — jet pumping stub; *10* — compressed air nozzle; *11* — slotted partition.

3.5 Industrial exhaustors

A special category of filters are those incorporated in the dust and fume exhaustors that serve various types of process equipment, like saws, grinding or polishing machines, metalworking and woodworking machinery, plastics processing equipment, etc. Similar exhaustors are also used to remove the dust deposits which are apt to build up in industrial premises. Most of these devices employ specially adapted kinds of cloth filters. The variety of exhaustors now on the market is such that only a few of them can be described here by way of an example.

Medium-performance exhaustor

The unit shown in Fig. 134 is intended for clearing up dust sediments in industrial premises by means of portable suction hoses, which are attached as the need arises to couplings on a network of fixed, stationary suction pipes. The dust-laden air enters cyclones (*1*), which divert the coarsest fractions into hopper (*2*). The air next passes into cloth filter (*4*), where the bags are supported on wire frames. The bags are cleaned, while filtration is still in progress, by a reverse flow of outside air admitted through solenoid valve (*5*); in view of the relatively high vacuum inside the filter, this air flows in spontaneously, and the pressure differential is sufficient to ensure an efficient cleaning action. The air stream is drawn through the unit by an eight-stage fan (*6*) downstream of the filter, which discharges through a third cleaning stage, filter (*8*) at the delivery side of the fan.

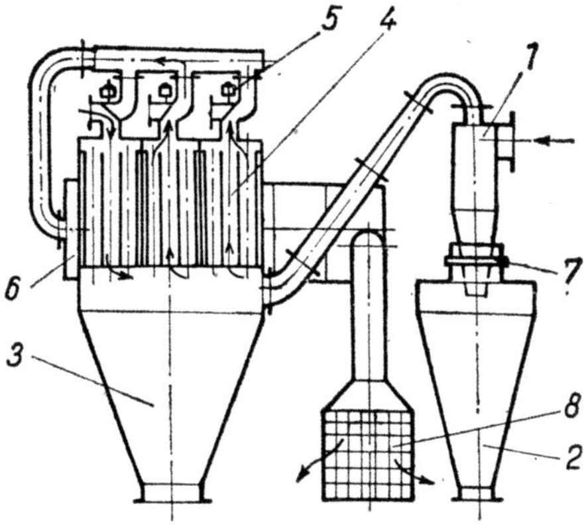

Fig. 134 The SOB-900 medium-capacity exhaustor:
1 — cyclones; *2* — cyclone hopper; *3* — filter hopper; *4* — cloth filter; *5* — solenoid valve; *6* — fan; *7* — frangible diaphragm; *8* — fan outlet-side filter.

Units of this kind, available in both stationary and mobile versions, are capable of cleaning any dry gas at inlet temperatures up to 30 °C. They have been widely adopted in metallurgical plants and foundries, ceramics factories, in the production of building materials, and in various other industries. The eight-stage fan induces 1000 m^3 of air per hour, and builds up a static suction pressure of 23,000 Pa at the unit inlet. With only the first two cleaning stages in operation, the outlet dust concentration is consistently below 100 mg m^{-3}; with all three stages engaged, it drops to less than 10 mg m^{-3}.

High-performance exhaustor

The unit sketched in Fig. 135 resembles the one described in the preceding paragraph, but is rated for an air intake of 1400 m^3 per hour at an inlet suction pressure of 25,000 Pa. The installation is always stationary, and is designed to exhaust to the free atmosphere. The dust-laden air is again induced through cyclones (*1*), which separate out the coarsest particles, into a cloth filter (*2*), which is cleaned by a reverse flow of air admitted through purging valve (*6*). The trapped dust accumulates in hopper (*3*), which is fitted with a frangible-diaphragm outlet (*8*) to jettison the blast wave in the event of an explosion, and with a dust level indicator (*7*). The pressure drop across these units ranges from 3500 to 9500 Pa, in dependence on the amount of dust retained in them. The unit is again powered by an eight-stage fan, and is intended for large-scale dust clearing in steelworks, foundries, and other premises where substantial dust layers are liable to build up between successive cleaning operations.

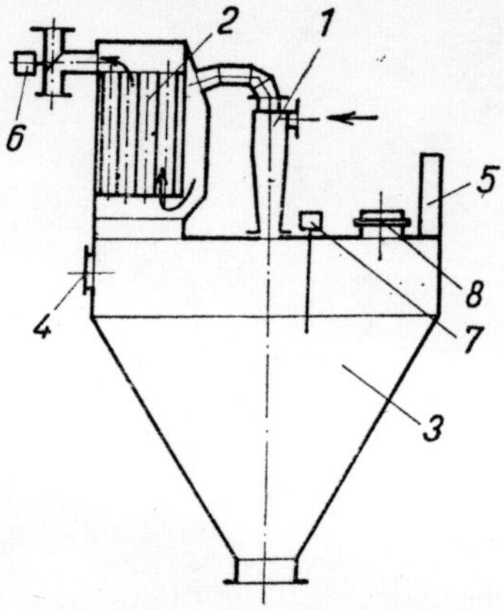

Fig. 135 The SOC-1400 high-capacity exhaustor:
1 — cyclones; *2* — cloth filter; *3* — hopper; *4* — inspection port; *5* — control panel; *6* — solenoid valve; *7* — dust level indicator; *8* — frangible diaphragm.

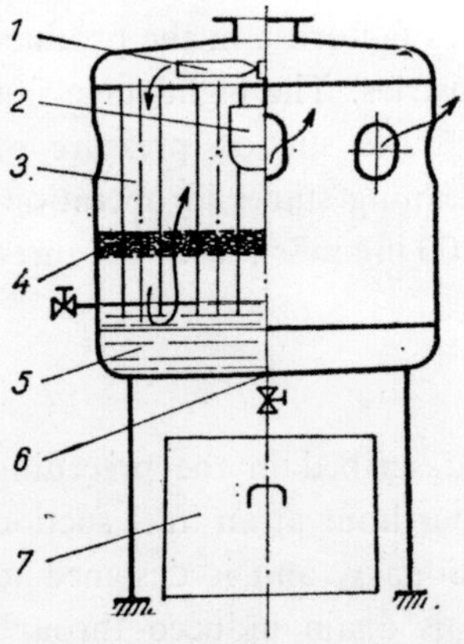

Fig. 136 The Velux 7 M wet exhaustor:
1 — fan impeller; *2* — fan motor; *3* — upright steel tubes; *4* — packing layer; *5* — liquid bath; *6* — drain plug; *7* — sludge settling tank.

Wet exhaustor

The exhaustor in Fig. 136 can deal with almost any non-fibrous industrial dust, including fatty or adhesive dusts which are difficult to trap in dry exhaustors. The dust-laden air is drawn in at the top of the casing by fan (*1*), which is mounted along with its drive motor (*2*) on a steel plate; the latter is perforated to open into the eight vertical steel tubes (*3*) that are attached to its bottom. The lower ends of these tubes are immersed in a liquid bath (*5*), normally made up of some 8 to 10 litres of machine oil to about 20 litres of water. Above the bath level, a grate (*4*) carries a layer of Raschig rings, or hollow cylinders about 1 to 2 cm high, 1 to 2 cm in diameter, with walls 2 to 3 mm thick. The air emerging from the bath has to pass through this packing layer on its way to the outlet, and deposits most of the entrained liquid on the rings. This exhaustor is rated for 750 m^3 of air an hour at a static suction pressure of 1000 Pa at its intake. The sludge and liquid are discharged, usually at weekly intervals, through drain (*6*) into settling tank (*7*).

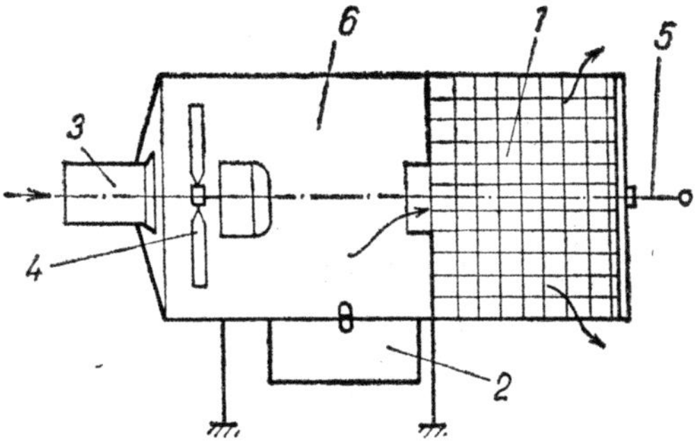

Fig. 137 The PO-2 portable exhaustor:
1 — cover with filter cloth; *2* — dust receptacle; *3* — intake stub; *4* — fan; *5* — manual rapping rod; *6* — cylindrical casing.

Portable exhaustor

The portable device in Fig. 137 is intended for sucking up dry non-adhesive dusts, including metal powders and filings. It is widely used for clearing up the swarf that arises in the machining of steel, light alloys, plastics, and other materials, but cannot cope with ceramic or fibrous dusts. The dust-laden air is induced through inlet (*3*) by fan (*4*). The bottom of the cylindrical casing (*6*) bears a dust receptacle (*2*), the clean air outlet is provided with a removable cover (*1*) that has a filter cloth stretched across it. The cloth is cleaned by means of a manually operated rapping rod (*5*). The unit draws in some 600 m^3 of air per hour, and generates a suction pressure of up to 1000 Pa at its inlet.

Stationary local exhaustor

Units of the type shown in Fig. 138 are meant to serve one particular location, and to be permanently installed in the immediate vicinity of the point where the dust is generated. The air enters the cylindrical casing through inlet (*2*) at its bottom, directly above the dust receptacle (*1*), and rises into the chamber (*3*) which houses 19 cloth bags with a total filtration area of 7 m^2. The rapping mechanism is actuated, with the unit shut down, by means of pedal (*4*). The fan (*5*) is surmounted by a silencer at the top of the unit. This exhaustor is designed for dry, non-fatty and non-fibrous dusts, is rated for a throughflow of 1500 m^3 of air an hour, and develops a static suction pressure of 2250 Pa at its inlet.

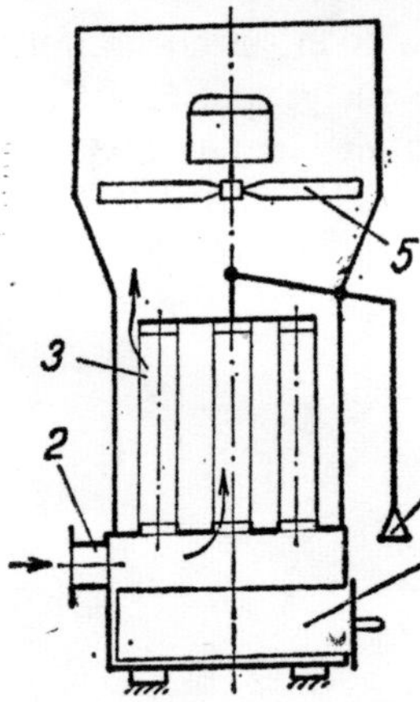

Fig. 138 The "Sajax" exhaustor for stationary local duty:
1 — dust receptacle; *2* — intake stub;
3 — cylindrical housing with filter bags;
4 — rapping pedal; *5* — fan.

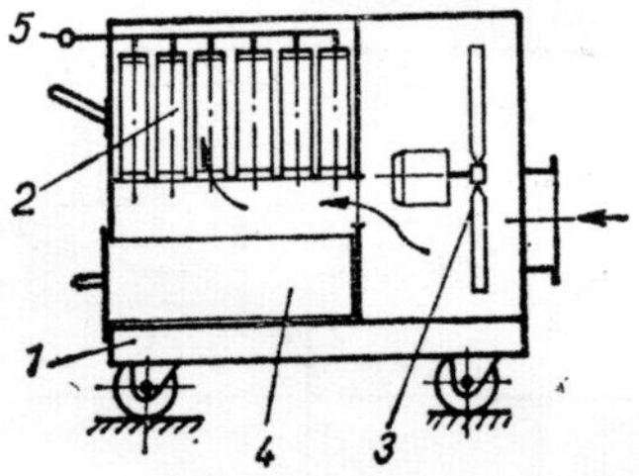

Fig. 139 The PUV mobile exhaustor unit:
1 — undercarriage; *2* — filter bags;
3 — fan; *4* — dust receptacle;
5 — manual rapping lever.

Mobile exhaustor

Wheelborne exhaustors, of the kind shown in Fig. 139, are useful for scavenging dust deposits in industrial premises and on plant equipment. The wheels (*1*) carry a box-shaped casing with a cloth filter (*2*) and an electrically driven fan (*3*). The rapping mechanism is operated manually by means of lever (*5*). The dust receptacle (*4*) at the bottom of the unit has a capacity of 0.08 m^3. The three-stage fan builds up a static suction pressure of 9000 Pa at the inlet port.

3.6 Coagulators and agglomerators

Coagulators and agglomerators are not meant to trap the liquid or solid particles, but merely to drive them together into larger drops or clusters, which are then easier to capture. The trapping itself is usually done in cyclones installed further downstream. Nowadays, the prevalent way of driving the particles together is by ultrasonic energy, in devices which are commonly but quite erroneously known as ultrasonic separators. This technique might at first seem incongruous, because standing waves of acoustic energy are so often used to stir, i.e. to dissipate the particles of, emulsions and liquid suspensions. However, their effect on aerosols or airborne suspensions is the very opposite. Several different mechanism are involved in this action, and only the chief of them is described below.

The sonic waves cause any droplets or solid particles less than about 10 microns in size to oscillate in the gas in which they are dispersed. That increases the number of collisions per unit of time, both of the oscillating particles with each other, and between them and the coarser particles which remain relatively unaffected by the acoustic waves. Now it is a peculiar feature of these very fine particles that the forces of adhesion which tend to hold them to each other are much stronger than the forces of repulsion which drive them apart, so that particle clusters, once formed, are unlikely to disintegrate again.

This process will obviously hinge on the magnitude of the particle displacement which we can induce. This quantity is commonly calculated on the assumption that we are dealing with standing waves arising by the summation of two mutually opposed wave emissions with identical wavelengths and amplitudes. That is more or less what happens when the sonic emission meets previously emitted waves which have been reflected back from an obstacle. Let us denote the amplitude of the wave A; its frequency f; its wavelength λ; let x be the investigated point on the horizontal axis linking two nodes; and let τ be the time interval in which the displacement u at point x is to be determined. This displacement is then defined as

$$u = -A \sin 2\pi f \tau \sin 2\pi \frac{x}{\lambda} \tag{90}$$

If we are considering a spherical particle of diameter a (in m) and specific gravity ϱ (in kg m^{-3}) in an environment with a dynamic viscosity of η (in Pa s), the displacement u_c of this particle will be

$$u_c = \frac{A \sin 2\pi \frac{x}{\lambda} \sin (2\pi f \tau - \varphi)}{\sqrt{(\pi \varrho a^2 f / g\eta)^2 + 1}} \tag{91}$$

where φ is the phase shift between the dust particle and the gas particles around it.

We can now divide equation (91) by the absolute value of expression (90) and, on inserting $\varphi = 0$ (i.e. assuming that the solid or liquid and the gas particles move in phase with each other), obtain

$$\frac{u_c}{u} = \frac{1}{\sqrt{(\pi \varrho a^2/(g\eta))^2 + 1}} \tag{92}$$

This relation indicates that, given constant values of ϱ and η, the ratio between the displacements of admixture particles and of the carrier gas particles will depend on the product fa^2. Thus, for any given particle diameter a, the $u_c : u$ ratio will depend solely on the frequency f. This frequency therefore determines the relative (i.e. effective) displacement of the admixture particles which we can attain in ultrasonic equipment.

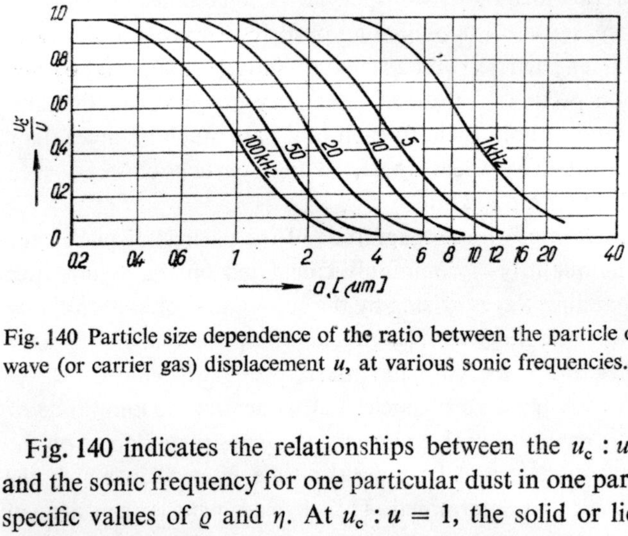

Fig. 140 Particle size dependence of the ratio between the particle displacement u_c and the sonic wave (or carrier gas) displacement u, at various sonic frequencies.

Fig. 140 indicates the relationships between the $u_c : u$ ratio, the particle size, and the sonic frequency for one particular dust in one particular gas, i.e. at certain specific values of ϱ and η. At $u_c : u = 1$, the solid or liquid particles will to all intents and purposes move in unison with the gas around them. Once the combination of particle size and sonic frequency exceeds a certain limit, the $u_c : u$ ratio will drop to less than unity, i.e. the admixture particles will in view of their greater inertia no longer be able to follow fully the motions of the gas particles around them. At a certain critical particle size, the $u_c : u$ ratio drops off to zero, which means that the solid or liquid particles will then remain more or less unaffected by the sonic emission.

Apart from inducing these particle oscillations, the sonic energy also aids coagulation and agglomeration processes by the acoustic pressure which the standing waves exert. This pressure arises as the gas is alternately thickened and rarefied in the course of its oscillations, and increases the dust particle concentration at the antinodes or vibration loops. The effectiveness of this mechanism grows with the intensity of the sonic emission, with the cube of the particle size, and with the time elapsed since the mechanism first came into play.

The frequency dependence of the main agglomerating or coagulating mechanism accounts for the fact that the frequencies employed for this purpose very often lie in the ultrasonic spectrum, which in theory is mostly defined as beginning at 20 kHz. In practice, of course, the boundary between the audible and the ultrasonic range is purely subjective, so that the threshold is sometimes quoted as 15 and sometimes even as low as 8 kHz.

Research and development work in this field has shown that, especially when the throughflows are fairly large, say 10,000 to 100,000 m^3 per hour, the success or otherwise of the ultrasonic treatment will depend essentially on three factors:

a) A well-selected acoustic field intensity,

b) The residence time of the gas in this field, and

c) A correct choice of the acoustic frequency.

Let us now examine these key factors one by one.

a) The least acoustic field intensity which will produce a detectable degree of coagulation or agglomeration in a practically acceptable time span is about 140 dB. In plant practice, however, the intensity must always be higher than 150 dB, because the human ear simply cannot tolerate intensities below this limit. The sonic energy is usually generated by what is essentially a siren or a hooter. Just how efficiently these devices will convert the energy of compressed gas into acoustic energy depends on their design, especially on the geometry of their horns and the adjacent spaces. The design of these components is complicated by the fact that their aerodynamics, their technological aspects and economics have to be respected no less than their acoustic performance. The literature variously quotes a 40 to 70 per cent acoustic efficiency as a realistic target figure for the design of these ultrasonic generators.

b) Since the agglomerating or coagulating effect depends on the number of collisions between particles, it seems obvious that it will be the greater, the more time the particles spend in the acoustic field. That, however, holds true only up to a certain limit, beyond which no further dwell in the field will enhance the effect any further. This limit varies with the aerosol and field intensity in question, but is always reached when agglomeration or coagulation have progressed to the point where each unit volume of gas contains only a relatively small number of particle clusters. A well-selected field intensity and sonic frequency will produce this state after a gas retention time which, as a rule, should not exceed a few seconds.

c) There is a certain optimum frequency range for every particular aerosol and every particle size. This is the range of frequencies which will most effectively overcome the inertia of the particles and displace them by the two mechanisms already explained – by creating a velocity differential between the admixture particles and the gas, and by setting up an acoustic pressure gradient. These two factors combine to exert different pressures on each of the mutually opposite faces of every particle; the resultant forces then set the particles into motion, or accelerate or decelerate them. Now, even given a sufficient field intensity, too high a frequency will fail to produce much effect on relatively massive particles: the resultant forces generated by the two mechanisms will not, within the short time span available at the high frequency, build up to a level adequate to overcome the inertia of such particles. Similarly, too low a frequency or too low an inertia of the particles will cause the latter to move almost in unison with the gas around them, thus rendering the main coagulating mechanism ineffective. In practice, we are of course always dealing with a mixture of various particle sizes; we must therefore select that frequency which will induce the maximum of relative motion as between particles of different sizes, as only that approach will ensure a maximum incidence of particle collisions. Experiments have shown that the optimum frequency range extends down to 1.0 kHz for 10-micron particles, is slightly higher for particles in the submicron range, lies above 5 kHz for ammonia and arsenic mists, and higher still for tobacco smoke.

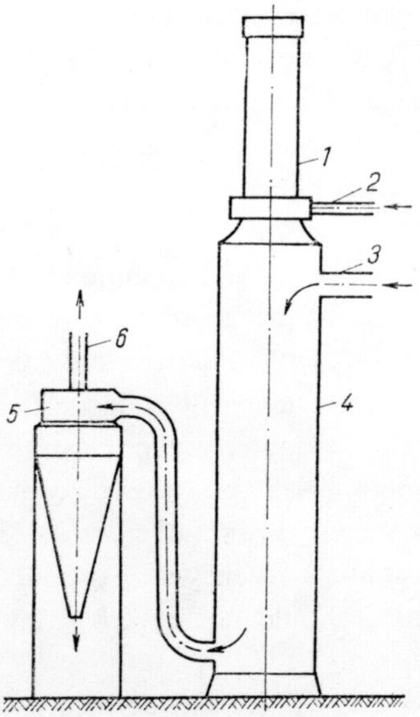

Fig. 141 Scheme of an ultrasonic separator unit:
1 — ultrasonic generator (siren); *2* — compressed gas piping; *3* — gas inlet; *4* — coagulation or agglomeration vessel; *5* — cyclone; *6* — clean gas outlet.

The main type of ultrasonic generator at our disposal for industrial applications are sirens. These typically need about 1 kW of input energy per 1000 m^3 of gas per hour cleaned in a smallish installation, but their specific power consumption drops to something like 0.5 kW per 1000 m^3 per hour in large units. These figures, of course, apply only to the actual generation of sonic energy, and do not cover the power expended on moving the gas through the unit or through the separator downstream of it.

A scheme of a complete installation of this kind is presented in Fig. 141. The ultrasonic generator is built into the top of the coagulating or agglomerating column, and the sonic waves pass down this column to be reflected from its bottom. The interaction between the emitted and the reflected waves then gives rise to the standing waves which drive the particles together into clusters. To support this process, the raw gas is admitted into the column tangentially, so that its swirling motion tends to pre-concentrate the admixture particles at the column walls. Some of the particles or clusters settle at the bottom of the column, the rest leaves it with the gas stream and passes on into one or more cyclones. It is vital that the gas velocity at the cyclone inlets should be kept low: a high inlet speed would be liable to decompose the particle clusters, and thus largely cancel the effects of the ultrasonic processing. The result would be a substantial loss of collecting efficiency in the cyclone stage.

Gas cleaning equipment incorporating ultrasonic agglomerators or coagulators has gained a measure of acceptance in several industries, and has become fairly common especially in the chemical industry, but has not so far been adopted on anything like a really large scale. A typical application is the trapping of sulphuric acid mist in the production of this acid by the contact process, where the droplet size ranges from 0.5 to 5 microns. A representative installation for this duty handles some 48,000 m^3 of gas an hour at an inlet temperature of 125 °C. Its primary stage comprises a column 2.4 m in diameter and 10.5 m high, with an ultrasonic siren mounted on its top, where the acoustic field intensity exceeds 150 dB. The column discharges into a bank of cyclones, which release the clean gas into the atmosphere. Trials with this equipment have demonstrated that a mere four seconds of ultrasonic coagulation will produce a state in which more than 90 per cent of the acid can be trapped, about half of it in the column and the rest in the cyclones, so that the latter discharge the gas with a residual acid content of only 20 to 30 mg m^{-3}.

A similar ultrasonic apparatus is used at a paper mill to trap fine NaOH particles, less than a micron in average size, from 100,000 m^3 of fumes per hour at 95 °C. In this case too, a four-second dwell in the ultrasonic field proved fully adequate. In another such typical application, carbon black is separated out from some 3000 to 3600 m^3 of gas an hour, by a sonic frequency of 3.5 kHz at a field intensity of roughly 160 dB. That agglomerates 96 per cent of the carbon black particles, a proportion of which is retained in the column while the rest is captured in two cyclones arranged in tandem beyond the column outlet.

Altogether, then, ultrasonic processing can greatly facilitate the subsequent mechanical trapping of fine solid or liquid particles, less than about 10 microns in size, provided that these particles display a natural tendency to agglomerate or coagulate. A distinct advantage of the process is that it remains virtually unaffected by the gas temperature, which can range all the way from 0 to 1000 °C, as long as the equipment is built of materials which will withstand those temperatures. Moreover, the process is not sensitive to the electrical properties of the particles, and can therefore deal even with admixtures which are difficult to handle in electrostatic precipitators. Finally, the process presents no fire or explosion hazards, and can consequently be applied even when the admixture or the gas itself is inflammable. That is a distinct advantage over electrostatic precipitators, which cannot possibly handle inflammable materials, for fear of an arc-over igniting them.

With all these advantages stated, it must be pointed out that this process is subject to certain constraints. The most important of them is the concentration of admixture particles per unit volume of incoming gas, which must be high enough to ensure a sufficient incidence of particle collisions. The actual concentration threshold naturally varies with the particle size, but, broadly speaking, for particles in the 1 to 10 micron interval the inlet concentration should not be lower than $10\ g\ m^{-3}$. When most of the particles are in the sub-micron bracket, reasonable results can be attained even with lower inlet concentrations. Besides, the concentration level can be artificially increased by such simple expedients as spraying water or other droplets into the gas, or enriching the gas to be treated with some other, more heavily dust-laden gas. An often convenient way of raising the concentration level is to cool the gas, causing it to contract, which will leave the same number of particles dispersed in a smaller volume.

Another potential limitation is the noise the unit emits. In these times of increasingly stringent public health regulations, there have been cases where ultrasonic installations have had to be shut down and dismantled on this account. But then the whole sphere of ultrasonically assisted particle separation is still at a very early stage of its development, where teething troubles are only to be expected. The noise problem is only one of them, and should prove no more insuperable than the others.

4. ANCILLARIES OF SEPARATORS

Ing. O. Štorch, CSc, Ing. J. Urban

4.1 Removal of trapped pollutants

No matter what type of equipment we use to capture the admixture particles, we cannot expect it to function efficiently and reliably unless we ensure the continual and fully dependable removal of the trapped material that accumulates in it. This material can either be recovered for further processing, or taken away and dumped at some usually distant site; sometimes, part of it is processed and the rest is dumped. Our choice of equipment for its removal will therefore depend not only on the physical properties of the material itself, but equally on the state in which we want to obtain it to facilitate its further processing or disposal. Material earmarked for further processing is nearly always wanted in the dry state. Material intended for dumping may just as well be wet; in fact, a certain moisture content is usually welcome, because it prevents the generation of secondary dust as the material is tipped onto the dumps. Moreover, material that is to be dumped should preferably have good bonding properties, so as to present a compact, reasonably dust-free surface even after drying out.

The removal of the trapped material breaks down into two distinct phases: the evacuation of the hoppers, chutes or throughs on the separating equipment itself, and the delivery of the evacuated material to its destination.

Hoppers and chutes are generally emptied by gravity alone, which imposes certain requirements on the shapes and slope angles of their walls. Fine and/or adhesive materials must be discharged continuously, to prevent them forming arches which would block the outlets (this is easily the most common cause of separator breakdowns and malfunctioning). Even coarser dusts sometimes call for continuous discharging, especially when their cooling in the hopper is likely to entail the condensation of moisture and consequent wetting of the dust, because moist dust is particularly prone to arching. In short, continuous emptying is by far the safest solution in all cases when we cannot be absolutely sure that the handled material will remain loose and flow freely under every possible set of conditions in the hopper. Under some circumstances, even continuous discharging provisions, adequately inclined chutes, and hopper walls insulated to prevent the condensation of moisture, will still not ensure a fully reliable removal of the accumulated material. In those cases there is often nothing for it but to heat the hopper walls. Another device often used, both to check the state within the hopper and to loosen any material that blocks the outflow, is a raking bar, which slides in and is sealed within a ball mounting set into the hopper wall.

In large sets of separating equipment, the tendency nowadays is to replace the conventional array of tapering outlet chutes or hoppers by a single trough that contains a built-in conveyor. This undeniably facilitates the discharging process: all the material in the trough flows along the line of maximum inclination, while in the conventional inverted-pyramid hoppers, part of the material inevitably finds its way into the corners, where the slope is gentler than that of the adjacent walls. The trouble about these troughs is that the conveyor is located within the separator, and is therefore accessible for inspection and maintenance only when the whole unit has been shut down. Obviously, any design of this kind calls for conveying equipment which will remain utterly dependable even under the worst of service conditions. Which is a tall order, because conditions at the outlet ends of separators can be extremely severe.

The hopper outlets are generally provided with both sliding gates and dosing closures. The former serve to stop the outflow when the latter are undergoing repairs. The dosing device usually serves two distinct purposes: it meters out the trapped material, preferably at the rate at which that material drops into the hopper; and it prevents or at least limits the ingress of outside air into the hopper, as the pressure inside the hopper equals that in the separator above it, and the separators are mostly run with a partial vacuum within them. The induction of outside air through the hopper would thus reduce the efficiency of the separating process. It could also cool the material in the hopper, and cause the condensation of moisture, which would impede the flow of material towards the hopper outlet.

The sliding gates are mostly fitted with flat, rectangular slide plates, which are displaced manually. Smaller slides are usually shifted by means of a lever linkage, larger ones by a lead screw and nut or a rack and pinion drive controlled by a hand-wheel or a sprocket and chain transmission. The main requirements on these slides are that they should dependably close and seal the outlet opening, and that they should be easy to operate. This calls for a slide plate sufficiently rigid to withstand, without undue deflection, the considerable forces that may act on it. Furthermore, the slide must be supported in a way which will forestall any tendency of the plate to jam in its guideways or frame in the course of its travel. Naturally, the drive mechanism must always be kept clean, free of dust, and properly lubricated. Since it often takes a good deal of force to shift these slides, their drive mechanisms are usually designed with large step-down ratios, so the resultant motion of the slide is necessarily slow. That is why these slides generally serve no other purpose than to block the outlet during equipment shutdowns. In these cases, however, they must be capable of bearing the weight of the trapped material, which may continue to accumulate until it fills the whole hopper.

Where it is essential that the slide should be capable of closing quickly, as is the case when it is to perform some regulating or other process function, it is best actuated by a pneumatic cylinder.

The dosing devices fall into two main categories—flap-type closures, and rotary air locks. The former can have either automatically opening or power-operated flaps.

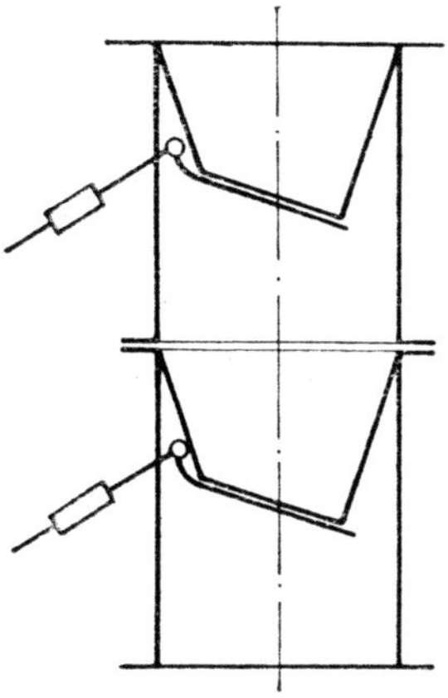

Fig. 142 A two-stage dust outlet gate with vertically superimposed flaps.

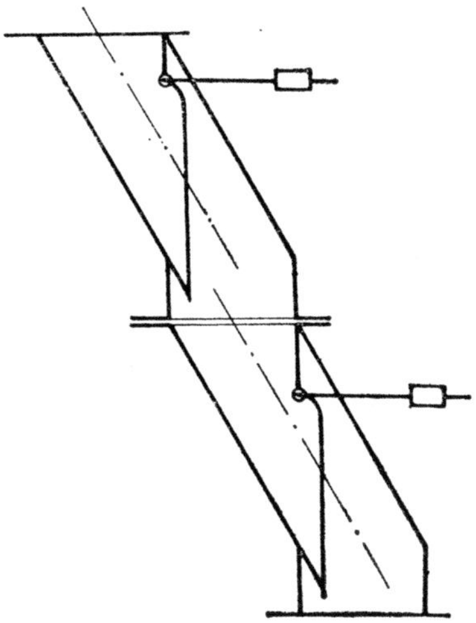

Fig. 143 A two-stage dust outlet gate with rubber flaps closed by atmospheric pressure.

The "automatic" or gravity-operated flaps open when the weight of the material resting upon them overcomes the resistance of the counterbalancing weights attached to them, and are closed by these weights when a proportion of the material has been discharged. It is essential that a certain layer of material should remain above the flap as the latter closes, because this layer acts as a seal to prevent the ingress of outside air. When the equipment is operated at much less than atmospheric pressure, a single air lock of this kind may not be capable of providing the requisite sealing in the face of the large pressure difference, so two of them may have to be installed in tandem. In all cases, the flaps must be carefully adjusted so as to permit the smoothest possible discharge, yet so as to limit their opening height or angle to a minimum. The flaps are normally made of steel, and in two-stage air locks are either superimposed vertically, as shown in Fig. 142, or staggered. In an alternative design, illustrated in Fig. 143, the flaps are made of rubber, are again opened by the weight of material on them, but are closed with the aid of the higher air pressure acting on their lower or outside surfaces. This design, however, imposes extreme demands on the mechanical properties of the rubber, which must not only retain its strength and resilience over long periods, but often also has to withstand abrasion and combined mechanical stressing at elevated service temperatures.

Power-operated flaps are always installed in pairs, and interlinked so as to operate alternately: one flap is always closed while the other one is open. They are opened by lever linkages or cam mechanisms, which are usually driven by electric motors acting through reducers, and are generally closed by return springs.

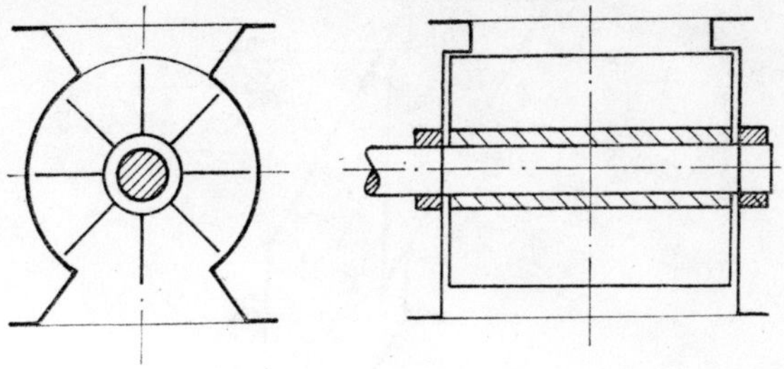

Fig. 144 A rotary air lock.

Flap closures afford reasonably efficient sealing, and possess the great virtue of simplicity. On the other hand, they are of necessity tall, which can often be a distinct drawback. Automatically opening flaps constantly have to retain a certain amount of undischarged material above them, which can be another disadvantage. Furthermore, the throughput rates attainable in these devices compare unfavourably with the sizes of the closures themselves.

These shortcomings are largely avoided in rotary vane air locks, as sketched in Fig. 144. In these units a rotor, radially partitioned into a number of segmental compartments and driven by a gear motor, fits fairly tightly in a casing provided with inlet and discharge openings. The rotor compartments are filled one by one as they come into alignment underneath the inlet opening, and empty one by one as they reach the discharge opening. The sealing efficiency of these units depends essentially upon the clearance between the rotor and the casing, and upon the number of rotor compartments.

The main advantage of these rotary air locks is their relatively small size for any given throughput rate. This is partly offset by the fact that they are liable to wear, which gradually impairs their sealing, and by their susceptibility to foreign objects, which can stop or even damage the unit. The rotors are therefore usually driven either through couplings fitted with shear pins that snap under overloads, or else by motors with thermal circuit breakers which cut the current when the motor starts overheating.

The second phase in the removal of the trapped material is its transport, either back into the process as a recycled substance, or else to the point of disposal. Our choice of means for this operation is limited by the finely pulverous or coarsely particulate nature of the material we are handling. As a rule, the whole delivery system must be sealed off from the outside atmosphere, to prevent excessive pollution of the environment. Moreover, our choice of equipment will mostly be governed by the distance to be covered and by the state in which the material is to be delivered to its destination.

Recycled material is more often than not required to re-enter the process in the dry state. It must therefore be conveyed either mechanically or pneumatically. Material which is to be dumped on open tips can be handled in either the dry or the wet state, but is nearly always required to be wet at the instant when it is dumped. The options in this case include hydraulic conveying systems, which obviate the need for spraying or otherwise moistening the material at the end of the delivery line. Very often, economic considerations dictate a combination of two or more different modes of transportation at various stages of the delivery route.

The mechanical handling devices in most widespread use at present are worm conveyors for horizontal and elevators for vertical or inclined transportation, and a variety of chain-operated bulk conveyors for both horizontal and inclined displacement of the material. Very often a whole train of similar or different conveyors is needed, with the material passing from one to the next down enclosed transfer chutes. Mechanical conveying systems are virtually impossible to seal hermetically: the dust raised both on the conveyors and at the transfer chutes will tend to escape through every chink and crevice in the hoods or shrouds. Consequently, there is often nothing for it but to connect the whole enclosed space of the conveying system to a central exhaustor unit, and to run the equipment under a partial vacuum which will preclude or at least minimize the escape of dust from it.

Mechanical conveying systems are mostly employed for small to medium throughflows, of the order of a few dozen tons per hour, and for delivery distances up to roughly 100 metres. These systems tend to be bulky, which is a serious drawback when space is short, and require extra space to afford access for their maintenance. They can move a given mass of material for a relatively low power consumption; but then they generally have to be supplemented with exhaustor systems, which increases both the first cost and the power consumption and other running costs of the equipment.

Fig. 145 is a scheme of a typical mechanical conveying system for returning trapped material to the process equipment, either directly or via a storage bin. Fig. 146 shows a set of mechanical conveying equipment typical of the installations fitted to small boilerhouses to remove fly ashes intended for dumping. The trapped fly ashes are extracted by a worm conveyor, discharged into a bunker, and picked up by a second worm conveyor, on which they are moistened before being dumped into the vehicles that take them to the tips.

The alternative mode of dry transportation is by pneumatic means. These take up much less space for any given throughput rate; simplify the selection of the material flow route; and are equally suitable for horizontal, vertical or inclined displacement of the handled material. Moreover, pneumatic conveying systems must of necessity be virtually leakproof, and therefore need no exhaustor ancillaries to avoid polluting their environment with dust. These systems are suitable for medium and large delivery distances. However, some of them need substantially

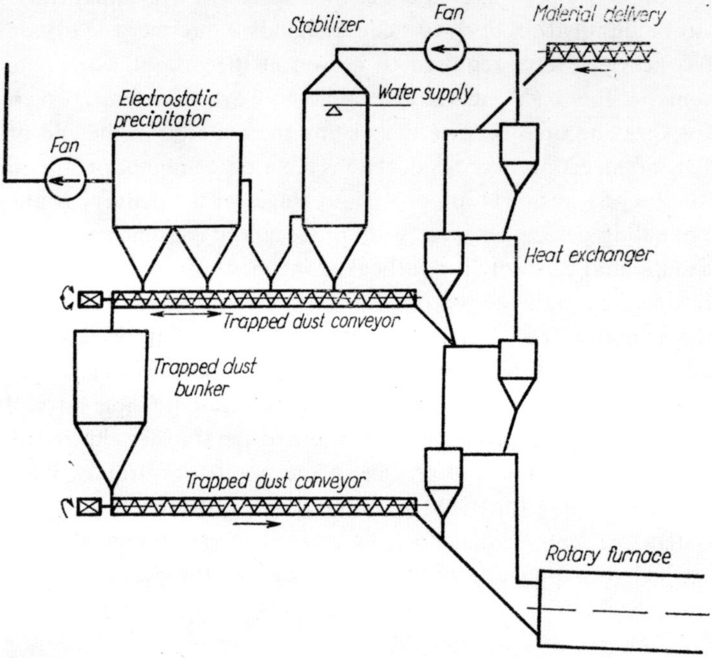

Fig. 145 A typical mechanical conveying scheme for recycling the material trapped in separators.

higher energy inputs than their mechanical counterparts. Furthermore, pneumatic systems are more prone to wear when handling abrasive substances, and less versatile than mechanical systems, in that they can handle materials with only a fairly narrow range of physical properties. They are also more susceptible to the effects of particle agglomeration, adhesion and abrasion, and extremely sensitive to the particle size of the handled material. Fortunately, the grain size distributions of trapped dusts are mostly well within the scope of pneumatic conveying systems.

Pneumatic conveying systems can exploit either of two basic principles. In the one case, the conveyed material is mixed into the air stream and moved through an enclosed piping by the air pressure. In the other, a fluidized layer of material moves down an inclined plane under the action of gravity. This latter alternative is sketched in Fig. 147. The layer of material rests on an inclined porous plate which admits air into it from below. The finely dispersed air in effect cancels the forces that normally act between solid particles, and thus "fluidizes" the material, causing it to flow down the slope like a liquid. This same stratagem of air-supported fluidization is also sometimes used in bins or hoppers to counteract any tendency of the material to arch, and to assist its smooth outflow under otherwise unfavourable circumstances.

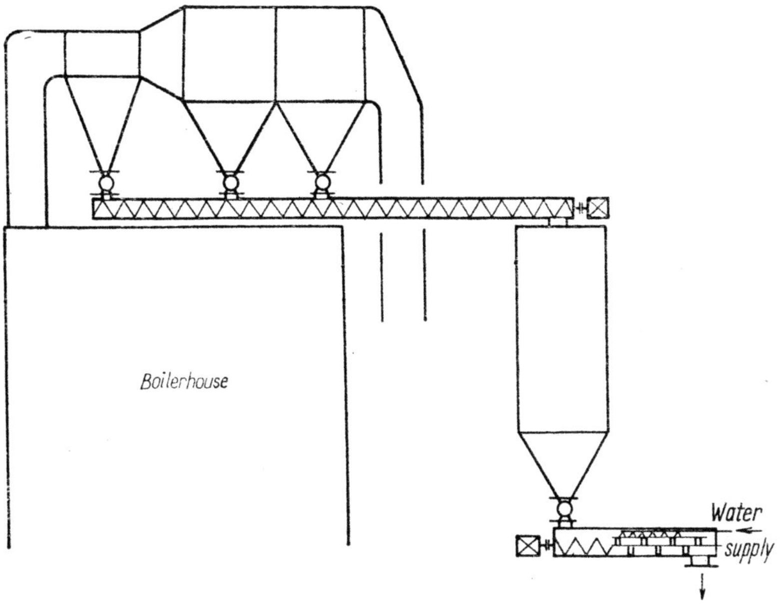

Fig. 146 A typical mechanical conveying system for removing trapped fly ashes intended for tipping.

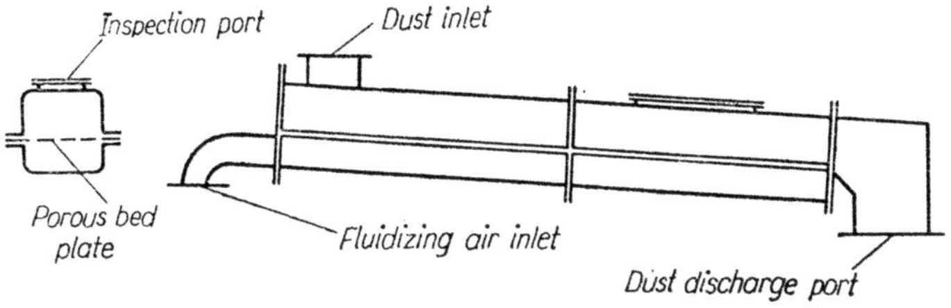

Fig. 147 An inclined fluid-bed conveyor.

Fig. 148 is a scheme of a pneumatic conveying system where the material is forced through pipings by positive air pressure. The compressed air, if necessary divested of water and oil droplets, is supplied to mixing chambers at the outlets of the various hoppers, where it entrains the dust particles. The chambers or closures must be designed to prevent any ingress of compressed air into the interior of the hoppers. The airborne dust then passes through pipings which, as the sketch indicates, may be branched to accept dust flows from two or more sources. At its destination, the dust is trapped in a separator and evacuated from the discharge

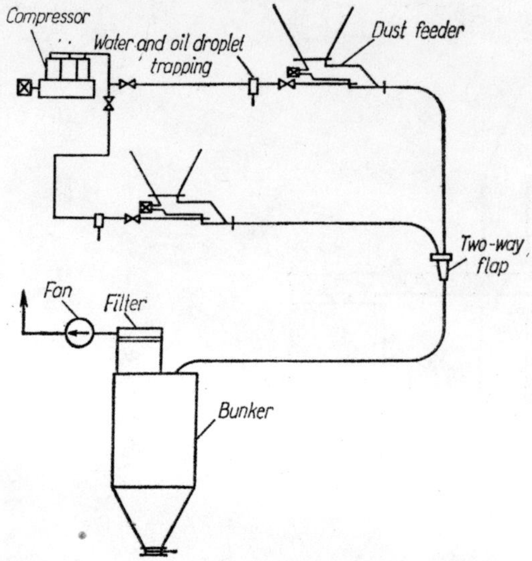

Fig. 148 Scheme of a pneumatic conveying system operating under gauge pressure.

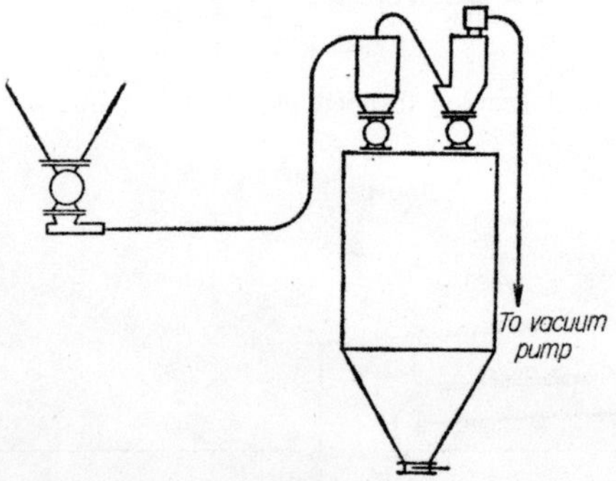

Fig. 149 Scheme of a vacuum conveying system.

hopper of this unit, while the air (or gas) is generally released into the atmosphere. An alternative system, shown in Fig. 149, employs vacuum pipings, through which the mixture of air and dust is drawn by sub-atmospheric pressure at its destination. The starting point of the delivery line is again a mixing stage, where the dust is entrained by the air or gas stream. The delivery end is again a separator, where the dust is retained in a bin or hopper, while the air or gas passes on to the vacuum pump or other vacuum generator which expels it into the atmosphere.

The gauge or suction pressure required in such systems depends both on the delivery distance and on the mixing ratio, i.e. the proportions by weight of dust and of air which are to pass through the pipings per unit of time. As a rule, the pressure rises with the throughput rate. The magnitude of the requisite gauge or suction pressure is a convenient criterion for classifying these systems into low-, medium- and high-pressure types, each of which employs its own specific kinds of pressure sources and feeder or mixer designs.

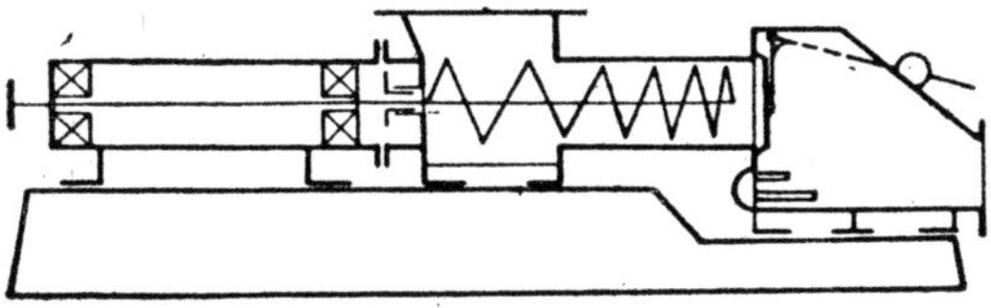

Fig. 150 Scheme of a high-speed worm conveyor.

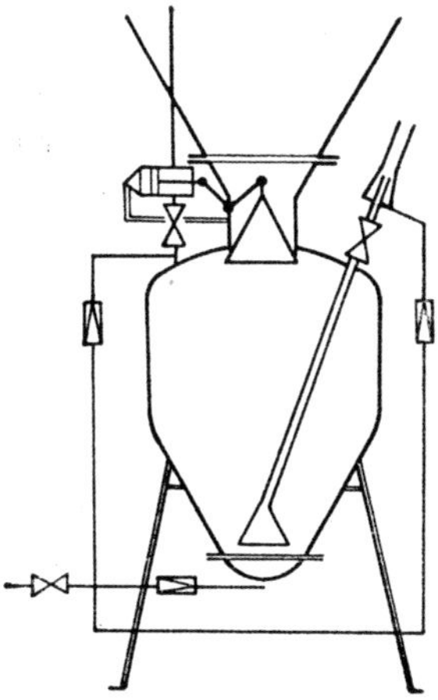

Fig. 151 A chamber-type feeder for intermittent dosing.

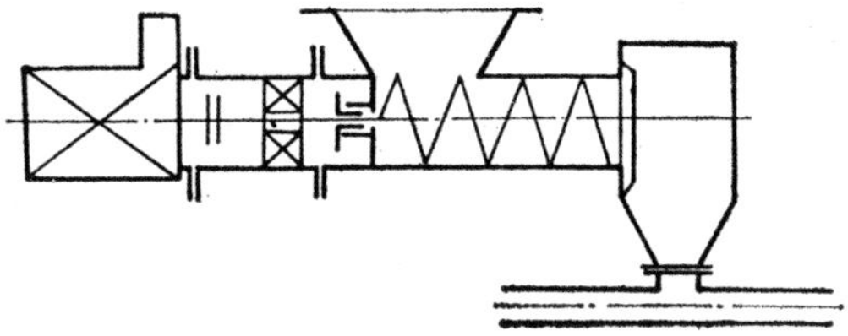

Fig. 152 Scheme of a slow-running worm conveyor.

Pneumatic conveying systems which employ a high gauge pressure are commonly fitted with either of two types of feeders: high-speed worm conveyors of the kind shown in Fig. 150, or dosing chambers as sketched in Fig. 151. The former operate continuously, while the latter work in cycles and hence need some form of automatic control of the sequence of cycle steps (filling, pressurizing, discharging, and air purging). Worm feeders, apart from functioning continuously, also have the advantage of being small, and particularly of a small overall height. However, they need relatively powerful drive motors, and are apt to suffer a good deal of wear, particularly when handling abrasive materials. Chamber feeders suffer very little wear, and permit higher operating pressures, but are bulky, and especially tall. Moreover, they function intermittently, and require relatively complex and costly control equipment.

Systems working at medium gauge pressures mostly employ either slow-running worm feeders, as shown in Fig. 152, or rotary vane mixing units. Alternatively, they are sometimes provided with the "ejectors" or jet pumping feeders more commonly used in low-pressure systems, as sketched schematically in Fig. 153.

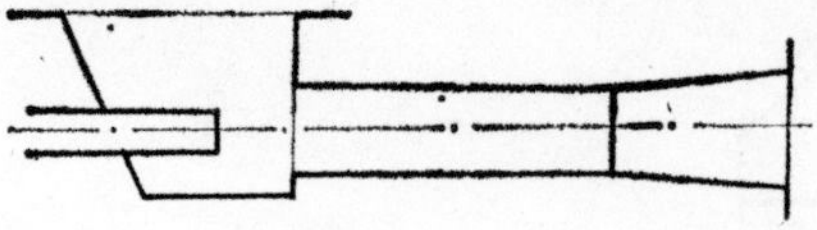

Fig. 153 An ejector (or jet pumping) feeder.

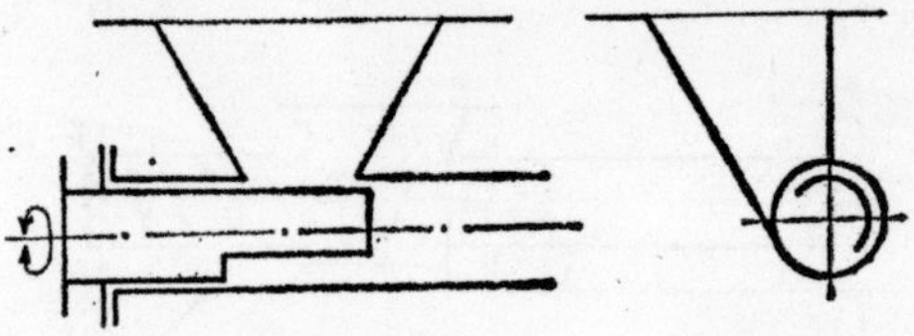

Fig. 154 A mixing chamber with an indexing inlet governor in the form of a cut-away cylinder.

In vacuum conveying systems, the solids are usually admitted either by a rotary vane metering device or from a hopper fitted with a discharge governor of the type seen in Fig. 154. In the latter, the throughflow is regulated by the indexing of a cylindrical closure which has part of its cylindrical surface cut away to present a choice of different effective inlet cross sections. The separators at the delivery ends of vacuum pipings must naturally have their discharge outlets sealed to retain the suction pressure. To this end, they are equipped with the rotary air locks or gravity- or power-operated flap closures which have already been described.

High-pressure conveying systems derive their gauge or suction pressures from compressors or vacuum pumps. Medium-pressure systems are in the main served by rotary-piston blowers, while low-pressure systems generally draw their operating pressure from fans.

The alternative fluid-bed conveying devices usually consist of enclosed troughs, mostly of rectangular cross section, where the interior space is divided into an upper and a lower compartment by partitions of some material that has microscopic pores. A fan delivers air to the space beneath this partition; the air passes through the pores, enters the layer of material on top of the partition, and fluidizes it so that it flows like a liquid. The partition is set at a slope angle of 4 to 10 per cent, depending on the type of material handled and on the material of the partition itself. The material flows down this slope to the discharge point, where the fluidizing air is evacuated through openings in the upper part of the trough and led off to dust trapping equipment; it is then often recirculated. These conveyors take up little space, and use up little power to shift a given material throughflow, but they restrict the choice of equipment layouts by requiring installation at a closely defined gradient. Furthermore, they are applicable only to certain kinds and sizes of dust particles, the ideal material for them being spherical grains of a virtually uniform size—and most dusts of course stubbornly refuse to meet either of these requirements.

The flow routes along which the trapped material is removed usually fall into two successive sections. In the first of them, the material discharged from various hoppers is brought together at a single common collecting point, along several flow routes or branches with a relatively small throughflow in each of them. In the second section, the material collected at this point is taken on to its final destination, which may be remote from that point, usually through a single high-capacity flow channel spanning a medium or large distance. The first section is generally operated at a medium gauge pressure, and fitted with slow-running worm feeders or jet pumping feeders, the second section is normally a high-pressure system provided with worm or chamber-type feeders. Fig. 155 shows a typical combination of such systems for the delivery of trapped fly ashes to a distant plant where they are utilized in the production of precast concrete blocks.

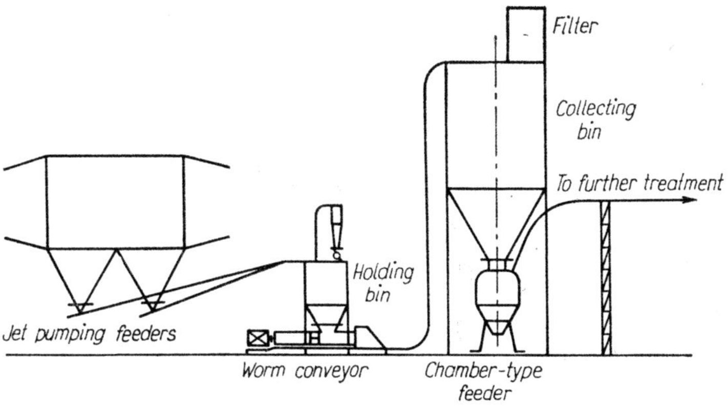

Fig. 155 A typical conveying layout for delivering fly ashes over long distances.

The first or collecting section is sometimes equipped with fluid-bed or worm conveyors, chosen on account of their relatively low power consumptions. The trouble with them is that the second-stage conveying system downstream of them often has to be set below ground level, and in the presence of a high ground water table this alone can offset all the advantages of these collecting systems.

The material brought together by the collecting system can be charged, in the dry state, straight into road or rail vehicles like those illustrated in Fig. 156. The containers on such vehicles are usually designed to be emptied by compressed air into storage bins or silos.

When the trapped material is to be dumped on tips, it may just as well be transported in the wet state by a hydraulic conveying system. Again, these systems are mostly split into a short-range primary collecting section and a longer-range secondary delivery section. The collecting system may employ either hydraulic

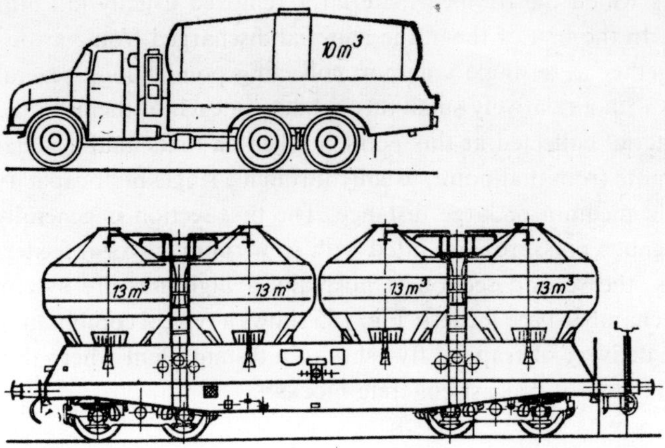

Fig. 156 Road and railborne bulk transporters.

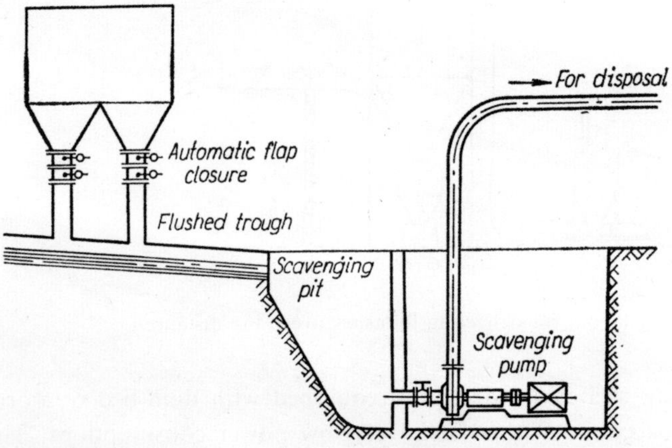

Fig. 157 Scheme of a hydraulic conveying system.

ejectors, or jet pumping devices, to drive the mixture through pipings, or else open troughs or channels served by flushing systems. In either case, it normally delivers into pits, from which the sludge is scavenged and pumped through pipings to the disposal area. Fig. 157 is a scheme of a representative installation, where the collector section consists of open troughs.

Whatever conveying system we adopt, it will always have to meet two principal stipulations: the utmost reliability, and the lowest possible maintenance and attendance requirements. Both are all the more difficult to achieve in view of the severe service conditions, which involve continuous duty and the handling of mostly abrasive materials. Consequently, the system must be robust, and must be backed by stand-by facilities for use while the main system is undergoing repairs. Even fully duplicated conveying systems are often considered justified, because the correct functioning of the main separating equipment depends on the constant availability of a sufficiently rated conveying system for disposal of the trapped material.

4.2 Sludge handling systems

Practically all the industrial countries of the world are short of water. Consequently, the users of wet separating equipment are almost invariably forced to recirculate their process water, using fresh water only to make up losses by evaporation and the loss of the water bound to the dust particles. How the water is recirculated, and how clean the water fed back into the separators has to be, depends upon several factors: on the type of separator installed, on the sedimentation characteristics of the dust, and on the possible further processing or utilization of the sludge.

In some types of separators, notably the Roto Clone N, the Tilghman, or the MHA type which is their Czechoslovak equivalent, the water recirculates spontaneously within the unit casing, with no need for any pump. The dust settles in a thick pasty slurry, which has to be evacuated from the bottom of the unit by raking or scraping. Other kinds of equipment in which water recirculation presents no special problems are those which impose no great demands on the cleanliness of the water, so that the settling tanks can be kept relatively small, and those to which the water need not be supplied under pressure, so that it can be recirculated by sludge pumps. These two stipulations must not be confused with each other: some units can accept water with a high solids content, but need it pressurized; that rules out sludge pumps, and any other kind of pump will suffer excessive wear when handling these dense slurries. Recent developments in gas scrubbing equipment indicate that the ease of water recirculation has finally come to be fully appreciated as a major design consideration: the newly evolved MSA unit seems to be a typical sign of the trend.

A totally different picture confronts us when the recirculated water has to be fairly clean, for instance so as not to clog spray nozzles. In those cases, the obvious approach is to cut down the specific water consumption, so as to minimize the amount of water which has to be subjected to costly cleaning.

No matter what kind of wet separators we install, we shall always be left with a greater or lesser quantity of sludge from which, as a rule, the water has to be extracted. From this aspect, it is almost immaterial whether the water is then drained off or recirculated: the first task is to separate it from the solid sediments. True, the centrifugal pumps generally used in high-pressure recirculation systems cannot cope with as high a residual solids content as is acceptable when the water is drained off into the sewers, but that is only a matter of degree.

By far the most common way of extracting the water from these slurries is by sedimentation of the solids in large settling ponds or circular or rectangular concrete-lined pits. These have the salient advantage of being simple: the solids settle by gravity alone, and no internal fixtures or fittings are needed. On the other hand, these ponds, pits or tanks are too large to be convenient, especially in the often overcrowded vicinity of industrial plants; are costly to build by reason of their sheer size; are apt to fill up with deposits very unevenly, so that large parts of their total area are not properly exploited; and also require fairly frequent cleaning.

Slurries which contain relatively large solid particles, such as are generated for instance in the dressing or beneficiation of minerals, are sometimes thickened in hydrocyclones. These harness centrifugal forces to separate solids from the liquid, just as ordinary cyclones separate solids from gases; which implies that they will be efficient only given solid particles with a considerable inertia and a specific gravity differing substantially from that of the liquid. It follows that they are of little use for treating the sludges or slimes obtained from gas scrubbing equipment, where the solids are mostly very fine and often of much the same specific gravity as the water that carries them.

All this points to the need for some item of equipment which will effectively dehydrate or thicken the sludges discharged from gas cleaning plants, but be far more compact than settling tanks. The equipment should not require the sludge to be delivered under pressure, and should incorporate mechanical facilities for the continuous removal of the thickened sediment. The units devised to fit into this niche will be described in the subsequent Section.

Before going into details, it must be admitted that sludge handling facilities are always one of the least pleasant corners of any plant. That is apt to obscure their salient advantage, that the trapped solids are removed without any of the secondary dust generation that is unavoidable when they are handled in the dry state. Furthermore, these sludges lend themselves to hydraulic conveying, which is so much of an advantage that even dusts captured in the dry state are often deliberately moistened to fit them for this mode of transportation. The only qualification to all this is that some kinds of dust are liable to cement. Especially dusts containing CaO or similar binders will quickly form deposits which can ultimately block up the whole flow channel. Sludges containing such dusts should therefore never be pumped, nor conveyed in enclosed pipes. They should rather be left to flow by gravity alone down inclined troughs, with removable covers affording access for inspection and cleaning. That is all the more advisable in view of the fact that sludge handling systems are inclined to freeze up in winter, a problem which will be examined in more detail further on.

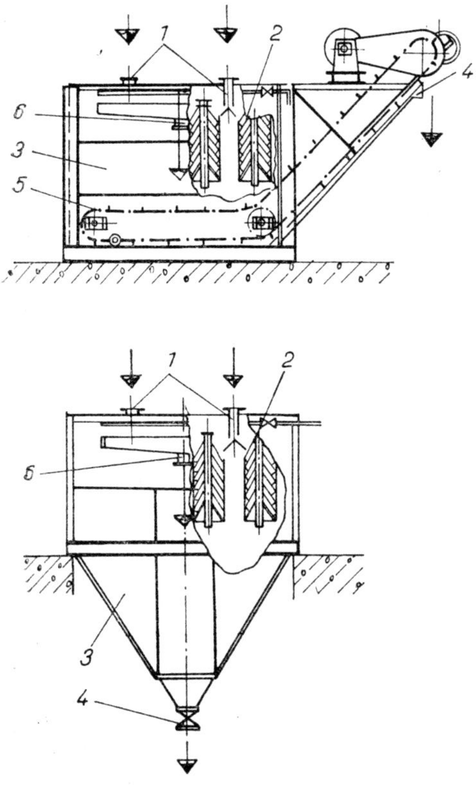

Fig. 158 The UNA and UNB settling tanks:
1 — sludge inlets; *2* — plate stacks; *3* — receiver vessel; *4* — thickened slime outlet; *5* — mechanical raker/elevator; *6* — clean water outlet.

4.2.1 Type UNA and UNB settling tanks

These two types of tanks have been evolved in Czechoslovakia for thickening the sludges arising in scrubbers of the MVA, MVB and MHC ranges, and in the bubble washers which are widely used in the ceramics industry, in foundries, etc. Both these types rely on the free, gravity-induced sedimentation of the solids in the sludge, and are designed to provide the largest possible settling area within the smallest feasible overall dimensions. As evident from Fig. 158, the two types differ from each other only by their provisions for the removal of the thickened sludge. In both these designs, a rectangular-section tank houses one or more identical cleaning units. Each unit comprises a rectangular-section receiver vessel for the cleaned water. The two longer sides of this vessel carry twelve vertically superimposed plates, 150 mm wide and spaced 50 mm apart, which slope downwards at an angle of about 50 degrees. Every unit is 1150 mm long. The raw sludge enters by gravity down an inclined inlet trough. As the bath level in the tank is slightly higher than that in the clean water receivers, the sludge flows slowly up the gaps between the adjacent sloping plates, depositing most of its solids content on the plate surfaces. The essential factors are the low flow velocity, of only about 0.6 mm s^{-1}, and the small height of the flow channel, which is 50 mm; these between them ensure a high degree of solids removal. The water then passes through holes in the receiver walls, facing the ends of the gaps between the plates, and is drawn off through a pipe near the top of the clean water receiver. The dense layers that build up on the plates gradually slide down them and drop to the bottom of the tank. In the UNA version, this is a funnel with a nozzle which continuously discharges the slurry, with an admixture of water, in the liquid state. In the UNB type, the thick pasty slime is scavenged by a scaper conveyor of the type fitted to the MHA and MHB washers.

In tanks containing more than one of these cleaning units, it is vital that all the units should be loaded equally, else much of the total effect would be lost. To this end, the individual receiver vessels are fitted with adjustable water outlets which control the throughflow through the unit upstream of them.

The range of UNA and UNB tanks now available includes types with 1, 2, 4, 6 and 16 cleaning units apiece. Each of these units is rated for a throughflow of 2.1 to 2.8 m^3 per hour; this is the flow rate at which, theoretically, all the solid particles or particle clusters with free falling velocities of 0.15 to 0.2 mm s^{-1} ought to settle on the sloping plates. Given particles with a specific gravity of 2 g cm^{-3}, these free falling velocities correspond to particle sizes of 16 to 19 microns.

The cleaning efficiency of these tanks is in practice usually higher than might be expected from the retained fractions curves of the sludges fed into them. This is explained by the coagulation and agglomeration processes which take place in the sludge. Unfortunately, these processes are totally unpredictable, and the way and degree in which they will affect the outcome can rarely be assessed in advance. These processes are largely governed by the size distribution and concentration level of the solids, but both these quantities are apt to vary continuously throughout the sludge circulation system. Moreover, coagulation and agglomeration are also affected by a number of other variables, such as the electrical charges and solubility of the particles, which are even less amenable to any calculations or forecasts. These processes can be intensified and accelerated by various additions, which may be either common substances like milk of lime, or special-purpose agents known as flocculants. The latter are usually specific in their effects, in other words, will produce the desired result only when carefully selected to suit the type of solids and the carrier liquid in question. So far, however, far too little is known about the efficacy of various flocculants as applied to the sludges generated by gas cleaning equipment.

The solubility or partial solubility of some solid particles, like those commonly arising in foundries, can create problems of its own. The dissolution of these solids raises the viscosity of the sludge, and thereby hampers the sedimentation of its solids content. This can reach the point where most of the solids pass clean through

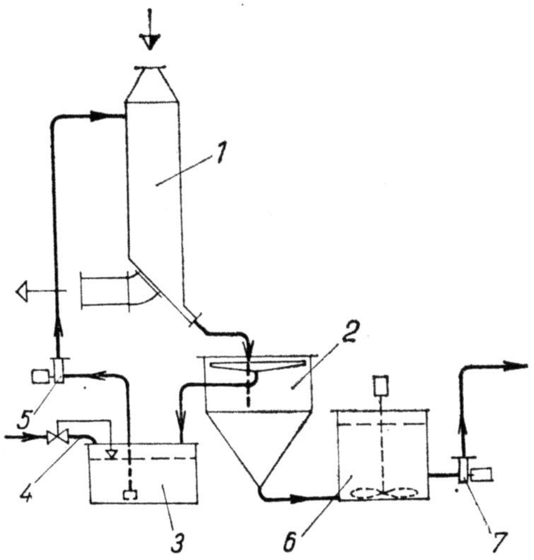

Fig. 159 A typical sludge thickening and water recirculation system incorporating an UNA tank: *1* — type MVA scrubber; *2* — UNA settling tank; *3* — recirculated water reservoir; *4* — float valve admitting make-up water; *5* — circulation pump; *6* — stirred sludge holding vessel; *7* — sludge pump.

the settling tank, and leave it in what was meant to be the clean water. Again, it is impossible to predict the viscosity variations and their effects on the sedimentation process, because too much depends on what is happening upstream of the gas cleaning plant. In foundry dust, for example, each of the various additions to the moulding sand, and especially any change in the bentonite ratio, can alter the whole picture. In the worst case, mechanically scavenged bath-type washers and/or the settling tanks that serve them will gradually fill up with a suspension of a gel-like consistency, which will ultimately put the units out of action. Depending on the dust concentration in the scrubbed air and on the bentonite content of the dust, it may take anything from a few dozen hours to several weeks for the equipment to clog up to the breakdown point, but even the longest of these time spans is still generally unacceptable. In such cases, the only preventive measure is to change the water in the scrubbing equipment at short intervals, and the only radical remedy is to replace that equipment with some other type of separators, for instance cloth filters.

The most common ways of incorporating UNA and UNB tanks in sludge handling systems are schematically represented in Figs. 159 and 160 respectively. In the system shown in Fig. 159, the sludge drained from an MVA scrubber flows by gravity into an UNA tank. The thickened slurry is discharged from this tank through a gate and a flow control valve, into a stirred holding vessel. Once this vessel has filled up to a preset level, the stirring mechanism is automatically engaged, and a sludge pump delivers the slurry to the disposal or further processing point.

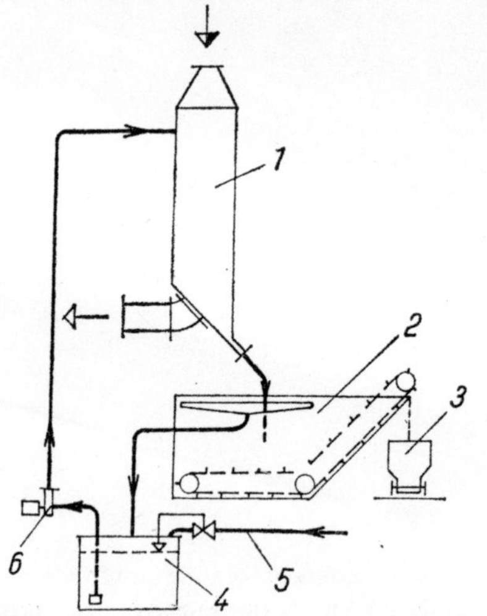

Fig. 160 A typical sludge thickening and water recirculation system fitted with an UNB settling tank:
1 — type MVA scrubber; *2* — UNB settling tank with mechanical raker; *3* — sludge disposal truck; *4* — recirculated water reservoir; *5* — make-up water piping with float valve; *6* — recirculation pump.

The cleaned water is led off to a reservoir, where it is topped up with fresh water before being recirculated to the scrubber. In the scheme in Fig. 160, the sludge again enters the UNB tank by gravity, but the caked slime is raked out of this tank by a mechanical scraper, and dumped into cars or trucks. The clean water is recirculated via a make-up reservoir as in the UNA installation.

The first of these schemes, employing the UNA tank, is usually preferable in cases when the thickened slurry is destined for further treatment at the plant which generated it, as is quite common e.g. in the ceramics industry. The alternative scheme, utilizing the UNB tank, is mostly used for the disposal of worthless wastes which are finally dumped. However, it is also applicable in cases when it is desirable for some reason to gain the caked slime as dry as possible.

Years of hard-won experience have brought to light some of the major advantages and shortcomings of the UNA and UNB tanks. For a start, neither of these types can reliably handle sludges containing dust which tends to cement or bond. That rules out for instance magnesite dust, or the fines which arise in ore sintering processes. They are also unsuitable for sludges containing large amounts of soluble solids, which on dissolution increase the viscosity of the sludge. If they do have to be used for such duties, it is advisable to drain and replace all the water in the system at intervals which may range from six weeks at the most to as little as a fortnight. And no matter what kind of sludge is handled, the vessels and chambers of the individual cleaning units need flushing out with a stream of water roughly once in every eight-hour shift of operation. Pressure distributor pipings for this rinsing are built into the tanks as a standard fixture.

On the other hand, both the UNA and the UNB tanks have proved capable of effectively dealing with sludges containing solids much finer than those for which the tanks are rated. The throughflow rating of these tanks and their vessels is based on the assumption that the solids will have free falling velocities of some 0.15 to 0.2 mm s^{-1}, but agglomeration and coagulation processes can be relied on to bring a substantial proportion of even much finer particles together into clusters which come within this bracket.

4.2.2 Sludge handling systems for sub-micron particles

At present there is only a single type of scrubber on the Czechoslovak market which is specifically designed for trapping particles in the sub-micron size range, and that is the MSA type. This unit can run on water containing up to 100 g of solids per litre, and consequently imposes no great demands on its sludge handling ancillaries. That leaves a choice of three alternative sludge thickening and water recirculation systems; which of them is adopted will depend on the specific site and operating conditions, on the size of the scrubber installation, on the dust concentration in the gas, and on the sedimentation characteristics of the solids in the resultant sludge. Briefly, the alternatives can be summed up as follows:

a) In plants equipped with central sludge treatment facilities and process water mains, there would usually be no point in providing a separate loop for the sub-micron scrubbers. These can constantly be fed with fresh process water, and can discharge their sludge through a reservoir and a sludge pump to the central sludge handling facility, as indicated in Fig. 161. The reservoir is set directly beneath the scrubber, and provides an air lock for its sludge discharge port. The bottom of this reservoir, a conical funnel with an apex angle of about 90 degrees, terminates in a flow control gate. The sludge level is kept constant by an overflow pipe set into the top of the reservoir, which delivers the overflow into the sludge pump intake piping downstream of the gate. In practice, normally about half the sludge

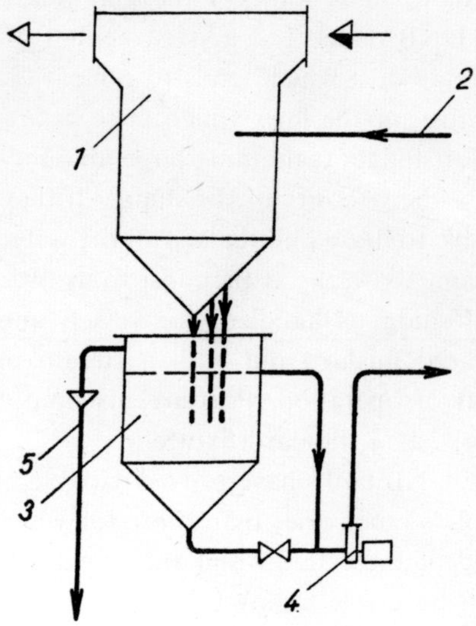

Fig. 161 Sludge handling system for predominantly sub-micron solids, in plants with central sludge treatment facilities:
1 — type MSA washer; *2* — feed water line; *3* — sludge reservoir acting as sludge discharge air lock; *4* — twin sludge pumps; *5* — emergency overflow pipe.

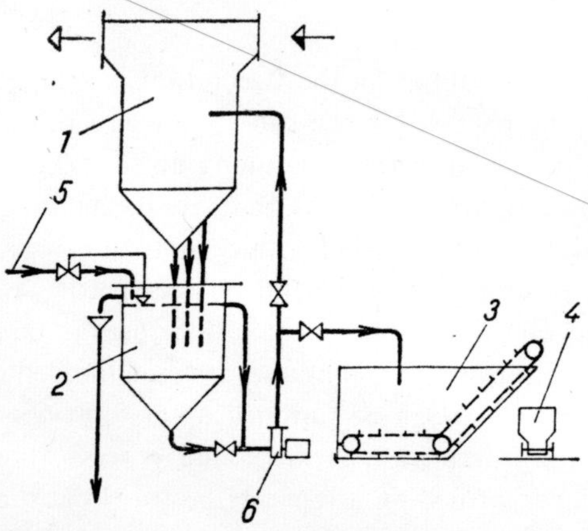

Fig. 162 Sludge handling system for mostly sub-micron solids, for intermittent thickening of the sludge (i.e. for low gas flow rates and/or dust concentrations):
1 — type MSA washer; *2* — sludge reservoir serving as air lock for washer discharge gate; *3* — mechanically raked settling tank; *4* — sludge disposal truck; *5* — make-up water supply controlled by float valve; *6* — twin sludge pumps.

is drained through the gate at the bottom of the reservoir and the other half through the overflow pipe. It is also customary for the discharge pipe from the scrubber to end at least 300 mm below the sludge level in the reservoir. The sludge pumps are generally duplicated, i.e. a stand-by unit is kept in reserve for use while the main pump is undergoing maintenance or repairs. Even so, it would be unwise to risk the consequences of a pump failure, or of a power cut stopping the pumps. A second or emergency overflow pipe is therefore set above the normal sludge level in the reservoir, to jettison the excess sludge when any such situation develops.

b) A different layout, sketched in Fig. 162, is commonly adopted when the dust concentration in the scrubbed gas and/or the total flow rate of this gas are small enough to permit intermittent rather than continuous thickening of the discharged sludge. In this case, the water is recirculated between the scrubber and the air lock reservoir beneath it until the solids content rises to the permissible limit of 100 g per litre. The reservoir must be able to accommodate enough water for this concentration level to be reached only after a considerable time span, because once the solids content of the water attains this level, all the water in the system must be drained to a further tank for sedimentation. Otherwise, the reservoir is identical in design and function with that described in alternative a). The fluid leaving it is induced by a sludge pump, which delivers it through either of two stop valves – either back to the scrubber, or into the settling tank. The latter is generally provided with mechanical scraping equipment for raking out the thickened sludge. The cleaned water emerging from this tank can be returned to the recirculation loop; it is made up, to cover evaporation in the scrubber and other losses, through a fresh water pipe with a float valve set into the reservoir beneath the scrubber. This reservoir must again be fitted with a dumping pipe for jettisoning the overflow in the event of a pump failure.

The economics of this system, and often its technical acceptability too, will obviously depend on the time for which the system can be run before the solids content in the recirculated water builds up to the 100 g per litre limit. Table 16 lists the number of running hours that will elapse before this limit is reached, for the seven standard sizes of the MSA scrubbers in conjunction with four sizes of the settling tanks, at six different dust concentrations in the scrubbed gas, on the assumption that the scrubbers retain a full 100 per cent of the solid particles. The seven scrubbers listed there afford unit throughflow rates ranging from 3750 to 90,000 m^3 an hour. The settling tanks to which the sludge is drained when the specified running time has expired are modified UNB tanks, stripped of their internal fixtures (the vessels, plates and overflow pipes), but of the standard dimensions and with the standard sludge evacuating scrapers. The net capacities of the four tanks listed are as follows: UNB 1 – 3 m^3; UNB 2 – 6 m^3; UNB 4 – 8 m^3; and UNB 6 – 11 m^3.

The times set out in Table 16 demonstrate that this system configuration is applicable only when the dust concentration in the gas and/or the total throughflow of this gas are low. The combinations of scrubber and settling tank which are of practical interest are naturally those which can be run continuously for at least eight hours before the water need be drained for sedimentation; in the Table, these combinations are marked by vertical lines.

c) Large scrubbers, and/or substantial dust concentrations in the incoming gas, call for a different approach. In such cases, only part of the sludge or solids-laden water can be recirculated. The rest must be continuously removed from the system – to settling tanks, sedimentation ponds, or central sludge treatment facilities. Obviously, the sludge thus withdrawn from circulation per unit of time must contain as much dust as the scrubbers trap within the same time interval.

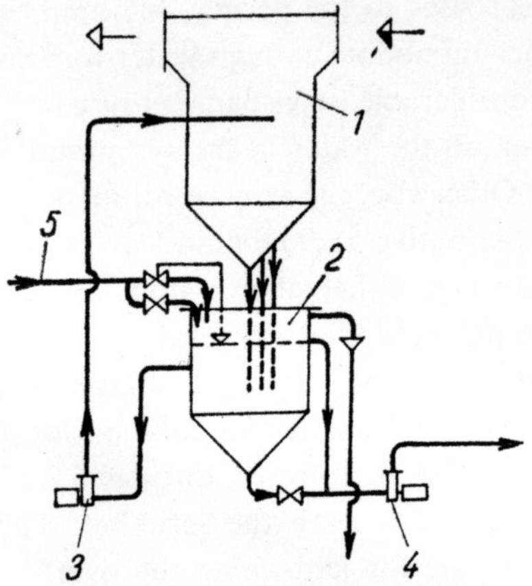

Fig. 163 Sludge handling system for prevalently sub-micron solids, with partial recirculation and partial dumping of the solids-laden water:
1 — type MSA washer; *2* — sludge reservoir providing air lock for sludge outlet of washer; *3* — twin sludge pumps (recirculation branch); *4* — twin sludge pumps (sludge withdrawal branch); *5* — make-up water line with float valve.

That will still leave a certain solids content in the recirculating water, and the proper target figure for this constant pollution level of the water is not at all easy to determine. It will largely depend on the type of dust handled, on its sedimentation characteristics, and on the capacity of the tanks, ponds or other facilities to which the surplus sludge is discharged. Similarly, the control of the solids content in the water, by regulation of the sludge withdrawal rate, is not nearly as straightforward a matter as it might seem. For instance, none of the existing sludge pumps can deliver at the extremely small flow rates which might sometimes be needed to counter slow, long-term increases in the solids content of the recirculated water.

TABLE 16

Maximum water recirculation times (in hours) in MSA Venturi washers fitted with various types of settling tanks*

Washer type and rating (m^3 $hour^{-1}$)	Tank type	Recirculation times (hours)** at inlet dust concentrations ($g\,m^{-3}$) of					
		1 gm^{-3}	3 gm^{-3}	5 gm^{-3}	10 gm^{-3}	15 gm^{-3}	20 gm^{-3}
MSA 1/2 (3750)	UNB 1	80	27	16	8	5.3	4
	UNB 2	160	53	32	16	10.5	8
	UNB 4	210	70	43	21	14	10.7
	UNB 6	290	98	59	29	19.5	14.5
MSA 1 (7500)	UNB 1	40	13	8	4	2.7	2
	UNB 2	80	26	16	8	5.3	4
	UNB 4	105	35	21	10.5	7.1	5.3
	UNB 6	145	49	29	14.5	9.8	7.3
MSA 2 (15,000)	UNB 1	20	6.7	4	2	1.3	1
	UNB 2	40	13	8	4	2.7	2
	UNB 4	53	17.8	10.7	5.3	3.5	2.7
	UNB 6	73	24.5	14.7	7.3	4.9	3.7
MSA 4 (30,000)	UNB 1	10	3.3	2	1	0.7	0.5
	UNB 2	20	6.6	4	2	1.4	1
	UNB 4	26	8.9	5.3	2.6	1.8	1.3
	UNB 6	36.5	12.2	7.3	3.7	2.5	1.8
MSA 6 (45,000)	UNB 1	6.7	2.2	1.3	0.7	0.4	0.3
	UNB 2	13	4.4	2.6	1.4	0.8	0.6
	UNB 4	18	5.9	3.6	1.8	1.2	0.9
	UNB 6	24.4	8.1	4.9	2.4	1.6	1.2
MSA 8 (60,000)	UNB 1	5	1.6	1	0.5	0.3	0.2
	UNB 2	10	3.2	2	1	0.6	0.4
	UNB 4	13	4.4	2.6	1.3	0.9	0.6
	UNB 6	17.5	6.1	3.7	1.7	1.2	0.9
MSA 12 (90,000)	UNB 1	3.3	1.1	0.7	0.3	0.2	0.16
	UNB 2	6.6	2.2	1.4	0.6	0.4	0.3
	UNB 4	8.8	3	1.8	0.9	0.6	0.4
	UNB 6	12.2	4.1	2.4	1.2	0.8	0.6

*Vertical lines mark the washer and tank combinations capable of running at least 8 hours without a water change.

**Running hours before the solids concentration in the recirculated water builds up to the maximum acceptable level of 100 g per litre, whereupon the water has to be replaced.

The scheme of such a system in Fig. 163 shows that the scrubber again surmounts a reservoir, of the same design and with the same functions as in alternatives a) and b). The discharge piping from this reservoir, to the settling tanks or ponds or central sludge treatment plant, must again be backed by an emergency overflow pipe to dump the surplus sludge when the normal discharge line is blocked. Pump priming water is supplied to the reservoir by a piping which terminates some 150 to 200 mm above the ends of the sludge inlet pipes that lead down from the scrubber. Make-up water, to cover the water losses in the sludge withdrawn from circulation as well as evaporation losses in the scrubber, can be fed in at a constant rate, which is adjusted by means of a flow control valve. True, it is always safer to provide a second feed-water inlet, controlled by a float valve, to cater for fluctuations in both the loss and the feed rates. But in a well-balanced system, where about half the sludge withdrawn from circulation reaches the sludge pump via the overflow pipe, and where this overflow is adequately matched to the inflow of sludge from the scrubber, the bath level will remain virtually constant in spite of all these fluctuations, and the float valve will seldom be required to open. As always, there is much to be said in favour of such a self-regulating system, where random fluctuations tend to cancel each other out and maintain an average state.

5. CRITERIA FOR THE SELECTION OF SEPARATORS

Ing. J. Kurfürst, CSc., Ing. O. Štorch, CSc.

All the technicalities of separators and their ancillaries should not blind us to the fact that the prime object of all these devices is to limit the emission of solid and liquid pollutants into the atmosphere around us. Wrongly selected separating equipment, or the malfunctioning of even well-selected equipment, can easily increase the pollutant emission by as much as two orders of magnitude, as against the minimum attainable by the best of contemporary equipment. And since this minimum is rapidly becoming the norm expected of industry by the legislators and the public at large, it will be as well to examine the criteria that govern the choice of separating equipment, and the factors which affect its reliability in routine service.

Apart from economic considerations, which will be dealt with later on in this Chapter, there are two main sets of requirements to take into account in the selection of equipment. One of them is compliance with the statutory regulations and other administrative measures adopted to combat air pollution. The other is conformance with those technical practices which are known to facilitate this task.

5.1 Legal considerations

Most industrial countries nowadays enforce legislation intended to keep air pollution levels within acceptable limits. The measures taken in this respect, which are tending to become increasingly severe, can be roughly broken down into:

a) *Legislative restrictions*, in the forms of laws, local government orders, court injunctions, or technical standards with mandatory force. As a rule, they either specify a maximum permissible overall level of atmospheric pollution, or restrict the concentration of airborne effluents emitted from any one source, or else place various constraints on the creation of new pollution sources.

b) *Enforcement measures*, such as spot checks or regular inspections of known pollution sources; investigations of atmospheric contamination levels by the pollutants, quantities and sources involved; systems of sanctions and incentives, ranging from fines, and subsidies for mandatory pollution control at the source, down to compulsory compensation payments for the inhabitants of the polluted areas; control of the fuel supplies in smoke-free zones; etc.

c) *Long-term programmes* for the conservation and improvement of the environment. These are increasingly becoming ambitious schemes with an integrated approach to all the factors and aspects involved, from research planning to publicity campaigns, from consistent monitoring of environmental changes, and the evaluation of the results achieved by various countermeasurs, down to the organization of large-scale field experiments. Though their direct impact may as yet be slight, they should not be underrated, because the sheer weight of public opinion behind these schemes will ultimately be hard to resist.

In most countries, the principles underlying all these repressive and preventive measures are written into various "Clean Air", "Public Health" or "Environmental Protection" laws. In practice, these lofty declarations of intent are less important than the mass of minor statutory and other regulations which specify the actual constraints on the release of airborne effluents, and especially the actual amounts of various pollutants that may be legitimately released.

There are several different ways of defining the limits on the pollutant emission from any given source. The one best fitted to the intentions of the conservationists would seem to be to restrict the emission rate to an amount (dependent on the height of the chimney stack) which will keep the pollutant concentration in the atmosphere, at ground level in the vicinity of that stack, within a specified tolerable limit. This is the procedure adopted in Czechoslovak Law No. 35/1967 on measures against air pollution. Under this law the emission limits are calculated, for every particular chimney height, from the maximum tolerable contamination level in the lowermost layers of the atmosphere. The actual maximum concentration level is taken to be the average of the values recorded at the point of measurement within a thirty-minute interval. For this calculation, the process by which the pollutants spread through the atmosphere is represented by a mathematical model, based essentially on the Sutton formula, which naturally involves a good many simplifying assumptions about the physical conditions. In particular, it applies only where the surrounding ground is flat, it averages out the wind speed, and it neglects all other pollution sources in the vicinity. Still, the procedure does yield a figure, in kg per hour or tons per year, which is directly applicable and can readily be checked in actual plant practice.

The shortcomings of this approach are obvious. New sources of pollution are only rarely built in areas free of all previous atmospheric contamination, and even more seldom in regions flat enough to fit the assumptions behind this calculation. The stipulations of this law result in a proliferation of very tall chimneys, and are unrealistic in areas where the height of the chimneys has to be restricted, e.g. because of the proximity of an airport. Finally, weather conditions often differ substantially from the average state on which the calculation is based. And unfavourable weather can cause the permissible local concentration levels to be exceeded even when the pollutant emission rate, as measured at the stack top, is well within the legal limit. That is perhaps the main weakness of this approach: no matter how carefully you control the emission rate, there will always be isolated localities or occasions where or when the atmospheric contamination at ground level will overstep the statutory limit.

It would of course be easy enough to formulate regulations which would take some of the other variables into account too. The rules could make due allowance for the presence of other pollution sources in the vicinity; for the residential or other character of the areas affected; for the configuration of the surrounding ground; for the type of contaminant in question, and the amount of harm it can do; etc. This more comprehensive treatment of the problem has been tried out in the German Democratic Republic. Unfortunately, the resultant regulations then tend to be too complicated to be easily interpreted, let alone easily applied and enforced.

Many industrial countries have tried to get around these problems by basing their emission limits not upon some universally valid and arbitrarily predetermined target figure, but rather on the minimum pollution level attainable by the best practicable means which are at present at our disposal. This approach fully recognizes that the outcome is not dependent on the separating equipment alone, but is decided mainly by the technology upstream of that equipment. The "best practicable means" cover not only the pollutant trapping facilities, but equally the selection of the process equipment, process materials, any special pre-treatment of those materials, the process control techniques, and other such factors which between them govern the type and amount of pollutants created. Emission limits determined in this manner are usually defined in either of two ways. They can be specified as the pollutant concentration in the gases released into the atmosphere, i.e. expressed say in mg m^{-3}; alternatively, they can be related to the quantities of raw materials entering the process, or the amounts of products or semi-products leaving it, or to the energy input or output, in which case they will be stated for instance in kg per ton, or kg GJ^{-1}.

A salient feature of emission limits established in this way is that they are specific to every particular process, and to every particular set of process equipment and conditions. That obviously complicates the legislative process, but makes its outcome far more effective. The aim is always to reduce pollution levels by as much as is currently feasible, rather than to impose some overall and not always realistic limit on them. The stack height is not treated as one of the initial, ruling factors, but is calculated for every specific case with due regard to the actual configuration of the ground, the vicinity of built-up areas, the presence of background pollution from other sources, etc. If this calculation then yields an unrealistic chimney height, or fails to arrive at an acceptable pollution level even with the tallest still practically feasible chimney, this indicates that the nature or size or location of the pollution source will have to be altered. That leaves the

[1]Details in Federal Register Vol. 26, No. 247 of December 23, 1971.
[2]Maximum values of two-hour averages.
[3]Smoke opacities of 40 % and 20 % correspond to Ringelmann scale fields 2 and 1 respectively; a 10 % opacity would correspond to 0.5, if the Ringelmann scale contained such a field.
[4]The SO_2 emission limits for Class I (steam generators) using mixtures of solid and liquid fuels are calculated as

$$\frac{0.8y + 1.2z}{x + y + z}$$

where x, y and z are the percentages of the total heat input provided by gaseous, liquid and solid fuels respectively.
[4]The NO_x emission limits for Class I (steam generators) using fuel mixtures are calculated by the formula

$$\frac{0.2x + 0.3y + 0.7z}{x + y + z}$$

where x, y and z are as defined in footnote 4.

TABLE 18

Simplified selection of emission limits valid in Sweden since 1971

Process, equipment, raw materials	Emission limits for	
	Newly built sources	Existing sources
A. *Combustion of liquid fuels*:	(per ton of fuel oil)	
1. SO_2 emission from power stations of 300 MW or more	20 kg	20 kg
2. Particle emissions from:		
— Power stations of 300 MW or more	1 kg	1 kg
— Gas turbines running 500 or more hours annually	Bacharach 3	
— Gas turbines running up to 500 hours annually	Bacharach 5	
— Boilers from 50 MW upwards	1.5 kg	1.5 kg
Max. combustibles content	1.0 kg	1.0 kg
	Bacharach 3	Bacharach 3
— Boilers of up to 50 MW	1.5 kg	2.0 kg
Max. combustibles content	1.0 kg	1.5 kg
	Bacharach 3	Bacharach 4
B. *Incinerating plants*:		
Particle emissions from plants with capacities of		
15 tons of refuse per hour or more	180 mg/m_n^3	
3 to under 15 tons of refuse per hour	250 mg/m_n^3	
under 3 tons of refuse per hour	500 mg/m_n^3	
C. *Cement and lime production*:		
Particle emissions from:		
Furnaces	250 mg/m_n^3	500 mg/m_n^3
Other dust sources (crushing, grinding, mixing, handling, filling, etc.)	250 mg/m_n^3	
D. *Bituminous gravel coating plants*:		
Particle emissions from:		
Plants further than 500 m from residential areas	250 mg/m_n^3	500 mg/m_n^3
Temporary plants more than 1 km from nearest houses	5000 mg/m_n^3	5000 mg/m_n^3
E. *Inorganic chemical processes*:	(per ton of product)	
Sulphuric acid production (from sulphur or pyrites):		
1. SO_2 emission	5 kg	20 kg
2. SO_3 emission	0.5 kg	8 kg
Chlorine production:		
Mercury emission	0.001 kg in extracted air 0.0005 kg in hydrogen	

(Table 18 continued)

Process, equipment, raw materials	Emission limits for	
	Newly built sources	Existing sources
F. *Iron and steel production*:	(per ton of product)	
Particle emissions from:		
Sintering plants	0.5 kg	1.0 kg
Blast furnaces	0.3 kg	0.3 kg
Open-hearth furnaces	0.5 kg	1.0 kg
Oxygen steelmaking convertors	0.3 kg	0.3 kg
Kaldo process convertors	0.15 kg	0.15 kg
Electric arc furnaces	0.3 kg	0.6 kg
G. *Grey iron foundries*:		
Particle emissions from:		
Cupolas with annual outputs (tons)	up to 2000	up to 3250
emission per ton	7 kg	8 kg
Cupolas with annual outputs T (tons)	up to 47,000	up to 47,000
emission per ton	$\frac{35}{T + 3}$ kg	$\frac{50}{T + 3}$ kg
Cupolas with annual outputs (tons)	over 47,000	over 47,000
emission per ton	0.7 kg	1 kg
Iron pouring, shaking-out and sandblasting operations		
at outputs over 2500 tons a year	150 mg/m_n^3	150 mg/m_n^3
at outputs under 2500 tons a year	300 mg/m_n^3	300 mg/m_n^3
H. *Ferroalloy production*:	(per ton of product)	
Particle emissions in:		
Fe—Si production	15 kg	30 kg
Fe—Cr—Si production	15 kg	30 kg
Fe—Cr production	5 kg	10 kg
Fe—Si—Mn production	0.3 kg	
Fe—Mo production	3 kg	3 kg
I. *Paper mills*:		
1. Particle emissions from the		
— Sulphate process:		
Recuperative boilers	250 mg/m_n^3	500 mg/m_n^3
Lime kilns and digesters	250 mg/m_n^3	250 mg/m_n^3
— Sulphite process:		
Recuperative boilers	250 mg/m_n^3	500 mg/m_n^3
2. SO_2 emissions from sulphite process (except for the combustion equipment)	20 kg per ton of pulp	30 kg per ton of pulp
3. Aromatics emissions are covered by detailed regulations		

options of relocating the plant, scaling it down, or reconsidering the choice of process equipment or materials or controls, or else improving the separating facilities.

This approach, where the yardstick are the results attainable by the best presently practicable means, has been adopted in Great Britain, Sweden, Switzerland, the U.S.A., West Germany, and some other countries. By way of an example, Tables 17, 18 and 19 present some of the emission limits currently valid in the U.S.A. and in Sweden. A close study of these Tables will both bring out the complications inherent in this treatment of the problem, and confirm that the results are well worth while the extra trouble.

The alternative technique employed e.g. in Czechoslovakia, of establishing the emission limits from a predetermined maximum atmospheric contamination level, rests upon the sort of figures exemplified in Table 20. These are the highest permissible thirty-minute and 24-hour averages of the pollutant concentration in the atmosphere, as laid down in Circular No. 34 of the Chief Public Health Officer for Czechoslovakia. The maximum tolerable dust concentration (of dust containing no more than 20 per cent of free SiO_2) is arbitrarily defined by a coverage of 12.5 g m^{-2} monthly, i.e. 150 tons per km^2 annually. No attempt is made to justify the choice of this particular limit. Moreover, the same document leaves regional Public Health Officers free both to impose stricter limitations, where these are justified by specific local environmental considerations, and to grant temporary local exemptions from the generally valid limits. Similarly, it is up to the regional public health authorities to specify the maximum acceptable concentrations of

TABLE 19

Particle emissions from Swedish metallurgical plants in 1967, and limits valid for the same equipment since 1975

Pollution source	Output (1000 tons)	1967 emissions (kg per ton of output)	1975 limits for existing sources (kg/ton)	Limits for new sources (kg/ton)	Assumed trapping efficiency (%)	
					1967	1975
Sintering pans	2436	3.0	1.0	0.5	80	93
Sintering strands	874	4.0	1.0	0.5	73	93
Pelletizing plants	4590	1.5	1.0		90	93
Blast furnaces	2250	0.3	0.3	0.3	99	99
Thomas convertors	96	25.0	—		0	—
Open-hearth furnaces	1336	5.2	1.0	0.5	0	81
Oxygen convertors	106	0.3	0.3	0.3	99	99
Kaldo convertors	1441	0.15	0.15	0.15	99.5	99.5
Electric arc furnaces	1820	3.9	0.6	0.3	13	87

TABLE 20

Maximum pollutant concentrations in the atmosphere permissible in Czechoslovakia

Pollutant	Max. short-term peak k_{max}* (mg m^{-3})	Max. daily average k_d* (mg m^{-3})
Dust**	0.5	0.15
Sulphur dioxide	0.5	0.15
Carbon monoxide	6.0	1.0
Nitrogen oxides (NO_2 equivalent)	0.3	0.1
Chlorine	0.1	0.03
Hydrogen sulphide	0.008	0.008
Lead (except tetraalkyl compounds)	—	0.0007
Carbon disulphide	0.03	0.01
Arsenic (inorganic compounds except for AsH_3)	—	0.003
Fluorine (gaseous inorganic compounds)	0.03	0.01
Carbon black and soot (as amorphous C)	0.15	0.05
Formaldehyde	0.05	0.015
Phenol	0.3	0.1
Manganese (MnO_2 equivalent)	—	0.01
Sulphuric acid (in H ions)	0.01	—
Nitric acid (in H ions)	0.01	—
Hydrochloric acid (in H ions)	0.01	—
Ammonia	0.3	0.1
Benzene	2.4	0.8

*At 0 °C and 101,325 N m^{-2}.
**Not soluble in body liquids; concentration limits for dust containing more than 20 % of free SiO_2 are fixed by regional public health authorities.

individual harmful pollutants in the atmosphere. In all these decisions, the regional Public Health Officers are expected to consult the appropriate local government departments and the national Inspectorate for Air Pollution Control.

Law No. 35/1967, the basic piece of clean air legislation in Czechoslovakia, essentially defines air pollution, identifies those responsible for it, and lays down the fines. It has since been supplemented by a directive of the Ministry of Forestry and Water Resources, issued in 1968, and by a further document of the same ministry, which in 1972 clarified the provisions of the basic law, the enforcement procedures, and the way the newly encoded draft regulations were to be tried out in practice. Law No. 35/1967 provides fines for anyone who fails to carry out pollution control measures ordered by the Inspectorate for Air Pollution Control, or by the public health authorities, or fails to do so within the specified deadline, or fails to ensure the satisfactory operation or proper maintenance of pollution control equipment. The basic fine is a flat rate of 100 crowns annually for every 1000 kg of contaminants released in excess of the respective legal limits; but this rate can be increased, up to twofold, in dependence on the nature of the area that is being polluted. Of the money thus collected, 60 per cent is allocated to the local government authorities in the areas polluted by that source, and 40 per cent to a central air pollution control fund administered by the Ministry of Forestry and Water Resources. The law further vested responsibility for all air pollution control programmes in this ministry, and set up a national Inspectorate for Air Pollution Control with sweeping supervisory powers.

The Appendices of this law define the maximum permissible pollutant emission rates; explain the formula for calculating the fines; and set down emission thresholds, with the stipulation that anyone exceeding those threshold levels must report to the relevant local authorities all the data needed for calculating the amount due in fines. For this purpose, two or more chimneys belonging the same owner and standing within the same circle of 1 km diameter are regarded as one chimney. For the cases when these chimneys differ in height, the directive of the Ministry of Forestry and Water Resources specified a procedure for calculating a fictitious equivalent chimney height. Pollutants (including dusts) other than fly ashes are dealt with in Appendix E, which lists the threshold levels for the compulsory reporting of their emission to the local authorities. These are the levels set out in Table 21. The Appendix also lays down the maximum tolerable short-term concentration peaks, as summarized in Table 20.

TABLE 21

Treshold levels for compulsory reporting of pollutant emissions

Pollutant	Emission rate (kg per hour)	Pollutant	Emission rate (kg per hour)
Dust (with up to 20 % SiO_2)	5.0	Carbon black and soot (as amorphous C)	1.5
Carbon monoxide	60.0	Formaldehyde	0.5
Nitrogen oxides (NO_2 equivalent)	3.0	Phenol	3.0
Chlorine	1.0	Metallic mercury	0.003
Hydrogen sulphide	0.08	Manganese (MnO_2 equivalent)	0.1
Lead (except tetraalkyl compounds)	0.007	Sulphuric acid (in H ions)	0.1
Carbon disulphide	0.3	Nitric acid (in H ions)	0.1
Arsenic (inorganic compounds except for AsH_3)	0.03	Hydrochloric acid (in H ions)	0.1
Fluorine (gaseous inorganic compounds)	0.3	Acroleine	3.0
		Ammonia	3.0
		Benzene	24.0

Appendix A restricts the air pollution that may be caused by fuel combustion processes to the values listed in Table 22, and Appendix B states that these same limits also apply to pollution generated by various technological processing operations. Appendix C applies to the production of clinker, cement and lime,

TABLE 22

Permissible emission rates for fuel combustion processes (thresholds for fine payments)

Chimney height (m)	Emission (kg/hour) Fly ashes	SO_2	Other pollutants*	Chimney height (m)	Emission (kg/hour) Fly ashes	SO_2	Other pollutants*
7	2.5	2	$4.0k_{max}$	85	290	257	$514k_{max}$
8	3	2.3	$4.6k_{max}$	90	325	295	$590k_{max}$
10	4	3.2	$6.4k_{max}$	95	360	335	$670k_{max}$
12	5	4.2	$8.4k_{max}$	100	400	375	$750k_{max}$
14	7	5.3	$10.6k_{max}$	110	490	900	$930k_{max}$
16	9	6.8	$13.6k_{max}$	120	580	1425	$1130k_{max}$
18	11.4	8.4	$16.8k_{max}$	130	675	1950	$1340k_{max}$
20	14	10	$20.0k_{max}$	140	785	2475	$1560k_{max}$
25	21	13.5	$27.0k_{max}$	150	900	3000	$1790k_{max}$
30	31	22.5	$45.0k_{max}$	160	1010	3555	$2060k_{max}$
35	42	32.5	$65.0k_{max}$	170	1130	4110	$2320k_{max}$
40	55	46	$92.0k_{max}$	180	1270	4665	$2600k_{max}$
45	70	60	$120.0k_{max}$	190	1400	5220	$2890k_{max}$
50	84	82.5	$165.0k_{max}$	200	1550	5779	$3200k_{max}$
55	110	100	$200.0k_{max}$	220	1820	6355	$3840k_{max}$
60	130	122	$245.0k_{max}$	240	2110	6930	$4500k_{max}$
65	160	145	$290.0k_{max}$	260	2400	7510	$5160k_{max}$
70	192	170	$340.0k_{max}$	280	2700	8085	$5820k_{max}$
75	225	195	$390.0k_{max}$	300	3000	8665	$6500k_{max}$
80	260	227	$455.0k_{max}$				

*For k_{max} values see Table 20.

and for chimney heights of 60 to 80 m ties the permissible solids emission to the total output rate as indicated in Table 23. For other chimney heights, and for output rates exceeding 150 tons per hour, a directive of the Ministry of Forestry and Water Resources imposes the emission limits stipulated in Appendices B and A, i.e. the emission rates set out in Table 22. Appendix D presents the formula for calculating the amounts payable in fines, which can be transcribed as follows:

$$A = (B - C)\,DEF$$

where A is the sum of the fine,

B – the actual emission rate (in kg per hour),

C – the permissible emission rate (in kg per hour),

D – the number of hours, in that year, during which the legal emission limit has been exceeded,

E – the basic annual fine, of 0.10 crowns per kg of excessive pollutant emission, and

F – a coefficient dependent on the land use in the contaminated area, as follows:

2.0 for the central areas of spas,

1.8 for other spa areas, holiday resorts, and specifically protected zones,

1.6 for residential areas,

1.2 for industrial areas, and

1.4 for all other areas.

TABLE 23

Permissible solids emission rates in the production of clinker, cement and lime

Total output rate (tons/hour)	25	50	100	150
Max. emission rate (kg/hour)	120	160	250	270

No fines are levied for emissions lower than the threshold levels listed in Table 21.

Clause 13 of Law No. 35/1967 establishes the national Inspectorate for Air Pollution Control, and defines its duties and powers. Briefly, this institution is to:

1. a) Inspect all devices intended to combat air pollution; ensure that they are installed and commissioned within the specified deadlines; and secure that adequate air pollution control measures are incorporated in the projects of all new plants.

b) Check and record the actual pollutant emission rates.

c) Assist all concerned with the technical problems involved in air pollution control; and, after consultations with the appropriate public health authorities, specify the technical and/or other precautions which individual pollution generators must take to prevent or limit contamination of the atmosphere.

d) Aid local government authorities in establishing the amounts due in fines, especially in checking the data supplied by the pollution generators and in the technicalities of the calculations.

e) Inform local government authorities of all pollutant emissions and sources which ought to be fined.

2. Cooperate with public health authorities in enforcing those provisions of the Public Health Law (No. 20/1966) which relate to air pollution, and particularly

in compiling the official Evaluation Certificates required by Clause 4 of that Law.

3. Have free access to all premises and areas which it needs to inspect in the course of its duties; have the right to conduct observations and measurements there, and to order the presentation of all relevant documents and data.

The main activities of this institution, obviously, are spot checks and detailed investigations of the more important pollution sources. The findings underlie the subsequent orders for mandatory countermeasures which the offender must take, and the deadlines set for him to carry out those orders. Typically, these orders call for the installation of particle or gaseous exhalation trapping equipment; for investigations to ascertain the pollutant types and amounts released; for specific improvements in the operation and/or maintenance of the trapping equipment; for the systematic recording of data on the operation of this equipment and/or on the actual pollutant emissions; for modifications of the process equipment, techniques or materials; or for limitation of the maximum output rates. In 1969 the Inspectorate was split into two separate institutions, one serving the Czech and the other the Slovak part of the country, but that in no way affected its basic functioning and standing.

Law No. 35/1967 leaves all the administrative routine work relating to air pollution fines to the local government authorities. They are also responsible for checking the data which the pollution generators report in their annual statements, though they can always request the assistance of the Inspectorate to cope with the technical aspects involved. In return, they retain 60 per cent of the money these fines yield, as an extra revenue earmarked mainly for environmental improvement schemes.

5.2 Technical considerations

If legislative and administrative measures are intended to make potential air polluters confine the amount of pollution they create to the minimum of what is presently attainable, then technical measures are what actually governs the limits of what is currently feasible in this respect. This is not just a matter of selecting the right equipment: even the best of equipment may malfunction, and in that state release many times the normal amount of pollutants into the atmosphere.

In the narrowest sense of the term, technical measures are understood to cover only the means and techniques used to separate, trap, and liquidate the pollutants. In the broader sense, however, this term also refers to all modifications of the process equipment, techniques and materials by which the pollutants are generated, and to all other changes which can affect the pollutant emission rates. At first, these measures used to be taken by the individual plants or industrial corporations on a largely voluntary basis, and this approach is still being maintained e.g. by the Swiss cement industry. As air pollution is becoming an ever graver, more widespread and more widely recognized problem, however, the amount of control which industry is ready to institute of its own free will is rapidly becoming inadequate. The initiative in the introduction of further controls is gradually passing into the hands of the legislators.

Technical measures to combat air pollution, in the broadest sense of the term, can be roughly broken down into the following categories:

a) Measures designed to reduce the absolute emission rates:

– The use of cleaner power sources, from nuclear power to gasified or variously pre-processed fuels;
– The replacement of the conventional raw or other process materials by substances with a lower content of pollutant-forming constituents; in the fuels field, this means mainly the substitution of liquid or gaseous for solid fuels;
– Treatment of the raw or other process materials to cut down the contamination they cause, e.g. reducing the sulphur content of fuels, the ash content of coals, coking, etc.;
– The trapping of solid, liquid or gaseous pollutants, including e.g. the sulphur content of flue gases.

b) Measures intended to spread the contamination more evenly in time, i.e. staggering of the combustion or other processes so as to shift part of the emission peaks into the trough periods; this is often vital when unfavourable weather conditions are apt to aggravate the effects of the peaks:

– Limitation of the peak output periods that entail excessive air pollution, whenever the output capacity thus lost can be made up by improved utilization of the remaining running time or from other sources;
– Continuous supervision of the combustion or other pollutant-forming process, and its regulation so as to avoid especially the transient or short-term emission peaks which in practice cause most of the outcry and sanctions;
– Temporary resorts to cleaner fuel or raw materials, timed to coincide with the peak emission periods, so as to alleviate the effects of those peaks;
– The installation of back-up equipment reserved for use during the maximum hazard periods. A good example are flue gas desulphurization plants where the facilities for processing the trapped material, or for rendering it harmless could not cope with the amounts that would accumulate in continuous operation. In such cases, the scrubbers are normally by-passed, and are brought into action only when weather conditions prohibit the release of the unscrubbed gas into the atmosphere.

c) Measures which spread the contamination over a wider area, so as to prevent the build-up of excessive local pollutant concentrations:

– Control of fuel supplies to ensure that the grades which produce the least

pollution will reach the areas where pollution problems are most acute;
- The relocation of pollution sources from already overcontaminated areas, and controls on the location of new sources, so as to ease the burden on the worst afflicted areas at the expense of hitherto relatively cleaner zones;
- The erection of tall chimneys to spread the emission more evenly over a larger region.

5.3 Reliability

Even the most efficient of separating equipment is little use if we cannot rely on it. Unfortunately, the dependability of such equipment is one of the most difficult factors to assess, and especially to evaluate numerically. Yet in the process of selecting such equipment, we cannot obviate an attempt to compare the degrees of reliability of various rival types, both in the abstract plane and under the envisaged specific service conditions. The former involves such questions as the inherent reliability of the moving parts, while the latter hinges on considerations like the physical and chemical properties of both the pollutant particles and the carrier gas, the resistance of the filter material to the intended service conditions, the likely maintenance requirements, etc.

A rational approach to reliability assessments calls first of all for a clear list of the principal criteria. For dry mechanical separating equipment, these can be summed up as follows:

a) The extent to which dust is likely to adhere within the equipment, either because it is adhesive or hygroscopic, or because it is extremely fine. Fines are a common pitfall especially in two-stage separating installations. Where heavy concentrations of abrasive dust have to be trapped with a high degree of efficiency, it is common practice to fit the second stage with small-diameter, high-efficiency cyclones, which necessarily work at high inlet velocities. To save abrasive wear and reduce the dust load on these units, the first stage is normally made up of larger-diameter cyclones with lower inlet velocities. This first stage then retains mainly the coarser fractions, and, to keep the second-stage inlet concentration within reasonable limits, has to do so fairly completely. That means the second stage is then fed almost exclusively with fines, which always tend to cling. All too often, deposits in the conical parts of the second-stage cyclones gradually put the units out of action, and result in the release of the fines into the atmosphere. Similar situations arise when high-efficiency cyclones are installed upstream of electrostatic precipitators. The latter then have to handle a proportion of fines, which can quickly coat their collecting and high-tension electrodes with dust layers too thick for the equipment to function properly.

b) The amount of abrasive wear to be expected. The wear rate affects not only the efficiency, but also the service life of the equipment, and is therefore crucial to the economics of the whole installation. Just how seriously efficiency will be impaired by a wear-induced puncture, in the cylindrical or in the conical part of a cyclone, depends mainly on the vacuum within the unit and the grain size distribution of the dust. Still, even if the effects are slight, the cyclone will soon have to be replaced, which is always a costly and often a difficult procedure. Therefore, the abrasive properties of the dust and the wear resistance of the equipment are primary factors to consider when choosing an installation.

c) The explosion and fire risks, if any. Dust presenting such hazards must never be allowed to accumulate in substantial quantities, and that naturally affects the choice of equipment and operating procedures. Moreover, the equipment must in this case be fitted with frangible diaphragms, and possibly other provisions to reduce the explosion risks, as outlined in Section 1.15.

d) Requirements on the dust discharge gate or closure. Each of these devices has its own inherent limitations. For example, rotary air locks should never be exposed to highly abrasive dusts, as the rapid wear of their sealing surfaces will soon cause air to leak past them from the hopper. Similarly, they are not suitable for dusts which are liable to form arches, as has been pointed out in Section 4.1.

e) The likelihood of water vapour condensing within the equipment. Condensation is liable to cause both clogging and corrosion, but can usually be forestalled by effective thermal insulation.

In wet scrubbing equipment, there are two further sets of criteria to take into account. One of them relates to the mechanical design aspects, e.g. of the sludge raking and conveying ancillaries or the flow control devices. The other can be lumped into three main points:

a) The bonding properties of the dust. This risk of bonding or cementation precludes the scrubbing of gas containing for instance sinter or magnesite dusts. True, in some types of scrubbers this bonding tendency can to some extent be countered by specially designed gas inlets with continuously rinsed walls, and by the use of fresh water only, but as a rule it is far safer to trap these dusts in equipment that is run dry.

b) The corrosive effects of the pollutants and of the carrier gas as combined with water. For example, SO_2 in the gas will form sulphuric acid vapours, which will subsequently condense as variously diluted sulphuric acid. Dry-running equipment will corrode rapidly only during the initial period of its operation, as later on a continuous layer of corrosion products will build up on its metal surfaces and protect them from further direct contact with the aggressive constituents. From then on, corrosion proceeds by diffusion only, at an incomparably slower rate. In wet scrubbers, however, the high-speed flow of particle-laden liquid is apt to flush away the corrosion products as they arise, and thereby prevent the formation of a protective deposit or incrustation. (On the other hand, when only non-corrosive materials are being handled, this same liquid cushion greatly reduces the rate of abrasive wear in wet as compared to dry separators). Anti-corrosive coatings and enamels are usually not much use: the surface finish of the base material is only rarely good enough for these layers to stay in place for any length of time, in the face of the erosion and abrasion to which they are exposed, so the coatings generally start peeling off at the sharp edges and corners. Corrosion-resistant steels do offer a solution, but also present an often insurmountable cost barrier.

c) The water supply problem, where the technical problems must usually be examined with their economic implications in mind all the time. In this age of ever graver water shortages, the process water mostly has to be recirculated, i.e. cleaned down to the residual solids content which the scrubbers can accept in their feed water, and fed back to them at the requisite pressure. Both the cleaning and the pumping can present technical problems which often translate into unacceptable costs. For example, scrubbers that trap ferrosilicon particles produce a slurry containing solids so fine that they take a prohibitive time to settle out. Even worse problems are encountered with soluble particles which tend to form gel-like suspensions, such as foundry bentonites. In these cases, there is often nothing for it but to dispose of the spent water, and run the scrubbers on fresh water alone.

In cloth filters, dependability hinges largely on the following factors:

a) The mechanical design of the unit, especially as regards the reliability of its rapping mechanism and of any control system used to govern the rapping.

b) The way the properties of the filter cloth are matched to those of the dust it will have to retain; the main questions are how quickly the cloth will clog up, how easy it will be to clean by rapping or blowing, and how well it will resist puncturing by any coarse or hot particles that may reach it.

c) The probability of water vapour condensing in the unit, and the risks this represents of either clogging the cloth with slime, or damaging it if the gas contains sulphur or other potentially aggressive substances. Sometimes, condensation can be avoided simply by adequate thermal insulation of the filter.

d) The explosion and fire hazards, which in cloth filters are always more serious than in cyclones. For one thing, it is practically impossible to prevent at least some build-up of dust deposits in a cloth filter; for another, the filter bags are far more susceptible to fire damage than the usually all-metal cyclones.

In electrostatic precipitators, the main considerations to take into account in reliability assessments are:

a) The robustness of the overall design, with special reference to the forces imposed by rapping, to the rigidity of the high-tension electrodes and their suspension systems, etc.

b) The risks of a gas explosion, especially when the gas is liable to contain a relatively high proportion of carbon monoxide.

c) The hazards represented by explosive or inflammable dusts, in view of the arc-overs that may occur in the working space.

d) The degree of efficiency to be expected from the rapping mechanisms serving the high-tension and collecting electrodes, especially when the dust is adhesive.

e) The likelihood of water vapours condensing in the unit, and of the condensate either interfering with the rapping action or corroding the equipment, particularly when the gas or the dust contain potentially aggressive constituents.

5.4 Economic considerations

The true capital outlay for a gas cleaning plant is always far more than the price tag on the items of which the installation is made up. Which means that low-cost equipment need not necessarily represent any true economy. To arrive at a realistic figure, we must take into account all of the following costs:

a) The project and design work on the installation.

b) The purchase of the equipment, meaning not only the separators, but also the ducting or piping runs, fans, supporting structures, the dust or sludge handling facilities, and any other ancillaries, such as feed stabilizers or rectifiers, controls, compressors, etc.

c) The installation of all this equipment on site.

d) The protection of all this equipment against the weather, which usually means roofing over or enclosing a good deal of space, and the attendant costs of the space thus enclosed.

e) All building and construction work entailed by the scheme.

These first costs must be written off over the envisaged service life of the installation, which is usually taken to be ten years. When the equipment has to handle abrasive or corrosive materials, however, its life expectancy is commonly much lower. Let us now examine these individual cost components one by one:

a) The project costs are, in Czechoslovakia at least, easy to work out in advance. Since the price of the project design is a fixed percentage of the value of the installation it covers, the procedure is simply to add up the cost prices of all the separating equipment, and multiply the sum by the relevant coefficient as listed in the project pricing instructions. The project costs N_1 per 1000 m^3 of cleaned gas then come to

$$N_1 = K_p N_2 \qquad \text{(crowns per 1000 m}^3\text{)} \qquad (93)$$

b) The cost price of the separating equipment, per 1000 m^3 of cleaned gas, works out as

$$N_2 = u' C_{oz} \frac{1000}{Qh} \qquad \text{(crowns per 1000 m}^3\text{)} \qquad (94)$$

c) The costs of site erection work are, in Czechoslovakia, again a fixed percentage of the price of the equipment that is being installed. Hence, the cost price of the equipment is simply multiplied by the appropriate coefficient quoted in the assembly pricing instructions, to obtain the erection costs N_3 per 1000 m^3 of cleaned gas as

$$N_3 = K_m D_2 \qquad \text{(crowns per 1000 m}^3\text{)} \tag{95}$$

d) The space enclosed, in m^3, is always understood to mean not only the amount of space actually occupied by the equipment, but equally the space needed for its operation and maintenance, including the usual facilities for the attendants. Some types of equipment, such as bubble washers, are mostly installed indoors, and in that case part of the costs of the building that houses them must be written off against them. Other types, like cyclones or electrostatic precipitators, are usually located outdoors, but even then the ground area which they cover must be included in their first costs. The costs of the building to house the equipment are

$$N_4 = u'' C_p V \frac{1000}{Qh} \qquad \text{(crowns per 1000 m}^3\text{)} \tag{96}$$

The costs of the ground occupied by the equipment are

$$N_4' = u''' C_u V' \frac{1000}{Qh} \qquad \text{(crowns per 1000 m}^3\text{)} \tag{97}$$

The running costs, as distinct from the capital costs, are largely dependent on the type of equipment we use, and break down into the following main components:

a) The cost of electricity for the fan which has to overcome the pressure drop across the separator.

b) The cost of electricity for the operation of the separator itself.

c) The cost of the water, if any, needed for the equipment.

d) The costs of spares and replacement parts.

e) The wage bills and associated overheads for the operation and routine upkeep of the equipment.

f) The wage bills and associated overheads for the replacement of worn or damaged parts.

g) The interest and capital servicing charges on the initial outlay, which should be spread evenly over the whole expected service life of the equipment.

With the principal cost components thus defined, we can take a closer look at them:

a) The power consumption of the fans that move the gas through the separators is sometimes augmented by further fan inputs, e.g. for the reverse blowing of cloth filters, or for the air circulation that is maintained in some cyclone installations. However, the major component is the one dependent on the pressure drop across the separator; this cost, per 1000 m^3 of cleaned gas, is established as

$$N_5 = EhC_e \frac{1000}{Qh} \qquad \text{(crowns per 1000 m}^3\text{)} \qquad (98)$$

b) The power consumption involved in the actual operation of the separators means not only the power fed to these units themselves, but also the electricity inputs to all their ancillaries, such as the compressors that serve cloth filters, or the sludge pumps of wet separators. The total electricity costs per 1000 m^3 of cleaned gas are determined as

$$N_6 = E'hC_e \frac{1000}{Qh} \qquad \text{(crowns per 1000 m}^3\text{)} \qquad (99)$$

c) Wet separators—scrubbers, washers, wet electrostatic precipitators, and gas stabilizers—also need a water supply. Except for the stabilizers, however, most of these units normally run on recirculated water, and the first and principal running costs of the recirculation system have already been covered in the respective paragraphs b). Consequently, the only water consumption to reckon with is that of make-up water to replace evaporation losses, water dumped in the sludge, and overflows from units like the MHB washers. The cost of water per 1000 m^3 of cleaned gas thus comes to

$$N_7 = AhC_a \frac{1000}{Qh} \qquad \text{(crowns per 1000 m}^3\text{)} \qquad (100)$$

d) As for replacement parts, the majority of the costs involved are caused, in the various kinds of separators, by the following items:

Cloth filters: filter cloths; components of the rapping mechanism; components of the rapping control system; and the dust discharge gates or air locks.

Cyclones: dust discharge gates and air locks.

Scrubbers and washers: nozzles; components of the sludge raking and evacuating mechanisms; pump impellers; automation components for control of the water supply; and the working parts of the separators themselves.

Electrostatic precipitators: high-tension electrodes; components of the rapping mechanism; and rectifier control components.

The cost of spares and replacements per 1000 m^3 of cleaned gas is ascertained as

$$N_8 = C_n \frac{1000}{Qh} \qquad \text{(crowns per 1000 m}^3\text{)} \qquad (101)$$

e) The wage bill for operation and routine maintenance, per 1000 m^3 of cleaned gas, works out as

$$N_9 = H_u M_u \frac{1000}{Qh} \qquad \text{(crowns per 1000 m}^3\text{)} \qquad (102)$$

The main servicing needed by the various kinds of separators can be summed up as follows:

Cloth filters: lubrication of the rapping mechanism; functional and leakage checks of all flaps and air locks; filter cloth inspections; checks on the dust evacuation facilities; and pressure drop measurements.

Scrubbers and washers: lubrication of the sludge raking mechanisms and pumps; inspections of the water supply and outlet systems; checks of the water pressure and of the gas pressure drop; and overall inspections of the unit.

Electrostatic precipitators: lubrication and inspection of the rapping mechanism; lubrication of the rotary rectifiers; checks of the rectifiers and their control systems; and regular examination of the high-tension and collecting electrodes.

f) The wage costs for the replacement of worn or faulty components, per 1000 m^3 of cleaned gas, are calculated as

$$N_{10} = H_d M_d \frac{1000}{Qh} \qquad \text{(crowns per 1000 m}^3\text{)} \qquad (103)$$

6. INDUSTRIAL APPLICATIONS OF SEPARATORS

Ing. O. Štorch, CSc.

6.1 Power generation

6.1.1 Power stations and district heating plants

The trapping of the fly ashes emitted by steam generator sets is generally regarded as the most important single task in the whole field of air pollution control. The sheer weight of these ashes easily exceeds that of all the solid pollutants released from all other atmospheric contamination sources rolled together. The task is made no easier by the widely varying grain size distributions, chemical compositions and concentration levels of the fly ashes. All these factors depend primarily on the properties of the fuel and on the combustion process itself, i.e. on the type of boiler employed.

The fly ash concentration per unit volume of the combustion products is as a rule inversely proportional to the calorific value of the fuel, and directly pro-

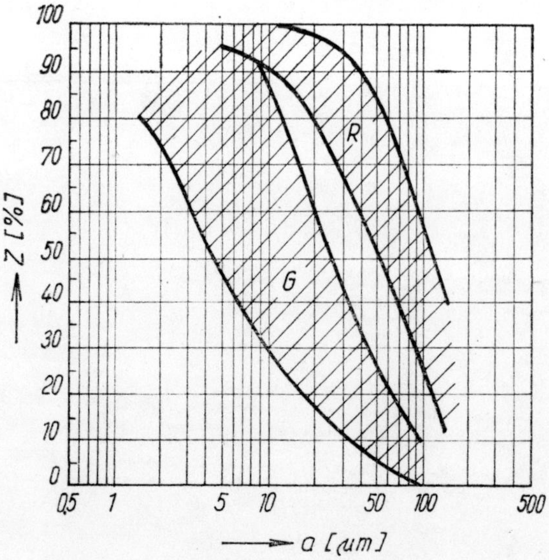

Fig. 164 The usual intervals of grain size distributions of fly ashes from stoker-fired boilers (*R*) and from slag-type boilers (*G*).

portional to the ash content of the fuel. Even stoker-fired boilers with travelling grate or reciprocating grate fireboxes can emit as much as 15 to 30 per cent of this ash content in the form of fly ashes. In boilers fired with pulverized fuel, especially in slag type boilers, this proportion can run as high as 80 to 90 per cent. In slag tap boilers this ratio drops to around 60 per cent, and in cyclone type fireboxes it is usually no more than 10 to 25 per cent. All this together yields a typical concentration range of some 2 to 50 g m_n^{-3}, with stoker-fired boilers occupying the lower end of the bracket, say 2 to 10 g m_n^{-3}, and boilers fired with powdered coal covering the 10 to 50 g m_n^{-3} interval. While on this subject, it must be borne in mind that apart from true fly ashes, the combustion products always contain a proportion of uncombusted though combustible solids.

The grain size distribution of the fly ashes, like their concentration level, is largely dependent on the boiler type. The coarsest fly ashes are produced by stoker-fired boilers: in Fig. 164, the zone marked *R* represents the size distributions of fly ashes from these boilers, as ascertained in various investigations and as reported in the literature. The diagram indicates that half the particles thus generated can be finer than 50 to 120 microns, and that 10 per cent of them will be smaller than the 10 to 30 microns size interval. Much finer fly ashes are obtained from boilers fired with pulverous coal, but in this case their size is also affected by the pulverizing itself, i.e. by the grain size distribution of the fuel supplied to the boilers. In Fig. 164, the fly ashes released by these boilers are represented by the zone marked *G*, which shows that half these solids will be smaller than 5 to 25 microns, and 10 per cent of them will fall below the 2 to 5 micron bracket. Even finer fly ashes are produced by slag tap boilers, but then the fly ash concentrations in the flue gases leaving these boilers are lower too.

As for the chemical analysis, the fly ashes created in the combustion of the brown and stone coals common in Czechoslovakia consist mainly of SiO_2 (46 to 55 %), Al_2O_3 (18 to 28 %), and CaO (2 to 6 %). Slag type boilers fired with powdered coal generate irregularly shaped fly ash particles with only a certain proportion of spherical grains among them, but slag tap boilers, and especially cyclone fireboxes, produce predominantly spheroidal particles with surfaces glazed by the high temperatures in the combustion chamber.

The combustion product temperatures range from some 170 to 300 °C in stoker-fired boilers to about 140 to 180 °C in slag type boilers fired with powdered coal, or roughly 200 to 220 °C in some of the older boilers of this type.

Stoker-fired boilers

These boilers generate relatively low concentrations (say 2 to 10 g m_n^{-3}) of fly ashes, which moreover have a fairly low content of fine fractions, and are generally rated only for low to medium steam outputs anyway. Consequently, the flue gases from these boilers are mostly cleaned in cyclones. In view of the grain size distribution of the fly ashes, the fact that cyclones are not very good at trapping the minus 5 micron fractions makes little difference. However, the parameters of the cyclones must be matched to the given concentration and size distribution of the fly ashes, and to the given chimney height, so as to ensure that the amount of solids released from the chimney will consistently remain within the legal limit. Another requirement on these cyclones is that they should take up relatively little space, because they are commonly added as an afterthought to existing boiler installations, and have to fit into what space happens to be available. Furthermore, they should be highly resistant to the abrasive effect of the coarse particles. With this much said, it must be admitted that the cyclone batteries and arrays now on the market all fall short of meeting these requirements in their entirety.

Both the running costs and the wear rate can be reduced by operating the cyclones at a relatively low pressure drop, in the region of 600 Pa. The centrifugal chambers of these cyclones should not be less than some 250 mm in diameter. Depending on their type and size, and on the grain size distribution of the fly ashes, cyclones in these applications normally achieve collecting efficiencies of 85 to 95 per cent.

Electrostatic precipitators are used for this duty only in the exceptional cases when the boilers pollute a residential area and emit a high proportion of minus 10 micron fines. Before even considering an electrostatic precipitator for any such application, the selectors must take into account two key complicating factors: firstly, precipitators are sensitive to any increase in the flow rate; and secondly, their efficiency diminishes abruptly when the ohmic resistance of the dust is abnormally high. Both these factors can severely limit the usefulness of electrostatic precipitators in this type of service.

In some particularly exacting circumstances, there may be nothing for it but to combine electrostatic precipitators with cyclones in a two-stage cleaning installation. These are the cases when the flue gases carry a high concentration of fly ashes with a large proportion of minus 10 micron fractions, plus a substantial amount of uncombusted fuel particles, and the siting of the plant is such that those gases must be cleaned with a high and consistent degree of efficiency. Under such special circumstances, there is much to be said for reversing the usual approach, and installing the precipitators as the upstream and the cyclones as the downstream stage. The precipitators in the primary stage would then trap not only some of the fines, but also most of the coarsest particles, if only by gravity-induced sedimentation alone. The wide mix of particle sizes reaching them would favourably affect their rapping, and hence the cleanliness of their electrodes, which is so vital to the efficiency of these units. The cyclones in the secondary stage would receive a reduced dust concentration, previously rid of most of the coarsest fractions, and would consequently suffer less wear.

On the face of it, a simpler solution would seem to be a two-stage layout with two cyclone stages in tandem. In practice, however, this arrangement has hardly ever proved satisfactory. For one thing, it entails a power consumption that is often unacceptably high. For another, high-efficiency cyclones in the first stage will feed the second stage with so great a proportion of fines that it will almost inevitably tend to clog up with adhesive deposits. And finally, this configuration will only seldom yield the high overall efficiency level expected of so costly an installation.

In small and medium-sized boilerhouses, with steam outputs up to about 50 tons per hour, the requisite efficiency could best be attained with wet scrubbers. For example, the MVB type described in Section 3.2.2.1, which affords efficiencies around 97 to 99 per cent, has already proved satisfactory in this application, but only when fabricated in corrosionproof materials. Moreover, the economics of any such installation depend heavily upon those of its sludge extraction and water recirculation systems, and these are areas where much remains to be done to reduce costs. Scrubbers are especially attractive for the boilerhouses of textile factories which produce alkaline liquid effluents: the latter can be used as cost-free process water, and will not only be neutralized in the scrubbers, but will also trap some of the sulphur dioxide in the flue gases.

Boilers fired with pulverized coal

Pulverized coal is used mainly in large steam generator sets, which are often grouped into plants comprising several such blocks. The 200-MW sets now in routine use generate flue gases at the rate of some 1 to 1.5 million m_n^3 per hour, depending on the boiler type and on the temperature of the gases. These gases are cleaned predominantly in electrostatic precipitators; sometimes, especially when the fly ash concentrations are substantial, these precipitators are preceded by primary stages made up of cyclones.

Electrostatic precipitators for these duties are almost invariably of the horizontal plate-type design; vertical plate-type units are employed only where space is critically short. The unit housings are made either of reinforced concrete or of steel. Concrete is often picked as being cheaper, saving steel, and dispensing with the thermal insulation which the steel walls require, but is seldom really leak-proof—and in units running with a partial vacuum inside them, this can offset all the advantages of concrete housings. The horizontal plate-type units are commonly split into two or three successive sections, each with its own power feed system. The number of sections depends on the concentration and size distribution of the fly ashes, and on the stipulated overall collecting efficiency. The first section, which handles the highest dust concentration, is normally operated at a slightly lower voltage than the sections downstream of it.

The gas velocity in the working spaces of electrostatic precipitators is nowadays mostly in the vicinity of 1.5 m s^{-1}, with peaks not exceeding 2 m s^{-1}. Attempts to raise this velocity to 2 m s^{-1} or more have lately been abandoned, because the resultant space savings have been outweighed by the penalties in the average collecting efficiencies.

Similarly, the gas temperature in precipitators must be kept down, in the interests of both dependability and efficiency. Trials performed so far have indicated that this temperature should not lie more than some 100 °C above the dew point which the combustion products would have if they contained their given content of water vapour only (as distinct from their actual dew point, which is raised by the presence of sulphur dioxide in them). At the average moisture contents of the coals fired in these boilers, this theoretical dew point, based on the water vapour content of the gases, is generally between 40 and 48 °C. Older boilers produce hotter flue gases, and consequently call for a flow velocity, in the working spaces of the precipitators, of no more than 1.5 m s^{-1} at the most.

Electrostatic precipitators will not function properly unless their collecting and high-voltage electrodes are kept reasonably clean. Just how clean they need to be depends largely upon the apparent specific resistance of the dust layers on the collecting electrodes, which in turn varies greatly with the temperature and humidity of the combustion products. Fly ashes containing a proportion of coarse particles (including uncombusted fuel residues) are easier to dislodge by rapping than a coating consisting exclusively of fines. Therefore, the first section of a horizontal plate-type precipitator is as a rule easier to keep clean than the sections downstream of it. The main exceptions are encountered in plants where cyclone primary stages are installed ahead of the precipitators. In these installations, the cyclones trap the coarser particles, and supply the precipitators only with strongly adhesive fines which tend to cling to the electrodes. Which means that when we decide on an efficient cyclone array upstream of the precipitators, we must also fit the latter with extremely effective rapping systems, or else expect trouble.

The risks inherent in a layout with two consecutive high-efficiency stages have recently led to the ever more widespread adoption of an alternative scheme, where the cyclones upstream of the precipitators are no longer high-efficiency tangential-inlet units. Instead, they are increasingly taking on the form of compact, space-saving batteries of axial cyclones with a straight-through gas flow, which serve to even out the flow rate reaching the precipitators as much as to reduce the solids concentration in that flow. A typical battery suitable for these duties is the SHA type described in Section 3.1.2.4. The efficiency required in this application is relatively low, so that the 50 to 60 per cent efficiency offered by these batteries is quite adequate. The fact that they are essentially non-selective is actually an advantage, as it means that the electrostatic precipitators will receive a reasonable proportion of coarser particles which will facilitate their rapping. However, any such arrangement deprives us of an important secondary advantage which in the past has often contributed to the decision in favour of high-efficiency primary-stage cyclones. The latter could, in the event of a precipitator failure or shutdown, trap a substantial proportion of the fly ashes, and especially of the coarser fractions, on their own. The axial-flow medium-efficiency batteries which are nowadays so often used in their stead cannot provide any such stand-by facility.

The progress made in this field over the past fifteen years is best exemplified by the following figures, which are representative of the state of the art in Czechoslovakia. As recently as 1960, it was customary for the combinations of primary-stage cyclones and electrostatic precipitators, fitted to newly commissioned boilers, to attain overall efficiencies around 90 per cent. Even this performance was naturally apt to deteriorate with the passage of time, with various changes in the mode of boiler operation or in the quality of the fuel, and especially during periods of equipment breakdowns or malfunctioning. Nowadays, well-selected precipitators with effective rapping provisions, powered by selenium rectifiers with automatic voltage control systems, reach efficiencies as high as 99 per cent, and are expected never to drop below the 97 per cent level. There is, however, one important qualification to all this: electrostatic precipitators do not readily accommodate significant changes in the flow rate. When they are to clean flue gases, we must always consider the likelihood of future changes in the quality of the fuel (its calorific value, moisture content, or content of ash matter) which might affect the average throughflow through the precipitators.

Installations consisting entirely of cyclones cannot clean the flue gases that arise in the combustion of pulverized coal with any high degree of efficiency. The best attainable with single-stage cyclone layouts is about 90 per cent, and two cyclone stages in succession will do only slightly better. Even that figure is strongly dependent on the size distribution of the fly ashes. The trouble is that in single-stage installations, we cannot use the small-diameter high-efficiency cyclone types, because they would be bound to wear at a prohibitively high rate. In two-stage facilities, the first stage is apt to trap all the coarser fractions, supplying the second stage only with fines which will rapidly clog it up. In practice, two cyclone stages in tandem are used only as a last resort in cases when gas cleaning equipment is being post-fitted to existing boilers, and where there is just not enough space to fit in electrostatic precipitators.

Newly erected power stations which burn powdered coal are now being provided with electrostatic precipitators large enough to dispense with any primary stage upstream of them. The makers of these precipitators commonly guarantee efficiency levels around 99 per cent.

6.1.2 Garbage incinerating plants

The piles of refuse generated by modern cities are difficult to dispose of by any other means than their incineration. Unless we want to eliminate one health hazard at the risk of creating another, however, the incinerating plants have to be fitted with highly efficient equipment for trapping the solids contents of the flue gases. The amounts of gaseous chemical contaminants released by these plants, such as chlorine or fluorine compounds, are so far too small to warrant any attempt to trap them.

Some of these plants burn garbage only, although their boilers are usually provided with auxiliary burners for ignition and stabilization of the flame. In other plants, the garbage supplements, or is augmented with, more conventional fuels like stone or brown coals, oil or gas. In either case, the flue gas temperature at the combustion chamber outlet must be kept to some 900 to 950 °C, as at gas temperatures below 800 °C the whole vicinity of the plant would be pervaded by objectionable smells. Which means that the gases must be cooled down to some 170 °C, or exceptionally to about 300 °C, before they are allowed to enter the gas cleaning equipment. Obviously, the most economical way of doing so would be to exploit the heat of those gases for steam generation, but in some small plants this would not be worth while. In such cases the gas temperature is brought down by the injection of water or induction of outside air. Some of the larger boilers are also fitted with water injection, downstream of the heat exchangers, but this serves only for damping major temperature fluctuations caused by the varying calorific content of the garbage, and/or for humidifying the gases prior to their admission to electrostatic precipitators.

Most of the incinerators in use at present burn some 10 to 20 tons of refuse an hour; the smallest of them burn as little as two, the largest now being installed up to 50 tons per hour. The quantities of flue gases they generate will naturally also depend on the calorific value of the garbage and on the air ratio. Given a calorific value around 4200 kJ per kg and an air ratio of 1.8 to 2.5, we shall be dealing with roughly 3.6 to 4.8 m_n^3 of flue gases per kg of incinerated refuse; at about 7500 kJ kg^{-1} and the same air ratio, the volume of flue gases will range between 4.7 and 6.3 m_n^3 per kg of refuse. All this, of course, applies only to the garbage itself, not to any supplementary fuels that may be combusted along with it.

The dust concentration in these gases varies with the type of combustion equipment, the process conditions and operating techniques, and with the kind of refuse that is being incinerated. Data published abroad quote values ranging from 0.5 to 15 g m_n^{-3}, with most of the figures lying in the 3 to 10 g m_n^{-3} interval. Higher concentrations are encountered when the specific loading of the boiler grates is high, and also in winter, when the domestic refuse that forms the staple diet of these boilers contains a higher proportion of ashes. Especially the very fine ashes that arise in the combustion of briquetted coals can markedly increase the solids content of the flue gases at incinerating plants. Another common cause of high solids emissions is the burning of garbage in equipment originally designed for coal firing.

The grain size distribution of the solids in the flue gas again varies a great deal from one installation to the next. In Fig. 165, the region covered by the typical retained fraction curves of these solids is denoted by the hatched area. The curves bounding this area indicate that half the solids are apt to lie in the minus 20 to minus 70 microns brackets, and 10 per cent of them will be smaller than 2 to 5 microns. This makes an intersesting comparison with the plots in Fig. 164. On the whole, the size distribution of the fly ashes from incinerators lies about mid-way between those for stoker-fired and for pulverized-coal boilers; but particularly in the region of the finer fractions it comes closer to the smaller particle sizes generated by the latter boilers.

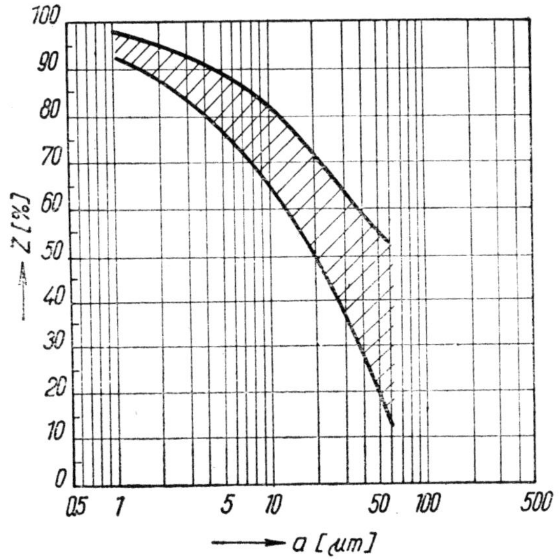

Fig. 165 The normal range of size distributions of the solids emitted by garbage incinerating plants.

The gas cleaning facilities at these incinerating plants are mostly cyclones or electrostatic precipitators. Cloth filters are ruled out by the high gas temperatures, the adhesive nature of the dust, and the risk of moisture condensation. Scrubbers or washers are employed only under exceptional circumstances – either when they can be harnessed to trap gaseous as well as solid pollutants, or else to serve small boilers, and deal with small throughflows, in cases where cyclones would not be efficient enough and electrostatic precipitators are ruled out by space or other considerations.

Cyclones are the usual equipment for the smaller boilers, which mostly emit not only only less gas per unit of time, but also lower solids concentrations in the gas. In these instances, cyclones are chosen for their economy and convenience. The fact that their efficiency in such applications is usually confined to some 80 or

90 per cent hardly matters, as long as the chimney is high enough to disperse the remnant of the (generally low) initial solids concentration. Another important proviso is that the cyclones must be readily accessible for cleaning, since in these installations they are particularly prone to foul up. The gas flow delivered by incinerators is apt to fluctuate far more widely and frequently than in most other applications, and every drop in the flow rate may cause some of the particles to adhere to the cyclone walls.

Medium-sized and large incinerators produce such a mass of solids that their flue gases must be cleaned with a very high level of efficiency, commonly around 99 per cent. This is the almost exclusive domain of electrostatic precipitators. In fact, small precipitators have been evolved for handling these gases at through-flows as low as 10,000 m^3 per hour. The precipitators employed at incinerating plants are mostly horizontal plate-type units with two sections in tandem. They should be fitted with deeply profiled collecting electrodes to facilitate the capture of flakes of charred paper, which are otherwise difficult to trap in precipitators on account of their high electrical conductivity: on contact with the collecting electrode, they tend to reverse their polarity and bounce off back into the gas stream.

The way an electrostatic precipitator will perform in actual service is known to be strongly dependent on three main factors: the electrical resistance of the retained dust layer, the humidity of the incoming combustion products, and their temperature. The trouble is that incinerators produce a mix of solids where the electrical resistance of the coarser fractions differs greatly from that of the finer ones. Even among the coarser sizes, there is a difference between the resistance of the minus 100 micron particles and the much lower resistance of the plus 100 micron fraction, which contains a large proportion of charred paper flakes. Sometimes the only way of preventing the escape of these flakes (or of other coarse particles which are not readily trapped in precipitators, and which would on escaping be likely to settle in the immediate vicinity of the plant) is to mount cyclones downstream of the precipitators. An alternative approach is to reduce the flow velocity in the working spaces of the precipitators to less than 1 m s^{-1}, and provide the precipitator inlets with internal fittings designed to shred up the larger flakes of paper. The smaller bits of paper are easier to capture, because on touching the collecting electrodes they are quickly covered by a layer of other particles, which tends to hold them down and prevent their re-entrainment.

The solids most likely to impair the functioning of precipitators at incinerating plants are the soot particles that arise in the combustion of rubber, plastics, bundles of paper, and oil remnants. The only known way of suppressing their effect is to combat the formation of soot at its source. That means mixing the soot-generating components intimately and evenly into the mass of other refuse; closely adhering to the specified combustion conditions, particularly as regards the primary and secondary air ratios; and maintaining a sufficiently high combustion temperature.

Ideally, the electrostatic precipitators should be big enough to do their job properly without any need for subsidiary measures. Unfortunately, practical considerations occasionally force plant designers to pick an undersized unit, and then improve its working by humidifying the combustion products so as to reduce the electrical resistance of their solids. The water must be injected at high pressures, of the order of $20 \cdot 10^5$ Pa, to ensure its complete evaporation before it actually enters the precipitator. This scheme does work, but is far better reserved for exceptional circumstances only. Particularly, plant designers should never allow themselves to be tempted into false economies: the countermeasures needed to remedy the situation often defeat the whole purpose of the exercise.

6.2 The iron and steel industry

6.2.1 Ore sintering plants

The world's iron and steel industries are nowadays increasingly reliant on fine-grained ores, which have to be sintered or pelletized into larger lumps before they can be charged into a blast furnace. The roasting pans and rotary furnaces formerly used for this process have now mostly given way to sintering strands, i.e. installations employing what is essentially a special-purpose conveyor belt. A homogeneous mixture of fine-grained ore, fuel (coke) and fluxing additions is moistened, deposited on the inlet end of the strand, and ignited. Turboblowers supply a blast of air to the moving grates of the strand, on which the mixture sinters into lumps at temperatures around 1150 to 1250 °C. At the outlet end of the strand, the granulated material is cooled, the fines are screened out, and the lumps are then delivered to the blast furnace. The input material is predominantly of minus 5 mm size, and has moisture contents of typically 10 to 20 per cent.

The air and gas stream emerges at the bottom of the strand at something like 1100 to 1200 °C, but this is brought down to roughly 100 to 400 °C by the induction of outside air before the stream reaches the foot of the chimney stack. This mode of cooling naturally increases the volume we have to handle, which is usually between 1000 and 4000 m_n^3 per ton of sinter, depending on the quantity of outside air drawn into the flow. The solids concentration in these gases varies with the grain sizes of the sintered materials; when these are extremely fine, it can run as high as 15 to 20 g m_n^{-3}. The size distribution of these solids differs widely from one plant to another, being dependent mainly on the particle sizes of the starting materials. Table 24 presents a typical grain size analysis of the dust in these gases, along with the percentages of minus 10 and minus 3 micron fractions ascertained

TABLE 24

Grain size distributions of dust in fumes exhausted from sintering strands

Sample No.	Particle size *a* (microns)					
	> 60	> 40	> 30	> 20	< 10	< 3
	Retained proportion *Z* (%)					
1	70	77	80	85	10	
2					22.7	
3					18.2	5.5
4					17.0	4.0
5					12.0	
6					9.1	
7					7.5	
8					8.2	

in some other measurements. This dust only rarely contains any sinter particles. Usually, it is made up entirely of the as yet unsintered raw materials, and its chemical composition is therefore governed by the analyses of those materials rather than by that of the sinter itself. A representative range of dust compositions is listed in Table 25. The chemical analysis of the gases which convey this dust is likewise dependent on that of the raw materials, but mostly lies within the interval set out in Table 26.

TABLE 25

Chemical composition of dust in fumes exhausted from sintering strands (% by weight)

Fe	approx. 50	Al_2O_3	2 to 8
SiO_2	9 to 15	C	0.5 to 5
CaO	7 to 24	S	up to 2.5
MgO	1 to 2		

TABLE 26

Chemical composition of gases extracted from sintering strands (% by volume)

O_2	10 to 20	SO_2	up to 0.4
CO_2	4 to 10	N_2	86 to 64
CO	0 to 6		

However, the strands themselves are just one of the many areas in a sintering plant where dust is generated in volumes large enough to call for countermeasures. The finished sinter is mostly taken to the blast furnaces on conveyor belts, linked by transfer chutes and holding bins, which create dust problems of their own. So do the sinter cooling, crushing and screening facilities, the storage bins or bunkers, and the transfer of the sinter into the furnace charging skips or conveyors. At all these points, however, the dust arises by the comminution and abrasion of the sinter itself, and therefore differs from the dust originating at the strands. The exhaustors serving these secondary dust generation areas will deliver air containing varying amounts of dust: the concentration level depends not only on the quality of the sinter (especially its mechanical strength), but equally on the design and layout of the suction hoods and the rating of the exhaustors themselves. Investigations at a number of sintering plants have yielded dust concentration figures ranging from 3 to 25 g m_n^{-3} at the exhaustor outlets.

As for the grain size distributions of these secondary or sinter-based dusts, the same measurements have proved that these dusts differ too widely for all of them to be regarded as one and the same material. However, the sieve analyses of these dusts do tend to fall into two distinct groups. Category *A* are the dusts arising at the sinter cooling, crushing and screening stations, while category *B* covers the dusts exhausted from the conveyors, transfer chutes and storage bins. The boundary between the two categories is indistinct, and there are frequent anomalies (like dust generated at the chutes coming squarely into category *A*), but at least no dust that does not fall into one or the other of these groups has ever been detected at any sintering plant, in Czechoslovakia or abroad. The range of size distributions

TABLE 27

Grain size distribution of dust formed at ore sintering plants

Particle size *a* (microns)	Retained proportion *Z* (%)	
	Class A dust (processing dust)	Class B dust (handling dust)
< 1	0.3 to 2	0.1 to 0.3
< 2	2 to 8	0.2 to 2
< 5	10 to 30	0.5 to 10
< 10	28 to 55	3 to 28
< 20	52 to 80	11 to 52
< 30	65 to 88	20 to 65
< 50	82 to 95	35 to 82
< 100	93 to 100	60 to 93

encountered in categories *A* and *B* is summarized in Table 27, and plotted in Fig. 166. Some of the sinter dust samples gained in this work have been subjected to chemical analyses, with the results set out in Table 28.

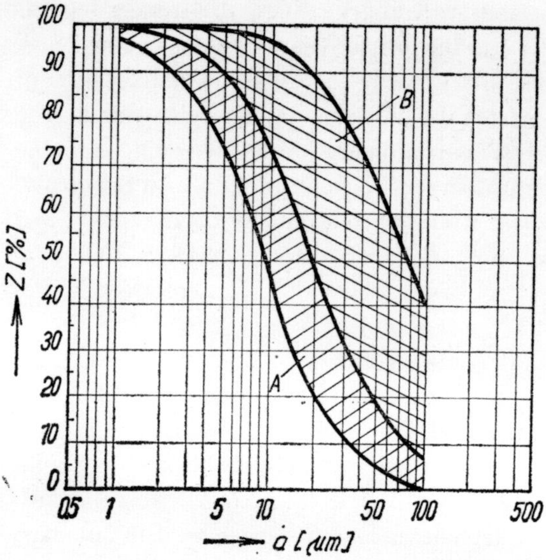

Fig. 166 Grain size distributions of the "hot" dusts (*A*) and "cold" dusts (*B*) trapped at ore sintering plants.

TABLE 28

Chemical composition of dust formed at ore sintering plants (% by weight)

Fe	26 to 40	Al_2O_3	2 to 6
Fe_2O_3	38 to 52	CaO	10 to 27
SiO_2	11 to 22	MgO	0.5 to 7

The dust exhausted at some sintering plants is liable to bond together and solidify on contact with water. This tendency is common to all the dusts which form at Czechoslovak sintering plants, and complicates the use of wet scrubbing equipment. Before examining the consequences, we must distinguish between the two different mechanisms which underlie this process.

The first of these mechanisms resembles the setting of ordinary lime mortar. The calcium monoxide which forms up to 20 per cent of these dusts can, in a sufficiently humid environment, combine with atmospheric carbon dioxide and turn into calcium carbonate, $CaCO_3$. The latter is not only very firm, but also capable of binding some other, normally inert constituents of the dust. The importance of this mechanism has been confirmed by analyses of the deposits removed from scrubbers at sintering plants; the scale layers were found to contain more than 50 per cent of $CaCO_3$. However, no such hard deposits ever build up e.g. in the water-flushed troughs into which some cyclone batteries discharge the trapped dust for raking out. The explanation is simple: in these troughs, the dust is completely covered with water, and hence inaccessible to the atmospheric carbon dioxide needed for this process to take place. That, unfortunately, is cold comfort for the equipment selectors, because in all the gas scrubbers and washers now in existence the dust is always wetted while exposed to the atmosphere.

The alternative mechanism is reminiscent of the bonding of cement. It again involves the joint action of water and carbon dioxide, but in this case upon a mixture of SiO_2, Al_2O_3 and Fe_2O_3.

In either case, the outcome will be strongly dependent on the grain size distribution of the dust. For one thing, there is a correlation between the particle size and the solubility of calcium oxide. For another, the smaller particles bond more easily. And finally, the chemical composition of the particles is apt to be size-dependent. Other factors which affect the solidification process are the properties of the gas (particularly its carbon dioxide content and temperature); the quantity, purity and chemistry of the process water; and the dust concentration in the gas.

Until fairly recent times, Czechoslovak sintering plants were fitted with cyclones to the virtual exclusion of any other gas cleaning equipment. These units, which had to clean both the combustion gases from the sintering strands and the dust-laden air exhausted from the secondary dust generation points, were mostly batteries or arrays of cyclones with diameters ranging from 300 to 630 mm. The collecting efficiency of these installations varied between 65 and 95 per cent, in dependence on the cyclone diameter, the grain size distribution and the origin of the dust. Factors detrimental to efficiency included very fine dusts; wrongly selected large-capacity cyclone units; puncturing of the cyclone walls by the highly abrasive sinter dust; and the induction of outside air through leaking air locks at the hopper outlets. These last two wear-induced factors often reduced the efficiency to even less than 65 per cent. In view of the relatively high dust concentration in the gases, these efficiencies were simply not adequate to keep atmospheric pollution in the vicinity of the plants within acceptable limits. The air exhausted at the conveyor transfer chutes, at the vibrating screens, crushers, and skip loading stations, is apt to contain some 10 to 20 $g\,m^{-3}$ of dust, so that a 65 per cent trapping efficiency will not reduce the final dust concentration to anything like the mandatory limit.

The only cyclones really fit for handling sinter dust are high-efficiency units capable of retaining particles down to some 3 microns in size. Depending on the actual grain size distribution of the dust and other specific local factors, these units can attain efficiencies of 85 to 98 per cent, but even that may in the least favourable circumstances leave a dust concentration of 1 to 2 g m^{-3} in the air released into the atmosphere. Therefore, when any such installation is contemplated, the concentration level and grain size distribution of the dust should always be ascertained in advance. That will not only facilitate the selection of the cyclones, but also indicate the efficiencies and outlet dust concentrations to be expected of them. Another aspect that calls for close consideration is the volume of air to be exhausted, per unit of time, from each of the secondary dust generation points. Too low an intake will fail to keep the workplace atmosphere reasonably clear of dust, while too large an intake will turn the whole proposal into economic nonsense, and bring in quantities of dust that will be difficult to handle. The cyclones must either be fabricated in thickish plate, or made of cast iron. The only cyclone equipment presently available in Czechoslovakia which more or less meets these requirements are the SGA arrays of units 630 mm in diameter, which are made of plates up to 8 mm thick and cleared for service temperatures up to 400 °C.

Wet scrubbers or washers, which could consistently afford efficiencies in the 97 to 99.5 per cent range, are unfortunately not suitable for these applications. Owing to the bonding properties of the dusts generated at sintering plants, very hard deposits would form in these units at the rate of 0.1 to 0.2 mm a day, and would mostly prove difficult to dislodge. Consequently, the units would either prove unreliable or impose excessive maintenance requirements. The only way of operating them at all satisfactorily would be to feed them constantly with fresh water, close them down for cleaning every few days, and remove the hard deposits by rinsing with dilute acids. That is not the sort of procedure which plant operators can be expected to accept with enthusiasm. Moreover, the same sort of trouble would be encountered in the sludge handling systems of these units. If the water were recirculated, its content of soluble dust constituents would grow rapidly, and would soon coat the interiors of the pipes, fittings and pumps with layers too thick to permit their further operation.

The only area of a sintering plant where scrubbers are worth considering is its raw materials stockyard, with its conveyors, bins and bunkers for ore, coke and limestone. The types best suited to the conditions in these stockyards are the MHB washers with mechanical sludge raking provisions. However, water-filled equipment does not take kindly to sub-zero winter temperatures, so these washers must always be set up in enclosed and heated spaces, and their water feed systems must be protected against freezing up.

Since wet scrubbers are virtually ruled out, and cyclones have not always rendered satisfactory services, ever more sintering plants are installing electrostatic precipitators to handle all the various dusts they produce. Sometimes the load on the precipitators is eased by cyclones mounted upstream of them. These are mostly medium-efficiency cyclones, with efficiency levels around 60 per cent, operated at a low pressure drop across them. The precipitator equipment is best split into two distinct installations, one to handle the combustion products from the sintering strands and the other to clean the air exhausted from all the other pollution sources. Since this equipment configuration obviously entails long and complicated piping runs, from the various dust generating points to the precipitators, the layout of this piping or ducting must be designed with extreme care. If any dust were allowed to settle in those pipings, the condensed moisture from the often highly humid gases would be apt to bond it into a hard coating that would be virtually impossible to remove.

The air or gases will reach the precipitators at widely varying temperatures and humidities, and with a huge variety of dust contents. It is therefore always advisable to provide the precipitator inlets with mixing chambers to ensure a more or less homogenous inflow, both as regards the dust concentration, the grain size distribution and electrical resistance of the dust, and in respect of the temperature and moisture content of the gas or air.

6.2.2 Blast furnaces

Blast furnaces produce pig iron from a charge consisting of ores with iron contents in excess of 25 per cent (which are lately being largely supplanted by sinter or pellets), fluxing additions like limestone, scrap additions, and coke. Blast air is blown in at the bottom of the furnace, at temperatures commonly ranging from 550 to 900 °C, to assist the combustion of the coke. The process generates large volumes of blast-furnace gas with a high content of carbon monoxide. The coke and gas reduce the ores, and melt the resultant iron at temperatures of 1500 to 1800 °C. The limestone, SiO_2 and other admixtures are molten down into slag. Contemporary blast furnaces mostly have capacities in the 2000 to 3500 m^3 bracket, though some of the latest large units are in the 4500 to 5000 m^3 range, and produce thousands of tons of pig iron daily.

The blast-furnace gas is cleaned of its initially very high dust content, and then used as fuel gas, mostly within the steelworks that produce it. It must be handled with care, since in view of its carbon monoxide content it is toxic as well as inflammable. A typical composition and some other properties of the raw gas are summarized in Table 29, some typical data on cleaned blast-furnace gas are listed in Table 30. The cleaning process of course does not significantly alter the chemical analysis or specific gravity of the gas. Blast furnaces generally produce something like 4000 m_n^3 of this gas for every ton of pig iron they turn out.

The dust content of this gas is of mixed origin. Some of the particles are entrained from the charge materials before these are molten down, others arise by the sublimation of the slag or metals. The quantity, chemical composition and grain size distribution of these dusts are all governed by the physical and chemical properties of the charge materials. Moreover, their chemical and size analyses vary with the distance from the blast furnace, owing to sedimentation in the pipings and to the selective action of the dust trapping equipment. The chemical composition ranges set out in Table 31 should therefore be regarded as no more than a rough outline of what can be expected under average conditions.

TABLE 29

Properties of raw blast-furnace gas

Amount per ton of pig iron	approx. 3900 m_n^3
Temperature at furnace outlet	100 to 300 °C
Dew point	35 to 50 °C
Calorific value	3140 to 4606 kJ m_n^{-3}
	750 to 1100 kcal m_n^{-3}
Dust concentration	10 to 50 g m_n^{-3}
CO_2 content	10 to 16 volume %
CO content	25 to 30 volume %
H_2 content	0.5 to 4 volume %
CH_4 content	0.2 to 3 volume %
N_2 content	50 to 60 volume %
S, H_2S, SO_2 and O_2 contents	Traces
Moisture content	50 to 60 g m_n^{-3}
Specific gravity (at 9 % CO_2)	approx. 1.3 kg m_n^{-3}

TABLE 30

Properties of cleaned blast-furnace gas

Moisture content	15 to 35 g m_n^{-3}
Dust concentration	under 20 mg m_n^{-3}
Calorific value	4186.8 kJ m_n^{-3}
	1000 kcal m_n^{-3}
Combustion ratio	0.9 to 1.2

TABLE 31

Chemical composition of dust emitted by blast furnaces (% by weight)

Fe	5 to 40	Al_2O_3	9 to 15	C	5 to 10
SiO_2	9 to 30	Mn	0.3 to 1.5	Zn	0 to 35
CaO	7 to 28	P	0.3 to 1.2	Pb	0 to 15
MgO	1 to 5	S	up to 0.1	Alkalis	0 to 20
				Cu	Traces

Blast furnaces normally generate anything from 20 to 200 kg of dust for every ton of pig iron they produce, in other words, their dust output equals 2 to 20 per cent of their iron output. And if these figures are frightening, then non-sintered low-grade ores can easily double even this dust rate. The dust concentration in blast-furnace gas mostly varies between about 10 and 50 g m_n^{-3}, but equipment failures, unfavourable process conditions, or negligence on the part of the operators can raise it to as much as 200 g m_n^{-3}.

The cleaned gas is burned mainly in the blast preheaters which serve the blast furnaces, in coking plants, steam generator sets, open-hearth furnaces, and, more recently, in the gas turbines that power some of the plant ancillaries. None of this equipment will operate efficiently when it is fouled up with dust deposits, and burner nozzles easily clog up when fed with dust-laden gas. The gas cleaning facilities must therefore be carefully picked to provide the cleaning efficiency demanded by the gas-consuming equipment.

Easiest to meet are the requirements imposed by boilers when the fuel gas only supplements pulverized coals. In these cases, inlet concentrations as high as 1 g m_n^{-3} are still fully acceptable, but the gas must be cleaned in dry equipment: if the residual dust content were moist, it would soon clog up the gas pipings and burner nozzles. Coke-oven batteries and blast preheaters need gas cleaned down to no more than 20 mg m_n^{-3}. Usually, the dust concentration in the gas fed to them is kept to something between 5 and 15 mg m_n^{-3} to avoid the risk of dust layers building up on their internal fixtures. Dust with a high alkali content must be separated out much more thoroughly than the more inert dusts.

To ensure compliance with the requirements, blast-furnace gas is mostly cleaned in several successive stages. The first pre-cleaning is followed by cooling, accompanied by further preliminary cleaning, and then by final cleaning down to the permissible dust content. The preliminary cleaning is always done in dry equipment, and is expected to reduce the dust concentration to preferably about 5, but never more than 10 g m_n^{-3}. The object is not only to ease the load on the further cleaning stages, but also to trap the fractions which are rich enough in iron to be re-usable in the furnace. At this stage, therefore, efficiency is not nearly as important as selectivity. The fractional collecting efficiency of the equipment should ideally be such as to ensure that the ferrous particles will be separated from all the other dust. The fractions recycled to the furnace will then be reasonably free of useless or even harmful components, like the non-ferrous fines which are apt to contain zinc and alkaline substances. This naturally calls for full and accurate data on the grain size distribution of the dust at the separator inlets, and on the specific gravities of its individual fractions, before we can select or design the equipment for this pre-cleaning stage.

A simple device often employed to trap the coarsest fractions directly at the furnace top is to evacuate the gas through ascending pipes, either vertical or inclined. These pipes often tower high up above the furnace top, to give the heaviest particles plenty of space in which to lose their inertia before they drop back by gravity.

For a long time, the usual pre-cleaning facilities used to be settling chambers or inertial dust arresters. To do their job at all effectively, they had to be huge – typically say 10 metres in diameter by 15 metres high. As demands on the efficiency of the pre-cleaning stage increased, these units were gradually supplemented with cyclones downstream of them, or even replaced by cyclones. These had to handle anything from 40,000 to 100,000 m^3 per hour apiece, and had to be correspondingly large. Nowadays, the trend is to use smaller cyclones, some 1.5 to 2 metres in diameter, or even cyclone batteries, so as to reduce the dust concentration by as much as is feasible before the gas reaches the final cleaning stage. When the final dust concentration does not have to be particularly low (for instance, when the gas is burned in boilers), this final stage can again consist of cyclone batteries, but these will wear rapidly unless the coarser fractions are trapped in settling chambers or large-diameter cyclones ahead of them.

Plant experience indicates that cyclones downstream of the dust arresters are justified only when the inlet dust concentration reaching them exceeds some 5 g m_n^{-3}. At lower concentration levels the dust generally consists of fines, which the cyclones will trap with efficiencies of only about 50 per cent, at the cost of a pressure drop of 400 to 800 Pa. True, the dust is gained in the dry state, and may be re-usable, but that is simply not worth the capital outlay, the space, and the power consumption required for the cyclones.

The gas coolers downstream of the preliminary cleaning stage provide a further pre-cleaning facility, where the collecting efficiency ranges from 30 to 80 per cent. They usually bring the dust concentration in the gas leaving them down to the 1 to 5 g m_n^{-3} bracket. These coolers used to be fitted with a variety of internal fixtures, baffles etc. to enhance their collecting efficiency; but these all too often merely collected dust which then bonded into hard deposits, raising the pressure drop across the unit without bringing any substantial benefits in return. Consequently, present-day coolers are mostly upright cylindrical vessels containing only two to four vertically superimposed spraying rings. The water is injected at pressures of 3 to 4 . 10^5 Pa or even more, at a rate of 0.3 to 0.5 m^3 per 1000 m^3 of gas.

Gas cooling practices have been radically altered by the now almost universal adoption of blast furnaces operating with top pressure. These emit the gas at a gauge pressure of usually 0.5 to 2.2 . 10^5 Pa. The gas can then be cooled in large-diameter Venturi scrubbers, at pressure losses of only some 2000 to 3000 Pa. Alternatively, water can be injected ahead of the flow throttling devices, which then also serve to cool and pre-clean the gas.

The final cleaning of blast-furnace gas used to be the domain of disintegrators and wet electrostatic precipitators. The former are no longer in use, but the wet precipitators generally used for this duty are now being joined by a few cloth filter installations, and even a few isolated dry precipitators. The gases from top pressure furnaces are sometimes finally cleaned in Venturi scrubbers, which do very well in these applications. The old disintegrators reduced the dust concentration in the gases leaving them to anything between 15 and 50 mg m_n^{-3}; they also in the process raised the gas pressure by some 500 to 1000 Pa, but required a droplet separator, usually of the baffle-flight type, immediately beyond them. The electrostatic precipitators, cloth filters and Venturi scrubbers used for these duties at present attain efficiencies mostly in excess of 99 per cent, and keep the dust concentrations at their outlets to generally less than 20 mg m_n^{-3}. Typically, these concentrations range from 5 to 15 mg m_n^{-3}.

Dry electrostatic precipitators offer the twin advantages of saving a lot of scarce water, and releasing dry clean gas suitable for the gas turbines which are now coming into vogue in iron- and steelworks. Therefore, a lot of work is now being done to overcome the obstacles that have so far prevented their adoption for the cleaning of blast-furnace gas. The key problem is one of temperature and humidity control. The requisite cleaning efficiency is attainable only if the gas enters the precipitator humid enough for its temperature to lie close to its dew point. So far, the most promising solution seem to be the stabilizers mentioned at the end of Section 3.3.1, which cool the gas down to a constant and narrow temperature interval and moisten it by complete evaporation of the water injected into it. Obviously, if the state created by the stabilizer is to be preserved, the precipitator itself will require thorough thermal insulation.

Theoretically, then, there would seem to be no reason why dry precipitators should not become the standard type of equipment for cleaning blast-furnace gas. In practice, however, it is usually very difficult to maintain their initially high efficiency level for any length of time. The collecting and high-voltage electrodes are easily coated with dust layers which tend to bond into hard deposits; the separating process is disrupted by increasingly frequent arcing; and it is not always possible to control the moisture content of the gas with the precision needed to accommodate the conflicting requirements of efficiency and dependability. Consequently, the efficiency of these units, dependent on a fine balance of so many mutually unrelated factors, is apt to fluctuate unpredictably. That has discouraged potential users, and has so far kept dry precipitators an exception rather than the rule in the gas cleaning installations of blast-furnace plants.

Wet electrostatic precipitators, which predominate in these installations, clean the gas while it is practically saturated with water vapour, and thus in the state best suited for the electrostatic separation of its solids content. Apart from being both highly and consistently efficient, these units also score by working out reasonably compact, since there is no need to restrict the gas velocity within them to less than some 1.2 m s^{-1}. Moreover, the gas is cleaned at very low temperatures, mostly no more than 25 to 35 °C, and the water consumption, at some 0.5 to 0.7 m^3 per 1000 m^3 of cleaned gas, is acceptable even in this age of water shortages. These wet precipitators are prevalently of the upright tubular type, and share their housing with an array of cooling sprays. An alternative arrangement that has gained some measure of acceptance in the U.S.A. is a horizontal plate-type precipitator which incorporates a cooler and consists of three sections in parallel. The advantage of this layout is that the cleaning, shutdown or breakdown of any one section has only a slight effect on the overall efficiency of the whole unit. On the other hand, this configuration will always be more expensive, and take up more precious space, than the vertical tubular units preferred in Europe.

Cloth filters were for a long time considered incompatible with blast-furnace gases. The first cloth filters tried out for cleaning these gases were cleared for service temperatures of only about 100 °C; the gas had to be cooled down to some 40 or 50 °C ahead of them, and then reheated to evaporate the droplets, since otherwise vapour condensation was likely to clog up the filter cloths. The gas thus passed through the filters at roughly 60 to 70 °C, and then had to be knocked down to about 25 °C by water sprays, so as to induce the water vapour to condense in the cooler rather than in the pipings downstream of it. The complications of this procedure were apt to offset all the inherent advantages of cloth filters. More recently, however, filter cloths made of glass fibres have raised the range of service temperatures to around 250 °C, thus obviating the need for preliminary cooling of the gases. The gas still has to be cooled beyond the filters, to prevent the condensation of moisture in the piping runs, but that has nothing to do with the filters themselves.

Top pressure furnaces, which deliver the gas at a gauge pressure of about 0.5 to $2.2 \cdot 10^5$ Pa, are increasingly being fitted with Venturi scrubbers. These units are relatively cheap, yet efficient, require no coolers ahead of them, and take up very little space, as they are generally installed in the upright position. Their main drawback, a high pressure loss, makes no difference in this application, since the gas pressure would have to be reduced anyway. Furthermore, Venturi scrubbers cost very little to run. Their power input is small, as the only item that has to be powered is their water supply system, and their water consumption rate is comparable to that of an equivalent wet electrostatic precipitator plus the cooler upstream of it.

The Venturi scr bbers presently installed at blast-furnace plants are mostly large-diameter units, which attain the requisite efficiency levels only because of the more than ample pressure heads available to them. In principle, however, blast-furnace gas can be cleaned in any type of Venturi unit, including the MSA units described in Section 3.2.5.1, as long as we are prepared to tolerate a pressure

TABLE 32

Survey of data on the individual stages of blast-furnace gas cleaning processes

Separator type	Max. overall efficiency (%)	Usual inlet concentration ($g\ m_n^{-3}$)	Usual outlet concentration ($g\ m_n^{-3}$)	Pressure drop (Pa)	Application
Dust arresters	50 to 80	10 to 50	5 to 10	150 to 400	Coarse pre-cleaning of uncooled gas regardless of inlet dust concentration
Cyclones	50 to 80	5 to 10	1 to 5	300 to 800	Coarse pre-cleaning of uncooled gas at medium and high inlet dust concentrations
Cooling towers	30 to 70	1 to 10	0.7 to 2.0	50 to 200	Coarse cleaning of pre-cleaned gas and cooling to 30 to 80 °C
Cloth filters	over 99	1 to 5	0.005 to 0.015	1500 to 2500	Final cleaning of gas at 60 to 80 °C pre-cleaned down to inlet levels of 1 to 5 $g\ m_n^{-3}$
Wet electrostatic precipitators	over 99	0.7 to 2.0	0.005 to 0.015	50 to 100	Final cleaning of gas at 30 °C previously cleaned down to 0.7 to 2.0 $g\ m_n^{-3}$
Venturi scrubbers	over 99	1 to 10	0.005 to 0.015	10,000 to 20,000	Final cleaning of gases not previously cooled, but precleaned down to 1 to 10 $g\ m_n^{-3}$

drop of never less than 10,000 Pa, and sometimes as much as twice this figure.

In a conventional three-stage cleaning procedure, the dust content of the gas leaving the furnace top is captured roughly as follows: 70 to 80 per cent in the pre-cleaning stage, 10 to 20 per cent at the coolers, and 3 to 10 per cent in the final cleaning phase. Modern installations are expected to reduce the dust concentration from the initial 10 to 50 g m_n^{-3} to no more than 5 to 15 mg m_n^{-3}, which represents an overall cleaning efficiency of more than 99.9 per cent. This figure will be hard to improve on, no matter what other advances are made in this field in such respects as economy, serviceability, etc. Table 32 surveys the typical performances attainable at the various cleaning stages with correctly selected, operated and maintained equipment.

6.2.3 Open-hearth furnaces

Open-hearth furnaces, though considered obsolescent, are still in widespread use in the world's steelworks. The earliest of them were commissioned in the times when no one ever bothered about air pollution. In fact, gas cleaning facilities began to be added to these furnaces only when their capacities started growing, from the once customary 30 tons or so to 400 tons or more, and especially when oxygen blowing was introduced. This development happened to coincide with the rise in the public outcry for cleaner air. The dust emitted by these furnaces, which consists almost entirely of very fine particles, was found particularly objectionable.

Open-hearth furnaces are mostly fired with producer gas, coke-oven gas, or a mixture of coke-oven and blast-furnace gases, although oil-fired units are no exception. A conventional melt in these furnaces lasts about eight hours, and the solids content of the gases leaving the unit varies widely throughout this period. Mostly the furnaces emit a greyish white smoke, but during the oxygen lancing periods the smoke turns reddish brown. It consists chiefly of iron oxides (Fe_3O_4, FeO, and sometimes Fe_2O_3 too), but also contains varying amounts of other admixtures, mainly compounds of silicon, manganese, zinc, aluminium, phosphorus, calcium, magnesium, sulphur, copper and lead. Which of these compounds will be present, and in what amounts, depends primarily on the composition of the charge materials.

The volume of combustion products generated per ton of steel differs, within very wide limits, in dependence on a variety of factors. The chief of them are the type of fuel; the specific heat input; the scrap ratio (i.e. the ratio between the molten iron and the solid constituents in the charge); the size of the furnace; and the air ratio in the combustion process. In general, the specific volume of fumes will be the greater, the lower the calorific value of the fuel, the smaller the furnace, and the lesser the proportion of liquid iron in the charge. For instance, furnaces of less than 50 tons capacity, burning gas with a calorific value lower than 8500 kJ m_n^{-3}, will usually generate some 3500 to 3800 m_n^3 of fumes per ton of steel they produce, while in large modern oil-fired furnaces the fume rate can be as low as 2100 m_n^3 per ton. Ideally, therefore, the actual flow rate of the combustion products should be ascertained by measurements before any attempt is made to decide on the gas cleaning equipment. Since measurements will often be out of the question, an acceptable alternative is to compute the theoretical flow rate from the properties and consumption rate of the fuel, and from a chemical analysis of the fumes. Table 33 reviews the composition ranges of the combustion gases produced by various fuels, as quoted in the literature.

TABLE 33

Chemical composition of fumes from open-hearth furnaces fired with various fuels

Fuel	Composition of fumes					
	CO_2 (vol. %)	O_2 (vol. %)	$CO + H_2$ (vol. %)	N_2 (vol. %)	H_2O (vol. %)	SO_2 (g m_n^{-3})
Producer gas	12 to 18	0 to 8	0 to 2	60 to 70	10 to 15	0.1 to 8.0
Mixed coke-oven and blast-furnace gas	8 to 14	0 to 5	0 to 4	65 to 75	12 to 18	0.1 to 1.0
Coke-oven gas only	6 to 8	0 to 10	0 to 6	60 to 70	18 to 25	0.1 to 2.0
Fuel oil	12 to 16	0 to 10	0 to 5	—	6 to 12	up to 5.0

What interests us, of course, is not so much the gas as the dust it conveys. This dust content will naturally be governed by what is momentarily taking place in the furnace—repairs, charging, melting down, liquidus heating, deoxidation, slag tapping, slag formation, finish refining, or steel tapping. Data are available on the dust concentrations created by each of these phases, and on the total dust emission in the course of one melt, but unfortunately the figures published by various authors differ so substantially as to be of little practical use. One fact on which all the sources agree, and which has been corroborated by recent measurements, is that throughout most of the melt period the dust concentration remains below 1 g m_n^{-3}. Oxygen lancing, and especially oxygen blowing at the bath surface, can raise it to several times that figure; in fact, the latter procedure can produce concentrations of several dozen g m_n^{-3}. Table 34 lists the range of concentration levels most frequently encountered in the various phases of a melt, in open-hearth furnaces of the Siemens—Martin and März—Böhlen types. These figures apply to the outlet of the furnace itself; the concentrations at the foot of the chimney stack are always somewhat lower, because much of the dust settles in the heat recuperators and flue ducts.

TABLE 34

Dust concentrations in open-hearth furnace fumes (at the furnace outlet)

Furnace activity	Dust level ($g\ m_n^{-3}$)	Furnace activity	Dust level ($g\ m_n^{-3}$)
Furnace repairs	0.15 to 0.25	Oxygen blowing at bath surface	up to 35
Charging	0.2 to 0.8	End of melt	0.5 to 1.5
Melt-down of solid charge	0.3 to 0.8	Finish refining	0.6 to 4.5
Charging of hot metal	0.5 to 0.9	Average for whole melt	0.4 to 0.7
Oxygen-assisted melting	up to 7.5		

The total weight of dust escaping from an open-hearth furnace is roughly estimated as 1.5 to 2 per cent of the weight of steel tapped from it. That figure is not quite as alarming as it might seem, because part of the dust is retained in the slag compartments, recuperators and flue ducts. Even if no gas cleaning facilities are provided, the amount actually released into the atmosphere will range approximately from 0.8 to 1.2 per cent of the weight of the steel output. Even so, a 400-ton furnace will, in the course of a single eight-hour melt, foul the air with 3.2 to 4.8 tons (or an average of 0.4 to 0.6 tons per hour) of a very fine dust which tends to stain anything it touches.

Oddly enough, in view of the magnitude of the problems created by these dusts, very little is known about their grain size distributions. The one thing on which there is general agreement in the literature, and which has also been confirmed by recent investigations, is that nearly all the particles which reach the chimney stack are in the minus 5 micron range, and half of them or more are smaller than one micron. The explanation is that the coarser fractions are heavy enough to settle out before they reach the chimney. Oxygen lancing produces an even finer dust, with half the particles or more in the 0.2 to 0.4 microns interval. The specific gravity of the dust varies with the chemical composition of the individual particles, but usually lies between 3.5 and 4.5 $g\ cm_n^{-3}$, with the isolated particles ranging from 0.8 to 5.3 $g\ cm^{-3}$. The bulk density of these dusts is mostly within the 1800 to 2500 $kg\ m^{-3}$ bracket.

The technical problems involved in the cleaning of these fumes, from open-hearth furnaces in general and from oxygen-blown furnaces in particular, were long considered insuperable. This is no longer the case: these fumes can be cleaned, and in a good many steelworks are being cleaned. The problems nowadays are economic rather than technical. The volumes of gas to be handled are huge, and vary sharply in the course of each melt; so do the dust concentrations in the gas; and then, the dust is extremely fine. For a fuller grasp of these problems, just consider the requisite cleaning efficiency. The mean dust concentration, averaged over the duration of one melt, is usually only about 0.4 to 0.7 g m_n^{-3}. At that rate, an acceptable residual dust content at the chimney top, say 150 mg m_n^{-3}, could be achieved with an overall collecting efficiency no greater than some 70 to 80 per cent. Yet there are peak periods, such as the oxygen blowing interval, when the dust concentration rises to several, or even to several dozen, g m_n^{-3}. At these times, it will take a 97 to 99 per cent efficiency to keep the dust emission into the atmosphere within the same limit. To make things worse, the maximum dust concentration in the gases occurs at the times when the heat input to the furnace, and hence the volume of fumes generated, are also at their highest. To round it off, the gases normally leave the furnace at 500 to 700 °C, though this can be brought down to about 200 or 300 °C in a waste heat boiler. The equipment selectors are thus faced with an enormous volume of very hot gases, and naturally have to aim for the efficiency level required during the peak emission periods. Consequently, the equipment that will meet these requirements will always be very expensive.

With that much said, it is time to review the options that actually face the equipment selectors. At present, these are confined to dry electrostatic precipitators, high-temperature cloth filters, and Venturi scrubbers.

Electrostatic precipitators tend to be too large to fit into the usually limited spaces in the vicinity of existing furnaces. Moreover, the gas must always be cooled down in a waste heat boiler before it enters them. The alternative mode of cooling, by the induction of outside air, would enlarge the throughflow and thereby necessitate an even bigger and costlier precipitator. Furthermore, the gas emerging from this boiler must be humidified and cooled down to about 150 °C in a stabilizer before it is admitted to the precipitator. All this is a high price to pay for the roughly 99 per cent efficiency which a well-selected and properly operated electrostatic precipitator is expected to render.

The same sort of efficiency, around 99 per cent, can be obtained from cloth filters with glass fibre filter cloths. These units can keep the dust concentrations at their outlets down to less than 50 mg m^{-3}, but have to be supplemented with coolers effective enough to ensure that the gas temperature at the filter inlet will under no circumstances exceed the temperature rating of the cloths. However, the chief reason why these filters have not been more widely adopted for this particular duty is that an installation capable of handling such large throughflows would be almost prohibitively expensive. That applies not only to their first cost, but to the running and maintenance costs as well.

Venturi scrubbers are much cheaper than either electrostatic precipitators or cloth filters, and are equally capable of keeping the dust emission into the atmosphere within tolerable limits. The trouble is that during the oxygen blowing periods, they can do so only at the price of a pressure drop in the region of 7000 Pa, which is bound to show up in their running costs. Nevertheless, the number of

Venturi scrubbers fitted to open-hearth furnaces has recently been increasing. The economics of these units can be improved by employing rectangular-section Venturi tubes with variable throat widths, which can be adapted to the fluctuations of the flow rate and dust concentration. At small throughflows and dust concentrations, the constriction can be widened to reduce the pressure drop; when the load on the unit is heavy, the throat can be narrowed down to achieve the most intense cleaning effect possible. Since these scrubbers are almost invariably run on recirculated water, and since the gases contain a high proportion of sulphur dioxide, special precautions must be taken to protect the equipment from chemical corrosion. Either the process water must be dependably neutralized in the course of its recirculation, or else the scrubber, and probably its sludge handling ancillaries too, will have to be made of special corrosion-resistant materials.

6.2.4 Oxygen steelmaking convertors

Oxygen convertors, the most modern type of steelmaking equipment, nowadays range in size from small 5-ton units, used mainly for the production of high-alloy materials, to the large vessels for often 350 tons or more which form the principal equipment of up-to-date steelworks.

These units are not to be confused with the Bessemer and Thomas convertors that preceded them. There, air was blown in at the bottom of the vessel to rid the melt of carbon and silicon in the acid-lined Bessemer units, or of phosphorus in the basic-lined Thomas vessels. The rate at which air was blown in, and the rate at which fumes were generated, depended both on the capacity of the vessel and on the composition of the charge. The Bessemer process usually yielded about 400 m_n^3 of combustion products per ton of steel, the Thomas process could produce up to twice this volume. The whole of the Bessemer or Thomas cycle (i.e. the preparation and charging of the convertor, air blowing, sampling, final refining, and tapping) lasted roughly 40 minutes, with the air blowing phase taking up some 15 to 18 minutes of this total. The gases generated in these convertors were mostly diluted with atmospheric air, but initially consisted in the main of nitrogen, carbon monoxide, and carbon dioxide; the proportions of those three components varied in the course of the cycle. The rate at which the gases developed also fluctuated; at its peak, after 10 to 12 minutes of air blowing, it would commonly attain about 150 per cent of the average rate for the whole of the blowing period. The gases emerged from the vessel at 1100 to 1600 °C, but were quickly cooled down by the induction of sometimes as much as several times their own volume of outside air.

Even a quiet air blowing period would load the combustion gases with large amounts of solids, chiefly particles of slag, lime, iron oxides, and other metal oxides. Unstable running used to raise the solids emission rate to alarming proportions. True, much of the dust would adhere to the convertor outlet or settle in the exhaustor piping beyond it, as the particles were mostly in a plastic state, but even so the volume of fines released into the atmosphere was, by present-day standards, quite unacceptable. The dust concentration and grain size distribution figures quoted by various authors differ a good deal (which is to some extent attributable to the differing amounts of outside air drawn into the flow), but, very roughly, the undiluted gases were likely to contain some 25 to 50 g m_n^{-3} of solids. This means that 1 to 2 per cent of the entire charge was apt to escape into the atmosphere. The gas cleaning equipment and procedures used for these convertors were essentially similar to those employed for modern oxygen convertors, as described in the following paragraphs.

The term oxygen convertor (or basic oxygen furnace) is nowadays applied to units blown with oxygen-enriched air or pure oxygen, or, sometimes, a mixture of oxygen with superheated steam. Oxygen-enriched air can be blown into the vessel either at its bottom or from the side. Pure oxygen (which has to be at least 95 per cent pure, and in some cases up to 99 per cent pure) can be blown either into the bath or at its surface. The design in most widespread use at present is the LD or Linz–Donawitz convertor, where pure oxygen is lanced at the bath surface.

In these basic oxygen processes, lancing produces clouds of brown fumes which are composed predominantly of iron oxides. Recent investigations have proved that, once the decarburization stage is completed, the flame hovering above the vessel consists almost entirely of very small incandescent smoke particles borne in a stream of nitrogen. Electron microscopy has revealed that these particles are spheroidal, with a mean diameter of about 0.05 microns when air is blown, and roughly 0.1 microns when oxygen is blown into the vessel. They are made up prevalently of iron oxides, with a small proportion of manganese oxides. Any larger solids detected in this work, which sometimes were several microns across, were invariably extraneous to the brown smoke; they were mostly small CaO and slag particles entrained in the gas.

Closer examination of the processes by which the brown smoke is evolved has revealed varying degrees of oxidation of the iron oxides at various locations. Within the convertor, most of the particles are made up of FeO. After leaving the vessel, i.e. upon contact with atmospheric air, they oxidize further to Fe_3O_4 and then to Fe_2O_3. Electron microphotography has demonstrated that isolated particles occur only in the hot part of the flame. In the colder upper zone of the flame, and in the exhaustor piping or ducting, the primary particles tend to cluster into long branched chains. That is ascribed mainly to the magnetic effect of the ferromagnetic phases of Fe_3O_4 and Fe_2O_3 once these oxides have been cooled to below the Curie point.

The chief reason for the current widespread adoption of oxygen convertors is their sheer productivity: in these units, the decarburization process takes only 18 to 20 minutes. The convertors are usually set up in pairs, or three abreast, with one unit free for maintenance or repairs while the other two operate in staggered cycles: while one unit is melting, the other one is being tapped and prepared for the next melt. This pattern of operation simplifies both the materials handling arrangements and the oxygen supply, and also facilitates the utilization of the combustion products in a waste heat boiler by evening out their flow rate.

The gases leave the convertor at a temperature that rises from some 1200 to 1500 °C at the beginning of the blowing period to 1600 or 1800 °C at the end of that period. They contain mainly carbon monoxide, which forms about 50 per cent of their volume when blowing is commenced, up to 95 per cent after ten minutes of blowing, and something like 85 per cent at the end of the blowing interval. The remnant is chiefly carbon dioxide. When pure oxygen is being blown, the gases evolved during the first roughly seven minutes of blowing contain a certain surplus of oxygen, so that the carbon monoxide is largely burned up before it leaves the convertor. The next five minutes, approximately from the seventh to the twelfth minute of the blowing interval, are the critical time as regards the demands on the gas cleaning plant. During this period, both the rate at which gas is evolved and the carbon monoxide content of the gas reach their maxima. The flue gases are mixed with outside air and combusted, producing stack temperatures as high as 2400 °C (unless the temperature is brought down by a substantial excess of air, over and above the amount needed to burn the carbon monoxide).

The gases usually contain about 0.8 to 1.6 per cent of the weight of the charge in the form of a very fine dust, roughly 80 to 85 per cent of which are iron oxides. The latter break down into generally 60 to 70 per cent of Fe_2O_3 and around 15 per cent of FeO. In the uncombusted and undiluted gas, the dust concentration can run as high as 150 to 200 g m_n^{-3}. This drops to roughly 15 to 40 g m_n^{-3} in the combusted gas, the actual reduction depending on the conditions of the combustion process, and particularly on the amount of outside air with which the gases are diluted. The particles are mostly spherical or spheroidal; there is no general agreement in the literature on their size distribution, typical data quoted being for example:

a) 80 per cent of all particles in the minus 0.8 micron range and 20 per cent in the 0.1 to 0.3 micron bracket, with a substantial proportion of them smaller than 0.05 microns;

b) 15 per cent of the particles in the 1 to 15 micron bracket, 65 per cent between 0.5 and 1 micron in size, and the remaining 20 per cent smaller than 0.5 microns.

The gas cleaning ancillaries of these convertors have in recent years undergone a process of rapid development, thanks to which the technical problems involved can now be considered as essentially resolved. The cleaning techniques are commonly classified into dry and wet ones, but that refers only to the state in which the trapped dust is obtained, not to the procedure itself. In both cases, the first step is to bring the gases down from the 1600 to 1800 °C at which they leave the convertor to something between 250 and 500 °C. Since this cooling strongly affects the economics and efficacy of the particle separating process downstream of it, the first step in our review of cleaning techniques must needs be to examine the various alternative modes of cooling. In principle, the gas can be cooled by any of the following procedures:

a) By inducing a large surplus of air (with an air ratio λ up to 6) and the direct injection of water. This simplest of the alternatives was applied in conjunction with the large-diameter wet Venturi scrubbers which were the earliest type of gas cleaning equipment used at convertor steelmaking plants. It nowadays ranks as totally uneconomical; the gases were released into the atmosphere with a dew point of 73 °C, i.e. with a water vapour content in the region of 350 g kg^{-1}.

b) By a water-cooled stack, resembling a radiant heat boiler, in combination with an air ratio λ around 4 and with direct water injection. The large excess of air used for burning the carbon monoxide content is necessitated by the inefficient combustion system. This arrangement again failed to survive the early large-diameter Venturi scrubbers for which it was devised, and released the saturated gas into the atmosphere at much the same temperature as the system described in a) above. Still, since much less air is drawn into the gas flow, this system has to handle only about half the throughflow entailed by the larger air ratio in alternative a).

c) By a waste heat boiler, at an air ratio λ of 2 or less, which is the system in general use at present. This approach cuts down both the volume of gas that has to be handled and the water consumption, since the boiler cools the gas, indirectly, down to some 250 or 350 °C. Moreover, it is now just as reliable as conventional coal-fired boilers have come to be. The boiler can discharge the gas either directly into wet cleaning equipment such as Venturi scrubbers, or into dry electrostatic precipitators, though in this latter case the gas must first be moistened and cooled down to about 150 °C in a stabilizer.

A more recent trend is to process only the gases that actually arise in the convertors, without any admixture of outside air and any combustion of the gas upstream of the cleaning equipment. This technique offers substantial savings: for one thing, the separators receive a much smaller volume of much cooler gas, and therefore work out smaller and cheaper; for another, the gas remains available for further use as a fuel. Most of the work done on this technique follows either of the following two approaches:

a) To prevent the induction of outside air, between the convertor mouth and the intake opening of the exhaustor pipe, by enclosing this space within an annulus supplied with some inert gas.

b) To limit the induction of outside air to a small and strictly controlled amount, which is always much less than the theoretical amount of air needed to burn the carbon monoxide in the gases; usually, the air ratio is resticted to $\lambda = 0.3$.

In either case, the gases are first pre-cooled in a specially designed length of piping or ducting so as to reach the gas cleaning facilities at about 1000 to 1200 °C. When these facilities are Venturi scrubbers or wet electrostatic precipitators, the gas can then be cooled down to an acceptable inlet temperature in a simple water-spray cooler. Gases that are to be processed in dry electrostatic precipitators must first be cooled to roughly 250 °C in a specially adapted stabilizer. At present, oxygen steelmaking plants are fitted with dry electrostatic precipitators or wet Venturi scrubbers to the virtual exclusion of any other types of separators, and our further considerations will therefore be limited to those two types.

Dry electrostatic precipitators are eminently suitable for this duty, because both the gases generated in convertors (once cooled down and humidified) and the dust contents of those gases display exactly the physical properties needed for electrostatic precipitation processes. The dust is as a rule non-adhesive, and is easily removed by rapping. However, the high concentration levels, small particle sizes, and the high collecting efficiencies stipulated in these applications all call for a very low flow velocity in the working spaces of the precipitators. Usually, this velocity is confined to the 0.6 to 1.0 m s^{-1} range. Furthermore, the carbon monoxide must be combusted more or less completely before the gas is admitted into the precipitator, as the fibriform streams it forms in the gas flow could otherwise cause an explosion. A carbon monoxide detector must be placed in the inlet duct well upstream of the precipitator, and must be complemented by provisions for cutting off the high-voltage feed to the unit whenever the CO content of the gases exceeds the danger point. All this, however, is a small price to pay for the advantages which precipitators have to offer in this application: an outlet dust concentration that can be well below the 100 mg m_n^{-3} mark; a relatively low pressure drop across the separator; and a trapped dust which accumulates in a thoroughly dry state. On the other hand, precipitators are apt to be costly and bulky, need highly proficient attendants and maintenance staffs, and are better not even contemplated when the gases are liable to contain any significant residues of carbon monoxide.

Venturi scrubbers capable of releasing only reasonably clean gas into the atmosphere fall into two categories: conventional circular-section scrubbers of small diameters, and rectangular-section units with narrow throats. Venturi scrubbers do best on moist gases, and should therefore preferably have water sprays installed upstream of them; these also reduce the effective volume of the gas entering the unit, and hence the pressure drop needed for the scrubber to perform efficiently. Modern Venturi scrubbers with water sprays at their inlets can work effectively at pressure drops as low as 7000 to 3000 Pa, are cheap, compact, highly reliable, easy to operate and maintain, and just as efficient as electrostatic precipitators. Still, there is no arguing away the fact that the pressure losses they entail, and the sludge handling ancillaries they need, are often highly unwelcome complications. Venturi scrubbers are at their best when the carbon monoxide content of the gases is not combusted, so that the volume of gases reaching the scrubber is relatively small.

6.2.5 Twin-vessel (tandem) furnaces

Twin-vessel furnaces are a fairly recent innovation in the field of steelmaking equipment. Czechoslovakia was one of the first countries to adopt these furnaces on any scale, largely because they offer a relatively high productivity and superior heat utilization for the comparatively slight cost of converting pairs of existing open-hearth furnaces into tandem units. These consist of two Siemens–Martin or März–Böhlen furnaces which stand side by side, are interlinked by their gas ducting, and operate in staggered cycles designed to make the most of the heat generated in the course of each melt. In principle, the hot gases evolved while oxygen is blown into one vessel are led off to the other vessel, where they preheat the charge and lining, before they are drawn off. The two vessels thus function alternately. A melt in one vessel lasts only 40 to 50 minutes, as against eight hours in a conventional open-hearth furnace. Most of the existing tandem furnaces tap 50 to 200 tons of steel at a time.

In view of the intensive oxygen blowing, it is hardly surprising that these units should resemble oxygen convertors in emitting huge amounts of reddish-brown fumes. The solids in these fumes are mostly micron or sub-micron particles of iron oxides, mainly Fe_2O_3, and attain concentration levels approaching those common in the fumes of oxygen convertors. Short-term measurements on a 75-ton twin-vessel furnace have shown that the solids concentration varies from about 30 to 50 g m_n^{-3} at the beginning of the melt to roughly 20 g m_n^{-3} half-way through its duration, and drops to some 5 to 10 g m^{-3} in its concluding phase. The average over the whole melt period is commonly between 15 and 20 g m_n^{-3}. The average chemical composition of the combustion gases, as established by several sets of measurements, is set out in Table 35.

The combustion products from twin-vessel furnaces thus contain high concentrations of predominantly micron and sub-micron particles, which means that the equipment used to clean them must attain an efficiency in excess of 99 per cent in the trapping of even these fines. That narrows the choice of equipment essentially to dry or wet electrostatic precipitators or wet Venturi scrubbers. Since most of the heat of the gases is abstracted by the charge in the second hearth of the unit,

TABLE 35

Chemical composition of fumes emitted by twin-vessel furnaces (% by volume)

CO	1.06	N_2	62.5
CO_2	26.0	H_2O	5.0
O_2	5.33		

and the rest is largely dissipated by the induction of a certain amount of outside air, there can be no question of incorporating a waste heat boiler in the scheme. The gases usually reach the separator inlet at something between 200 and 500 °C, the exact temperature depending on the stage which the melt has reached, on the amount of oxygen lanced in, the degree of preheating of the charge in the second hearth, and the amount of outside air drawn into the stream. These same variables also govern the rate at which the fumes are evolved. On average, a 75-ton twin-vessel furnace will generate about 30,000 m_n^3 per hour, or roughly 400 m_n^3 an hour for every ton of steel it produces.

Dry or wet electrostatic precipitators cannot handle these gases unless coolers are installed upstream of them. Gas to be treated in dry precipitators must be humidified and cooled to the specified inlet temperature range around 150 °C in a stabilizer, but gas intended for wet precipitators can conveniently be cooled down either in spray towers or in bubble washers.

Venturi scrubbers, which are the prevalent type of gas cleaning equipment for twin-vessel furnaces, can attain efficiencies in excess of 99 per cent only if the fumes are adequately cooled by water sprays upstream of them. These can be located either within the separator housing, just ahead of the Venturi tube inlet, or else in separate coolers close to the separator entry port. An alternative to water sprays is a bubble washer upstream of the Venturi unit. Measurements at an installation made up of type MSA Venturi scrubbers have ascertained an overall efficiency which, at various stages of the melt, ranged from 99.1 to 99.7 per cent. They have also established that the efficiency depends to some extent on the dust concentration and on the phase and course of the melt in progress. A particularly gratifying finding was that the collecting efficiency actually grew with the dust concentration. This work further revealed that efficiencies higher than 99.5 per cent are feasible only at a total pressure drop, across the bubble washer as well as the Venturi scrubber itself, in the region of 9500 Pa; about 1000 Pa of this total is accounted for by the bubble washer.

6.2.6 Electric arc furnaces

Arc furnaces, mostly reserved for the production of high-grade steels, range in size from small units which tap only 0.1 ton of steel at a time to very large ones which turn out 180-ton melts. The fumes evolved in these furnaces are normally exhausted in either of two ways:

a) The entire furnace is enveloped by suction hoods which shroud the electrodes and the charging door too. The hoods must not interfere with the tilting of the furnace or the thermal expansion of the various components, and must be joined to the exhaustor piping in a way which will preserve a sufficient vacuum at all points where dust is generated.

b) The furnace lid is provided with a suction stub, to which the exhaustor piping is attached via a variable-section annular orifice, so that only the interior of the furnace is evacuated.

In either case, the system is expected to exhaust several distinct types of gases:

a) Fumes emerging from the melt, at a rate dependent on the oxygen blowing rate and on the size of the furnace. These are mostly formed by the combustion of the electrodes and of carbon in the charge materials, but the melting of oily scrap can generate inflammable hydrocarbons, and moist scrap will give off water vapour which can under certain circumstances dissociate and release hydrogen. Carbon monoxide and dioxide are formed chiefly during the oxygen blowing period. The carbon monoxide, emerging from the bath at 1700 to 1800 °C, is only partly combusted by the excess oxygen above the bath surface.

b) Outside air leaking into the furnace past the electrodes, and into the space between the furnace and the suction hoods. The amount of air thus induced depends largely on the design and state of the furnace shrouding, but in general grows with the furnace diameter.

c) Outside air which, in furnaces exhausted through suction stubs in their lids, is deliberately drawn in through the variable aperture between this stub and the exhaustor piping, so as to complete the combustion of the inflammable gases. The intake of this air must be carefully controlled so as to forestall any explosion hazards without upsetting the process conditions in the furnace.

Table 36 surveys the average amounts of gases thus exhausted from arc furnaces of various sizes, as quoted in the literature. It indicates that the quantities extracted from the interior of the furnace only are substantially smaller than those induced in the alternative arrangement, where the furnace is provided with external hoods. It is further evident that the volume that has to be exhausted per ton of steel is much lower in large than in small furnaces.

The dust content of these fumes consists mainly of iron oxides formed by sublimation and oxidation, and is for the most part extremely fine. Table 37 lists the data reported by various authors on the grain size distribution of dusts drawn

TABLE 36

Amounts of gases exhausted from electric arc furnaces

Furnace capacity (tons)	Extraction through external hoods: Total amount ($m_n^3 h^{-1}$)	Extraction through external hoods: Amount per ton of steel ($m_n^3 h^{-1}$)	Extraction through furnace lid: Total amount ($m_n^3 h^{-1}$)	Extraction through furnace lid: Amount per ton of steel ($m_n^3 h^{-1}$)
2.5	13,600	5400	—	—
4.5	25,500	5650	7100	1580
9	39,000	4300	—	—
11	—	—	32,000	2900
26	51,000	1950	—	—
45	61,500	1360	—	—
50	68,000	1360	—	—
67	—	—	47,500	700
75	73,000	970	—	—
80	—	—	65,000	810
91	—	—	14,000	154
115	76,500	660	76,500	660
185	—	—	24,000	130

from arc furnaces which were not blown with oxygen. Oxygen blowing produces a dust where some 85 per cent of the particles are smaller than 4 microns. A charge which contains silicon, calcium or magnesium additions emits a very fine white smoke. Greasy or oily scrap (such as swarf) in the charge will generate an oil mist and a variety of hydrocarbons, which precludes the use of cloth filters in the gas cleaning plant.

TABLE 37

Grain size distribution of dusts exhausted from arc furnaces not blown with oxygen

Particle size *a* (microns)	Retained proportion *Z* (%) in measurements No.: No. 1	No. 2	No. 3
< 5	18 to 80	67.9 to 72.5	71.5
5 to 10		6.8 to 11.5	8.3
10 to 20	7 to 71	3.0 to 9.8	6.0
20 to 44		7.0 to 9.0	7.5
> 44	7 to 16	6.0 to 6.5	6.3

TABLE 38

Chemical composition of dust exhausted from arc furnaces (% by weight)

Fe_2O_3	35 to 42	SiO_2	2 to 10
FeO	4 to 8	MgO	2 to 8
CaO	6 to 15	C	2 to 8
Al_2O_3	3 to 13		

Table 38 presents the normal range of chemical compositions of this dust, as acertained by several different authors. Apart from the components listed in this Table, the dust is also apt to contain traces of tin, zinc, chromium, copper, and various other elements. A charge where the scrap is made up e.g. of bright strip or galvanized sheet trimmings can raise the contents of these metals, which are otherwise present in trace amounts only, to quite substantial proportions. The dust concentration is largely dependent on the rate at which the fumes are evacuated, but the average dust levels are commonly within the following limits:

a) In fumes drawn off through external hoods:
   when no oxygen is blown in   0.4 to 0.9 g m_n^{-3}
   when oxygen is being blown   2.3 to 5.3 g m_n^{-3}

b) In fumes extracted from the furnace interior:
   when no oxygen is blown in   1.7 to 2 g m_n^{-3}
   when oxygen is being blown   7.5 to 15 g m_n^{-3}

The former alternative, of inducing the fumes through external hoods at the chargings doors and electrode bushings, has been widely accepted for arc furnaces in the 15 to 25 ton bracket. At these smaller units, the total throughflow to be handled remains manageable even though this system necessarily entails a much larger intake than is the case when the fumes are extracted directly from the furnace interior. This latter mode of extraction now predominates in furnaces from about the 20-ton size upwards, in either of two versions:

a) The extraction stub is offset from the crown of the furnace lid, so that the gases which evolve during the melt keep the furnace atmosphere at a slight gauge pressure. Consequently, though most of the dust-laden gas is drawn off through this stub, a proportion of it always leaks out past the electrodes and around the charging doors. This leakage is most pronounced during the oxygen blowing period, when the electrodes are retracted until their tapering tips enter the bushings; that leaves relatively wide gaps around the electrodes open for the escape of gases. Another drawback of this design is that it creates a reducing atmosphere in the furnace. As a result, the brown fumes arising by sublimation of the iron then contain a *pyroforic dust* of pure iron or FeO particles less than 0.1 microns in size. The properties of this dust, along with the uncombusted remnants of carbon monoxide and/or hydrogen in the gases, create a serious explosion hazard, especially when the charge includes some oily or greasy scrap.

b) The more recent alternative approach is to maintain a slight vacuum within the furnace. That prevents any escape of fumes past the electrodes and charging doors, and thus makes for a relatively clean atmosphere within the furnace bay. It also makes for more complete combustion of the carbon monoxide and hydrogen in the course of the melt, and especially after the oxygen blowing period. The combustion of these gases within the furnace reduces the explosion risk in the exhaustor pipings, and also yields some useful heat which is transferred to the charge. The suction pressure required to prevent any leakage of fumes through the gaps and crevices, even during the oxygen lancing interval, ranges between 2.5 and 25 Pa. The vacuum actually maintained in these furnaces is usually about 5 to 10 Pa, as measured in the quiet zone at the top of the furnace vessel diametrally opposite the suction stub. To keep this vacuum constant, and for effective control of the extraction rate, the exhaustor system must be fitted with automatic controls.

Smaller arc furnaces which are not provided with oxygen lancing facilities, and from which the fumes are evacuated by means of external hoods, generally need no gas cleaning ancillaries. The fumes are merely exhausted through a short stack straight into the atmosphere. The large furnaces, where the fumes are extracted through a stub in the lid, and all oxygen-blown furnaces in general, do have to be supplemented with gas cleaning plant. The equipment currently employed to trap the exceedingly fine dusts emitted by these furnaces ranges from highly efficient Venturi scrubbers to cloth filters, and to both wet and dry electrostatic precipitators. No matter what type is installed, the fumes, which emerge from the suction stub at 1400 to 1600 °C, must first be cooled. The cooling requirements are dictated by the following considerations:

a) Cloth filters are restricted to the service temperatures which their cloths can withstand; for continuous duty, most synthetic fabrics are limited to temperatures between 130 and 180 °C, and fibreglass materials to 280 °C at the most.

b) Wet Venturi scrubbers will perform with the requisite efficiency only when the fumes fed into them are moistened and cooled (to prevent the evaporation of the finest droplets), preferably right down to their dew point.

c) Dry electrostatic precipitators do best when the gases are supplied to them in a humid state and at temperatures around 150 °C.

The cooling equipment must be selected to suit the parameters of the raw fumes and the other specific local conditions. Essentially, the options are as follows:

a) A long piping or ducting, the simplest and cheapest cooling device there is, offers the extra advantages of reliability and low maintenance requirements, and does not increase the volume of the fumes by the induction of outside air. Unfortunately, for the flow rates and temperatures in question, the piping would often work out impracticably long. In the limited space available at most steelworks, such pipings are often strung out along the walls of the hall, where they are out of the way.

b) A tubular heat exchanger, with forced air or water cooling, can also act as a preliminary dust arrester. The fumes pass through a cluster of upright tubes, and the coarser particles settle out into a receptacle at the foot of the unit.

c) Heat recuperators, with refractory linings and cold-air fans, afford a high cooling efficiency, but tend to be costly. These units generally consist of two chambers filled with refractory brickwork, which operate in an alternating heating and cooling cycle. While hot gas is passed through one chamber, to impart its heat to the refractories, the other chamber of the pair is cooled with fan-blown air. The chambers should be switched from one function to the other not at fixed intervals, but rather in dependence on the gas temperature at the outlet of the hot chamber. This mode of operation can keep the gas temperature e.g. at the inlets of cloth filters within a fairly narrow interval.

d) A stabilizer, as described in Section 3.3.1, serves to humidity and cool the gas by the injection of as much finely atomized water as can be fully evaporated inside the unit. The water injection rate must be regulated in response to the gas inlet temperature. Stabilizers are preferred especially upstream of dry electrostatic precipitators.

The cloth filters serving arc furnaces can be fitted either with lubricated fibreglass filter cloths, capable of sustained service at up to 280 °C, or else with synthetic fabric cloths. There are now many rival types of cloth filters intended for this application on the market, all of them aiming for the highest feasible long-term operating temperature rating. Glass fibre fabrics offer the highest temperature resistance, but are highly susceptible to attack by fluorine, hydrogen fluoride, and phosphorus. That rules out their use at plants where fluorspar is added to the melt in amounts exceeding about 0.5 kg per ton of the charge. A modern and well maintained cloth filter can usually keep the dust concentration in the gases leaving it down to something like 50 mg m_n^{-3}. Since the pollution limits imposed by current legislation mostly range between 100 and 150 mg m_n^{-3}, this affords an ample reserve for the inevitable gradual deterioration of the filter performance between overhauls.

High-efficiency Venturi scrubbers, which nowadays form the bulk of all gas cleaning equipment at electric steelworks, employ either small-diameter Venturi tubes up to about 120 mm across, or small rectangular-section Venturi tubes where the flow channel tapers down to a narrow slit. The units in use at present can clean arc-furnace fumes down to a residual dust content around 100 mg m_n^{-3}, or even less, but only at the cost of a pressure drop which typically ranges between 5000 and 8000 Pa.

Electrostatic precipitators, both wet and dry, can do equally well if the gas is suitably humidified and cooled upstream of them, and can do so for a very much lower pressure drop. However, these devices are so costly and bulky that lack of space alone often precludes their installation at arc-furnace plants. What precipitators have so far been utilized for this duty have usually been fitted only to the largest of arc furnaces.

6.2.7 Coking plants

The coking plants which feed blast furnaces with coke form an integral part of most ironworks. As far as air pollution is concerned, they are usually by far their most objectionable part. These plants cause intensive pollution especially of the lower layers of the atmosphere in their vicinity; emit a variety of toxic, aggressive or otherwise harmful substances; and present a host of major technical obstacles to any attempt to control their pollutant emission.

Present-day coke-oven batteries mostly comprise some 40 to 100 ovens arranged side by side. Each oven or chamber, typically about 10 to 16 metres long, 0.4 to 0.5 metres wide, and 3.5 to 8 metres high, is provided with doors at its front and rear ends. The side walls, which heat the chamber, house a system of channels in which gas is burned in the presence of preheated air. At each side of every chamber, the oven top is pierced by a hole through which the gas generated in the oven is led off, up an ascension pipe and through a flap valve, to the collector main that serves the whole battery. The gas is then piped to a by-product plant where various chemicals are extracted from it. The ovens can be charged in either of two ways:

a) With a rammed charge of coal, forced in through the pusher-side door, or

b) By gravity, in which case the oven top, about 1 metre thick, must be provided with four to five holes to accept the coal metered in by the charging lorry.

The ovens are charged with a suitable blend of stone coals having a specified size distribution (usually in the minus 10 mm range) and a moisture content of mostly 8 to 10 per cent. This charge, sealed off from the outside air, is then heated by the oven walls, which are kept at 950 to 1100 °C. As the heat gradually passes from the outer surfaces towards the interior of the charge, the moisture evaporates, and next, at 350 to 420 °C, the coal becomes plastic. In this state it exudes tar, so that pools of tar begin to permeate the charge from both its flanks and gradually spread towards its core. At about 450 °C the coal solidifies again in the form of semi-coke, which is then turned into coke by further heating. This process is concluded when the tar pools advancing from both sides meet in the middle of the charge, at which time all the coke in the oven is at a uniform temperature around 950 °C. It must be borne in mind, however, that the coking process, and hence the generation of vapours and gases, actually begins during the charging procedure, as soon as the charge contacts the hot walls and bottom of the oven. The finished coke is forced out of the oven by a pusher lorry, and transported to a spray tower for quenching. Once cooled down, it is dumped on a ramp, ready for conveying to the coke yard.

Nearly all the dust generated by coking plants arises at the coke-oven batteries. Most of it escapes during the charging and discharging operations, i.e. at times when the oven doors or the charging holes in the oven top are open. A rammed charge is inserted into the oven with the pusher-side door and the ascension pipe open, so the gas that previously occupied the oven space escapes through this pipe and past the entering charge material. It is joined by the volatiles which form as the charge contacts the hot oven walls and bottom – water vapour, tar mist, and a variety of toxic or evil-smelling gases like N, CO, CO_2, HCN, SO_2 and H_2S. The escaping gases entrain large amounts of coal dust and soot particles, forming a noxious smoke cloud with a high solids concentration. For gravity charging, the charging-hole lids must be removed, and the same mixture of dust-laden gases then emerges from these holes, through the ascension pipes, and leaks around the leveller bars.

The quantities of fumes thus released depend primarily on the oven size and charging time. The proportions of coal dust and soot, of water vapour, tar mist and other distillation products are governed by the properties of the coal, the oven wall temperature, and the design of the charging machinery. The amounts involved will probably never be known with any degree of accuracy, as measurements are precluded by the speed at which the whole pattern changes and by the rapid fouling of the instrumentation. At a rough estimate, a gravity-charged oven of 18 tons' capacity will, during its 3-minute charging cycle, emit about 60 m^3 of gases. This includes the 24 m^3 of gas at 1000 °C, corresponding to approximately 5 m^3 at 20 °C, which filled the empty chamber when charging was commenced; this gas consists, very roughly, of 15 % of carbon dioxide, 10 % of carbon monoxide, 75 % of nitrogen, and 1 % of oxygen.

The gases displaced during the charging process will inevitably entrain a lot of coal dust, the amount of which depends on the grain size distribution and moisture content of the coal, but can easily run up to 20 g m^{-3}. At that rate, the 60 m^3 escaping from our 18-ton oven during the three-minute charging interval will carry about 1 kg of dust. Moreover, the 8 to 10 per cent moisture content of the coal implies that the same oven will evolve some 7 kg of water vapour. Tar mist is present in the gases at an average of up to 5 g m^{-3}, so that the same oven will also release 0.3 kg of finely dispersed tar droplets.

While too many diverse factors enter the calculation for any hard and fast rules to be established, gravity charging mostly produces less pollution than rammed charging. This is due partly to the differences between the coals processed by each of these techniques, partly to the fact that gravity charging is as a rule faster.

There are several simple ways of reducing the gas and dust emissions during the charging period. The options are to increase the moisture content of the coals; to reduce their contents of ash matter; to eliminate the fines and/or screen out the smallest fractions; and to blend the coal mix to a high degree of homogeneity. Unfortunately, these are all palliatives rather than remedies. The only radical solution is to trap and exhaust the fumes as they leave the oven, and then process them so as to render them more or less harmless. This can be done by any of three principal techniques, which differ greatly in their degrees of effectiveness:

a) The exhausted gas can be dumped into the raw gas leaving the oven. Easily the most common way of doing so is to inject steam into the collector main. That generates enough suction pressure to draw the gases released during the charging period up through the ascension pipes into the collector main. This scheme has the advantage that these gases are made available for further utilization, but the drawback that they tend to degrade the by-products which are subsequently gained from the raw gas, especially the tar. Moreover, this admixture is liable to foul up the coolers and fans, and to impair the quality of the coke-oven gas, which is normally exploited as a fuel. Besides, the vacuum in the collector main is generally not sufficient to prevent the escape of a large proportion of the fumes into the atmosphere. This is not so much of a problem in gravity-charged batteries, where only the relatively small charging holes are open, but turns into a serious shortcoming when rammed charges are inserted through large oven doors. Finally, this system has in practice proved far from reliable: the fumes generated during the charging period are apt to clog up all the fittings and nozzles, and the steam pressure is difficult to control with the sort of accuracy which is needed if the system is to work as intended.

b) The gases can be burned. Generally, they are drawn off through openings in the oven tops by a special travelling exhaustor, and then ignited. That disposes of their inflammable constituents, but fails to reduce the overall pollution level to anything like the limits laid down in present-day legislation.

c) At gravity-charged batteries, the gases exhausted from the charging holes can be cleaned in a wet scrubber. A well-designed exhauster system can be so efficient that this has become the prevalent approach to the pollution problems arising in the charging of such batteries.

The usual practice is to exhaust the fumes emitted during the charging period through telescopic suction tubes that envelop the filler pipes. As the bottom end of the filler pipe is lowered into the charging hole, the telescopic sheath around it is extended to seat against the rim of this hole. The fumes drawn off through it are mixed with outside air and ignited, to burn their tarry constituents and combustible gases. A fan then draws the combustion products through a wet scrubber, generally of the Venturi type. The latter should be designed to operate on water with a high solids content, so as to keep the sludge handling and water recirculation system as compact as possible. Preferably, the whole system should be confined to a gravity-fed reservoir underneath the scrubber, and a pump to recirculate the water from this reservoir back to the scrubber.

The other main pollution source, apart from the charging process, is the discharging of the finished coke. The amount of fumes generated at this stage depends largely on the degree of carbonization achieved in the coal, i.e. on the degree to which the coke is of uniform quality. Since there is no hope of producing homogeneous high-grade coke unless the heating walls of the chamber are in mint condition, one of the chief (though often overlooked) contributory causes of excessive air pollution in the coke pushing phase is worn or damaged oven masonry. A poor state of the refractories is reflected in an uneven or insufficient carbonization of the charge; and that in turn greatly increases the emission of soot, carbohydrates and other volatiles, and the entrainment of coke dust. The exhaustor hoods and shrouds that have been fitted to the quenching cars or coke guides have up to now never proved capable of capturing anything like an adequate proportion of the fumes released at this stage.

Further fumes are generated as the coke is being quenched. These are best combatted at their source, by improvements in the design of the quenching facilities. For instance, the perforated tubes that were formerly used to sprinkle water over the coke have now given way to spray nozzles. These make for a more uniform wetting of the coke, form dense sheets of droplets which trap at least some of the solids, and also complete the quenching cycle faster, thus leaving less time for the emission of pollutants. Many quenching towers have also been fitted with wooden grids inside them, which are sometimes sprinkled or sprayed by further nozzles. The grids retain a proportion of the solid particles, by exploiting their inertia as the steam flow is deflected around the tower beams. Where such internal fixtures are desirable, however, they should preferably be incorporated at the design stage: when fitted as an afterthought to towers not designed to hold them, they can damp the updraught within the tower to the point where they may become a liability rather than an asset.

In these days of ever more stringent mandatory requirements, however, the only way to keep the pollutant emissions at the coke side of the battery within the legal limits is to alter the entire coke pushing and quenching procedure. A promising technique, now undergoing trials abroad, is to quench the coke in a rotating cylindrical vessel which travels on rails alongside the battery, and which is brought up against the oven to be emptied so that the face of the vessel almost touches the oven surface. The fumes entering the vessel, and those generated within it, are exhausted into a high-efficiency wet scrubber. This scrubber, usually a Venturi unit, and the fan that serves it, are again railborne, and are coupled up to the quenching car.

Another way of restricting the pollutant emission is by the dry quenching process, in which the heat of the coke is exploited for steam raising. So far, however, no practicable way has been found to overcome the secondary dust generation as the coke is dumped into the cooler, and in the subsequent handling of the dry coke after its quenching is completed. This secondary dust is liable to defeat the whole purpose of the exercise, as far as air pollution control is concerned; unless means are found of suppressing it at its source, dry quenching is not likely to make a coking plant any more acceptable as a next-door neighbour.

Altogether, then, we are still far from any really satisfactory solution of the air pollution problems created by the coke quenching process. Much the same applies to the dust that is raised as the quenched coke is screened and conveyed to its destination. The screens and the transfer chutes of the conveyor system must be provided with dust exhausting equipment; but coke dust is so adhesive that both the exhaustors and the separators to which they deliver are inclined to clog up very rapidly. That means the suction pipes must be kept short. It also means that the cyclones still sometimes used for this purpose can never reach the levels of efficiency and dependability which this application demands. Recently, they have been giving way to wet scrubbers or washers with sludge raking ancillaries, such as the MHB type described in Section 3.2.4.

6.3 Non-ferrous metallurgy

6.3.1 Aluminium production

The main raw material for the production of aluminium is bauxite. This complex mineral is made up of aluminium oxides, some of which are associated with bound water, and of oxides of iron, titanium, and silicon, as well as smaller amounts of other substances. It is turned into aluminium in a two-stage process: the bauxite is converted into Al_2O_3, and this oxide is then reduced to metallic aluminium electrolytically. Sometimes the two operations are performed at different plants.

The first steps are to crush the bauxite, grind it down to a powder, and dry it. The drying can be done either in rotary kilns right after the crushing, i.e. before the comminuted ore is ground down, or else in the course of the grinding process itself. All three of these operations generate so much dust that the equipment performing them must be fitted with exhaustor systems. Depending on the type of process equipment and the exhausting arrangements, the air evacuated from these plants can contain anything from 5 to 30 g m_n^{-3} of dust. The latter is mostly trapped in cloth filters; when the concentration levels are extremely high, it may be advisable to install cyclones upstream of these filters. The captured dust is generally recycled into the process. In cloth filters located downstream of the drying equipment, the gas temperature must be kept above the dew point to prevent the condensation of moisture; therefore the feed piping, and sometimes the filters too, are commonly provided with thermal insulation. These filter installations can attain efficiencies in excess of 99 per cent, and can keep the dust concentration emitted to the atmosphere down to the 100 to 200 mg m_n^{-3} interval.

Low-grade bauxites, with yields lower than some 30 per cent, as a rule undergo beneficiation treatment before they are ground down for further processing. This treatment normally takes the form of sintering, and produces gases which leave the equipment at 200 to 400 °C, with solids contents of 20 to 80 g m_n^{-3}. These gases are usually cleaned in horizontal electrostatic precipitators with bar- or rod-shaped electrodes; sometimes cyclones are mounted ahead of the precipitators to reduce the load on them. As long as the gas velocity within the precipitators is kept down to no more than 0.8 m s^{-1}, these units can capture 94 to 98 per cent of the dust the gases carry.

The ground bauxite is next processed chemically to yield aluminium hydroxide, $Al(OH)_3$, which collects in vacuum filters. Calcining in gas- or oil-fired rotary furnaces, at roughly 1300 °C, then converts this hydroxide into aluminium oxide, Al_2O_3. These rotary furnaces are a further source of pollution. The fumes exhausted from them contain typically some 20 to 30 g m_n^{-3} of dust, the exact concentration depending on the fuel, the combustion system, the calcination process and the structure of the product, and on the layout and performance of the exhaustor system. The fumes, which reach the separators at roughly 170 to 250 °C, are usually cleaned in dry electrostatic precipitators, either horizontal or vertical. As the precipitators are in this case required to work with an efficiency level around 99 per cent, the gas velocity within them must be restricted to about 0.4 or 0.6 m s^{-1}. To reduce the load on these precipitators, large-capacity cyclones or cyclone batteries are often installed ahead of them. All the trapped dust is recycled into the process. Apart from the dust, the fumes also contain some sulphur dioxide; when the rotary furnaces are fired with high-sulphur fuel, this content can be quite substantial. So far, however, next to no attempt has been made to trap or neutralize it. The usual device is to disperse the sulphur dioxide over a large area by means of tall chimneys. How much longer the legislators will tolerate this practice remains to be seen.

The second phase of the aluminium production process is the electrolytic reduction of the aluminium oxide into the pure metal. The oxide is molten down at about 950 °C in refractory-lined baths, where the aluminium is shielded against atmospheric oxidation by a supernatant layer consisting of a molten mixture of salts. The chief constituent of this layer is cryolite (Na_3AlF_6). The refractory lining is topped with a rammed layer of a carbon-based substance which serves to retain the shielding layer, and which also acts as the cathode, having ferrous conductors embedded in it. The anode, made of coke grains bonded with chemical binders, is lowered from above into the layer of molten cryolite and aluminium oxide that floats on the surface as the metal collects at the bottom of the bath. The supernatant layer, at roughly 950 °C, has a specific gravity of about 2.15 g cm^{-3}, while the metallic aluminium, which melts at 659 °C, has a specific gravity around 2.35 g cm^{-3}. What actually happens in these baths is that the direct current decomposes the oxide into aluminium and oxygen. The former sinks to the bottom, while the latter combines with the carbon of the anode to form carbon dioxide (or, if the process is not properly controlled, some carbon monoxide as well). Some of the cryolite inevitably decomposes too, and releases gases which contain fluorine.

The various types of electrolytic equipment that have been used for this process in the past are now beginning to give way to fully enclosed furnaces with tilting lids or covers. The primary gases arising in the process are exhausted through tubes set into the furnace casings, or through annular ducts surrounding the electrodes. The admixture content of the fumes is commonly reduced by burning the tarry constituents, along with the carbon monoxide, in burners mounted directly on the furnace, before the rest of the primary gases is drawn off to the absorption and particle trapping equipment. Unfortunately, none of the furnace or bath hooding layouts so far devised can preclude the escape of secondary fumes when the lids or covers are opened, e.g. for the charging of further aluminium oxide or cryolite. As the halls of aluminium plants often house arrays of up to several hundred tanks or furnaces, quite substantial amounts of gases and dust are thus released into the halls, and exhausted from them to contaminate the air in the vicinity of the plant.

TABLE 39

Data on gases and dusts generated in the electrolytic production of aluminium

Parameter	Unit	Primary gases from separators/absorbers	Secondary gases exhausted from furnace hall
Amount of gases (including false air)	$m_n^3 h^{-1}$	100 to 500	35,000*
Gas temperature	°C	100 to 200	35 to 45
Dust concentration	g m_n^{-3}	1 to 5	0.015
Aproximate dust analysis:			
Al_2O_3 content	%	40	
Na_3AlF_6 content	%	10	
Inflammables content	%	50	
Fluorine content of gases	g m_n^{-3}	0.5 to 2.5	0.001 to 0.002

*Per furnace, if the air in the hall is changed 30 times an hour.

Table 39 reviews the data so far available on both the primary fumes, which are drawn off from the tanks or furnaces into various absorption and particle trapping devices, and the secondary fumes, which are exhausted from the furnace halls and released straight into the atmosphere.

The dust content of the primary fumes could perfectly well be trapped in dry separating equipment. Cyclones could be expected to attain efficiencies as high as 90 per cent in this application, and dry electrostatic precipitators could keep the final solids content in the gases down to a mere 0.1 g m_n^{-3}. However, the gases also have to be divested of the fluorine contained in them in the form of hydrogen fluoride. That is why wet separating equipment is used almost invariably for the first stage, and more often than not for the final cleaning stage too. The most common types for this duty are spray towers where the gas flow is opposed to the droplet spray, and bubble washers, although some other types of scrubbers are employed in this role too. Hydrogen fluoride is readily soluble, and easy to absorb in the washing or scrubbing fluid, no matter whether this is ordinary water or a sodium solution. The wet scrubbing and absorption equipment now being installed at these plants can meet all normal requirements on its particle trapping efficiency. Only when particularly stringent demands are imposed on the cleanliness of the gases which are finally released into the atmosphere may it become necessary to add electrostatic precipitators downstream of this equipment.

The secondary gases drawn off from the furnace halls are much more of a problem, mainly because of the huge volumes involved. In the face of these quantities, most aluminium plants simply abandon all thoughts of cleaning those gases. Occasionally, when enough low-cost water is available, scrubbing equipment is mounted at or near roof level, but by and large these gases and their dust contents tend to be ignored. How much longer the public and the legislators will be prepared to tolerate this state, and what will happen when they clamp down on it, remains to be seen.

6.3.2 Lead production

The main materials from which lead is made are galena, lead oxides, wastes generated at lead-processing plants, and lead scrap. Pyrites, mostly fine-grained concentrates with average lead contents around 60 per cent, must be converted into oxides prior to their further processing. Lead ores usually contain traces of copper, zinc, arsenic, tin, noble metals, and other admixtures.

The first step in the production process is generally the roasting of pyrite ores, with various additions, to turn them into larger lumps of sinter. This exothermic process can be conducted in one or in several consecutive stages, but is always at least roughly described by the formula $2\,PbS + 3\,O_2 \rightarrow 2\,PbO + 2\,SO_2$.

The oxides are then reduced in shaft furnaces, where the other metallic admixtures are mostly trapped in the slag and can subsequently be recovered from it. Like the roasting process, this reduction again produces large amounts of dust-laden gases which have to be exhausted and cleaned. The dust separated from the shaft-furnace fumes is mixed with the dust trapped in the roasting stage, and usually recycled.

The lead gained in the shaft furnaces needs further refining, which is mostly done by remelting, as only some kinds of lead are capable of electrolytic refining. The refining process yields large amounts of intermediate products, which also call for further processing to exploit their valuable contents of metals. Altogether, then, the production of lead by this process, and the points where dust is generated in its course, can be described by the following scheme:

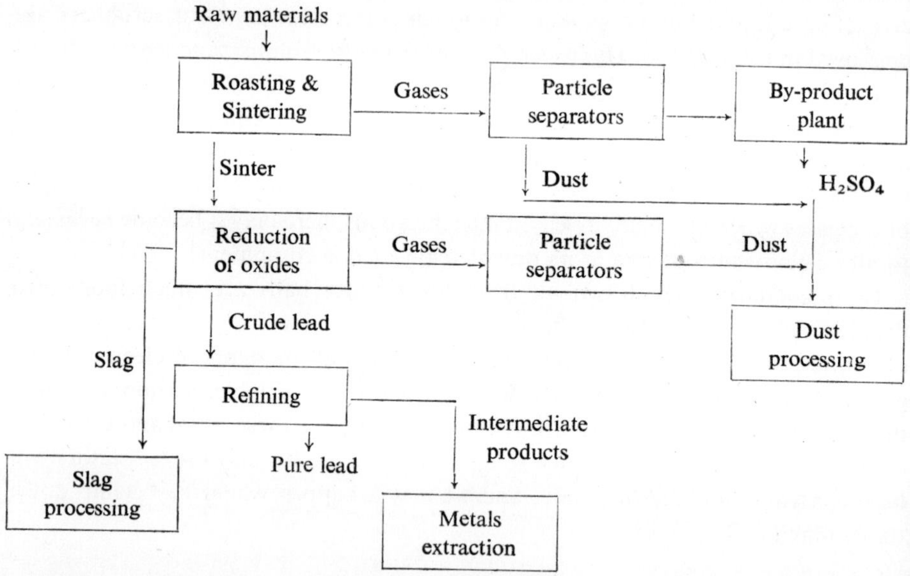

The gases that arise in the roasting and sintering processes contain some 1.5 to 5 per cent by volume of sulphur dioxide, and large quantities of dust. The dust concentration depends mainly on the type and grain size distribution of the charge materials and on the amount of combustion products exhausted from the process, but mostly lies within the 2 to 15 g m_n^{-3} range. This dust contains about 40 to 65 per cent of lead, 10 to 20 per cent of zinc, 8 to 12 per cent of sulphur, and other components which vary with the composition of the charge. It is made up of two distinct kinds of particles: one are the particles of ore, coke and other charge materials entrained by the gas flow, the other are condensed droplets of metals which evaporated during the process. The distinction is important, because the former particles are generally larger than 5 microns in size, while the condensed metal specks are very much smaller. Normally, this process yields some 3 to 3.5 m_n^3 of fumes per kg of sinter, at a temperature around 300 °C.

These fumes are commonly cleaned either in electrostatic precipitators or, after some preliminary cooling (e.g. in long piping runs), in high-temperature cloth filters. The precipitators employed for this purpose are mostly horizontal units, with collecting electrodes made up of corrugated or flat sheets or of bars, and attain efficiencies as high as 99 per cent. Sometimes, cyclones are used as a coarse pre-cleaning stage ahead of them. In view of their high sulphur dioxide contents, the cleaned gases are as a rule led off to a by-product plant, which turns out sulphuric acid.

The fumes drawn off from the shaft furnaces, at 100 to 200 °C, again contain the same two basic types of solids – particles abstracted from the charge materials, and particles formed by the condensation of metal vapours. The dust concentration is governed by the volume of fumes exhausted, their temperature, the composition of the charge, and the exact process conditions, but is usually inside the 5 to 15 g m_n^{-3} interval. The chemical composition of this dust fluctuates a great deal, as the proportions of pyrite, oxide and coke particles vary even in the course of one cycle, and so does the amount of fumes released, which can be anywhere between 15 and 50 m_n^3 per kg of coke charged into the furnace.

The shaft-furnace fumes are likewise cleaned in electrostatic precipitators or cloth filters, but are mostly of no further use, so the cleaned gases are released into the atmosphere. The chimney height must be selected with due regard to the sulphur dioxide content of these gases, which depends on the compositions of the fuel and charge materials, but as a rule ranges within much the same limits as in the combustion products from normal coal-fired boilers.

6.3.3 Ferrosilicon production

Ferrosilicon is produced in electric arc furnaces which, in view of the peculiar requirements of this process, cannot be fully enclosed. Consequently, the amounts of gases exhausted from these units are enormous. True, some fully enclosed furnaces with automatic charging provisions have recently been installed for the production of low-grade ferrosilicon, with useful contents of up to 75 per cent, but the bulk of the output still comes from the conventional open arc furnaces.

These units fortunately generate next to no gaseous pollutants. Carbon monoxide is present only in trace amounts, and the sulphur dioxide concentration, though dependent on the sulphur content of the coke in the charge, is only seldom large enough to present any problems. Besides, the gases evolving from the charge form no more than some 2 or 3 per cent of the total volume of fumes exhausted from these furnaces. This total volume thus consists essentially of dust-laden air, with a chemical composition not much different from that of the atmosphere around us. Its temperature, however, can be anywhere between 100 and 300 °C, depending on the configuration and effectiveness of the suction hoods and on the amount of fumes drawn off per kVA of furnace input. This latter figure is mostly in the vicinity of 8 to 10 m_n^3 an hour per kVA.

The dust content of these fumes consists of several components:

a) Dust entrained by the gas that passes through the layer of raw materials on the surface of the bath. This dust forms a mere 10 to 30 per cent of the total, and consists of relatively coarse particles, mostly of the plus 5 or plus 10 micron fractions. Some of these particles are present in the charge materials at the time of their charging, but others arise by the thermally induced decomposition of the charge.

b) Lower oxides arising by the chemical reaction of the quartzite in the charge with carbon. Normally, this reaction yields silicon, which collects in the bath, and carbon monoxide: $SiO_2 + 2\,C \rightarrow Si + 2\,CO$. However, wrong dosing or inadequate mixing can leave the charge materials too short of carbon for the quartzite to be reduced completely, in which case the reaction is modified into $SiO_2 + C \rightarrow$ $\rightarrow SiO + CO$. The gaseous SiO is relatively stable, at the high temperatures at which it forms, and escapes from the bath to oxidize upon contact with air into SiO_2. The latter then condenses in spherical particles, which may be several hundredths to several tenths of a micron across. The incidence of these lower oxides depends chiefly upon the nature of the charge materials.

c) Particles formed by the evaporation of the charge components. We must count with the evaporation of all those components which have boiling points lower than the bath temperature, and, depending on their vapour tensions, of some parts of the other constituents as well. The amount of metal thus volatilized varies strongly in the course of each cycle. The bath itself, at some 1900 to 2000 °C, is not hot enough to cause the sublimation of most metals, but the temperature of the electric arc is estimated at 3000 to 3500 °C, which is ample for this purpose. When a fresh charge is being melted down, the arc impinges directly on the solid charge materials; in view of their poor thermal conductivity, their spot temperatures can at this stage be close to the arc temperature itself. It is worth pointing out that this is an area where a lot depends on correct furnace operating procedures. The pollutant emission caused by evaporation can be greatly reduced by close control of the electrode voltage, and by a charge composition which aids a uniform distribution of the current intensity around the electrodes.

d) Particles generated by the reversible reaction $Si + C \leftrightarrows SiC$. In this high-temperature reaction, signalled by an intense white flame, silicon carbide particles arising at the electrodes dissociate and release silicon vapours. The latter then oxidize into SiO_2 upon contact with air. As the reaction is liable to damage the electrodes, every possible precaution is normally taken to precludeit.

The dust emitted by these furnaces consists prevalently of spheroidal particles with a pronounced tendency to form long particle chains. A typical grain size distribution of this dust, as quoted in the literature, is presented in Table 40. Similar measurements were performed by Silvermann, who reports a mean grain size of 0.3 to 0.4 microns. There seems to be no substantial difference between the particle sizes encountered in the production of FeSi 75 % and of the other grades, but there are marked difference between the ways the fumes arising in these processes respond to treatment in gas cleaning plants. The dust which is hardest to trap is that of the FeSi 90 % grade, which is almost pure silicon.

TABLE 40
Size distribution of FeSi particles in fumes exhausted from electric arc furnaces

Particle size a (microns)	Retained proportion Z (%)
10	93
3	88
1	78
0.5	62
0.2	30
0.1	15

Chemical analyses have shown that these dusts contain 85 to 93 per cent of silicon dioxide, which implies heating losses of 1.3 to 4.5 per cent. The remainder consists chiefly of Fe_2O_3, Al_2O_3, MgO, and some alkalis. Spectrographic analyses indicate that the silica content of this dust is in an amorphous state. What isolated quartzite grains were found in the samples would appear to have been entrained from the unreacted charge materials. This fact is vital, because while quartzite fines are known to cause silicosis, and therefore elicit a sharp response from the public health authorities, there is so far no evidence to suggest that amorphous silica constitutes any health hazard.

The dust concentrations in these gases are hard to represent by any single figure, both because they vary so widely in the course of every cycle, and because they are so strongly dependent on the grade of ferrosilicon we are producing and on the furnace operating conditions. The mean value, generally around 2 g m_n^{-3}, is apt to conceal variations as wide as 1 to 7 g m_n^{-3}. The bulk density of the dust that collects e.g. in the hoppers of cloth filters is very low, of the order of 150 to 250 g per litre. The amounts of dust generated are considerable – in the production of FeSi 75 %, for instance, we are commonly faced with 100 to 150 kg of dust per ton of alloy output, which means that 5 to 10 per cent of the quartzite charge leaves the furnace in the fumes.

This dust is far too fine to settle in the vicinity of the plant when released into the atmosphere. Despite the fact that it ultimately disperses over a wide area, however, it creates a clearly visible smoke plume even when its concentration at the chimney top is relatively low. Consequently, public opinion tends to react to this dust in a way out of all proportion to the actual degree of pollution it causes. Plants touchy about their public relations therefore find it worth while to clean the fumes down to a residual dust content of less than 100 mg m_n^{-3} whenever this is feasible. This last proviso applies to the technical problems as well as to the economics of the gas cleaning operation, because ferrosilicon and pure silicon particles are among the most difficult of all dusts to trap.

The outcome of the trials and experiments conducted so far can be summed up as follows:

a) The only wet scrubbing equipment that can provide the requisite degree of efficiency are the most effective of Venturi scrubbers. That means either the types with a number of Venturi tubes in parallel, like the Waagner Biró designs, or the units with a rectangular-section Venturi tube that tapers down to a narrow slit, like the Czechoslovak MSA types. However, the efficiency of all these units is highly sensitive to the pressure drop at which they are operated; and a collecting efficiency of 80 to 85 per cent would involve pressure losses as high as 9000 Pa.

Moreover, wet scrubbers in this application invariably run into trouble with their sludge handling ancillaries. The water must be recirculated, both to keep the consumption rate within acceptable limits, and because no one would tolerate the dumping of effluents containing FeSi particles into rivers or lakes. The trouble stems from the fact that FeSi particles take extremely long to settle out; and even upon finally sedimenting, they only form a gel-like suspension which contains a high proportion of water – sometimes several times the weight of its solids content. On top of it all, this suspension is easily raised or stirred up, whereupon the FeSi particles again disperse throughout the surrounding volume of water. Furthermore, since the coke in the charge inevitably contains some sulphur, the sludge leaving the scrubbers is apt to be acid, and the water extracted from it must be neutralized before it is recirculated. Bitter experience with even the most thoroughgoing of neutralization procedures has demonstrated that it is always advisable to construct the scrubbers and their entire sludge handling systems in stainless steel or some other corrosion-resistant materials.

The tricky sludge is best decomposed in large rectangular-section settling tanks, narrow enough for the grab of a travelling crane to reach any part of their bottoms. These tanks are operated in staggered cycles, with some of them receiving fresh sludge while others are left at rest for the solids to sediment out of suspension. Once the cleaned water is drained off, the deposit of FeSi particles can be scooped out. Obviously, a system of this description will take up a lot of space, and largely cancel out the cost advantages of the Venturi scrubbers. But then, in this application, the relatively low first cost of the scrubbers is apt to be offset anyway by their high running costs.

b) Electrostatic precipitators are a rare exception in this particular role, because the specific properties of FeSi particles, such as their high electrical resistance (sometimes of the order of 10^{13} ohmmetres), make them extremely difficult to trap by electrostatic processes. In view of those properties and of the small size of the particles, precipitators could achieve a satisfactory degree of efficiency only if the gas flow velocity within them were kept down to less than 1 m s^{-1}. That, however, would make the units unacceptably large and costly. Furthermore, FeSi particles, once settled on the electrodes, tend to defy all attempts to dislodge them again by rapping. Wet precipitators could on the face of it do much better than dry ones in this application, but in practice they always run into corrosion problems. Besides, their sludge handling ancillaries are just as much of a nuisance as those of the Venturi scrubbers described in the preceding paragraph.

c) Cloth filters are nowadays the prevalent type of equipment for this duty, but even they still have a long way to go before they will be capable of trapping FeSi dust without any major problems. For a start, the gas they are to clean must be cooled down to the temperature which their filter cloths can withstand. This cooling must be utterly dependable, and preferably controlled automatically. So far, only three cooling techniques have been considered promising enough to be tried out for this purpose:

i) The induction of outside air, which sounds simple, but can in fact be very expensive, because it increases the throughflow the filters have to handle.

ii) Heat exchangers, which have now been abandoned because no cheap and reliable way has been found to prevent the adhesion of electrically resistant dust to their walls; owing to its negligible thermal conductivity, any such dust deposit effectively precluded the abstraction of heat from the gases.

iii) Stabilizers in which water is sprayed into the gas stream and completely evaporated in it. This is now the usual mode of cooling gas ahead of cloth filters which cannot withstand the temperature of the raw incoming gas.

Cloth filters are susceptible to clogging by any moisture which condenses on their filter cloths. The simple way to avoid condensation is to run the filter at a sufficiently high temperature. That explains the present-day preference for synthetic fibre cloths, which are capable of permanent operation at 120 °C or more, and of sustaining short-term peaks as high as 140 or 150 °C. It also partly accounts for the growing popularity of fibreglass fabrics, rated for service temperatures in the 250 to 300 °C interval. Apart from this temperature resistance, however, the cloth must also absorb as little moisture as possible, and these two requirements often conflict with each other. For example, at a given atmospheric humidity level, tergal synthetics absorb only 0.4 per cent of their own weight of water, while woollen fabrics will soak up 10 to 13 per cent of their own weight. In view of the properties and small size of the FeSi particles, filter cloths for this application always need some form of surface treatment to delay their clogging and facilitate their reconditioning after use.

Cloth filters for this duty are usually operated at very low gas velocities, in the region of 1 to 1.5 cm s^{-1}. They need highly efficient reverse blowing systems supplemented by mechanical or, preferably, vibratory rapping provisions, as the FeSi dust deposits that build up on the cloths are not at all easy to dislodge.

6.3.4 Calcium carbide production

Calcium carbide is also produced in electric arc furnaces, but the properties of the solids in the fumes exhausted from these furnaces differ substantially from those of the particles generated in the production of ferrosilicon. No matter whether the furnaces are of the open or the enclosed types, the grain size distributions and chemical compositions of the dusts emerging from them are always much the same. Table 41 presents a typical size distribution of the dust emitted by an open arc furnace producing calcium carbide; Table 42 lists the prevalent chemical composition of this dust.

TABLE 41

Size distribution of solids in fumes drawn from open arc furnaces for calcium carbide production

Particle size *a* (microns)	Retained proportion *Z* (%)
plus 30	Traces
30 to 10	10
10 to 5	22
5 to 3	20
3 to 1	38
minus 1	10

TABLE 42

Chemical composition of dust emitted by furnaces for calcium carbide production (% by weight)

SiO_2	7.3	C	8.9
Al_2O_3	4.9	C_2Ca	1.3
Fe_2O_3	0.8	CaO	56.9
MgO	4.7		

The dust emission from these furnaces averages roughly 3 to 5 per cent of the weight of the calcium carbide they produce. In view of the very different amounts of gases drawn off from the two types, this represents a solids concentration of only 2 to 4 g m^{-3} in the fumes from open furnaces, but of some 80 to 150 g m^{-3} in those from enclosed furnaces. If the dusts from these two types of furnaces are essentially similar in their chemical compositions, then that does not apply to the gases which bear those dusts, as is evident from Table 43. The gases from the enclosed furnaces are potentially valuable as a fuel, since they contain some 11,000 to 12,000 kJ per m_n^3. Enclosed furnaces yield about 300 to 350 m_n^3 of gases per ton of calcium carbide produced in them, or about 100 to 120 m_n^3 per 1000 kW of their energy input, at gas temperatures typically between 600 and 700 °C. Open furnaces emit 8000 to 9000 m_n^3 of gases for every 1000 kW, usually at only 100 to 200 °C.

TABLE 43

Chemical composition of gases exhausted from furnaces producing calcium carbide (% by volume)

	Open furnace	Enclosed furnace
CO_2	0.6	1 to 2.4
CO	0.2 to 0.4	85 to 93
O_2	20	0 to 0.6
N_2	79	2 to 3.9
CH_4	—	0.2 to 1.2

The solids in those fumes are not toxic, but they have two properties calculated to arouse a public outcry against them. Firstly, they are so dark as to create an unfavourable visual impression even at relatively low concentration levels. Secondly, they are coarse enough for a sizeable proportion of this dust to settle in the immediate vicinity of the plant which generates it. Despite their relatively large size, however, these particles are usually very difficult to trap, both because of their chemical composition and because of their physical properties.

Conventional cyclones are generally not much use at these plants, because in this application their efficiency is limited to some 40 to 70 per cent. That would be acceptable only if the large amounts of untrapped solids could be adequately dispersed by very tall chimneys, but even then the properties of this particular dust would make the cyclones prone to frequent clogging.

Wet scrubbing equipment could on the face of it afford efficiencies as high as 90 per cent, and could do so for relatively low first and running costs, but its use is virtually ruled out by the cementing properties of the dust. The only way to prevent the formation of hard deposits would be to supply the scrubbers with a huge excess of water, calculated to flush away the residues before they can accumulate. And one look at the volumes of gas involved shows that a large excess of water would be prohibitively costly, even if enough water were available for the purpose.

Electrostatic precipitators are out of the question, since neither the gases nor the dusts generated in the production of calcium carbide can safely be exposed to the risk of an arc-over.

Cloth filters run up against their usual temperature limitations, which are aggravated by the fact that none of the normal methods of cooling the gas upstream of the filters are applicable in this type of service. The induction of outside air would increase the already large volume of gas to unmanageable proportions. The injection of water would cause the dust to cement. Heat exchangers would promptly clog up with thermally insulating deposits. Moreover, in this application the filter cloths would be difficult to recondition: neither conventional rapping nor vibration could dislodge the trapped particles without the aid of high-performance reverse blowing systems. Trials performed so far have indicated that the gas velocity in such filters would have to be kept well below 1 cm s^{-1}, which means that the filters would work out extremely bulky and costly. Finally, no filter cloth known at present can be considered really suitable for trapping the dust emitted by these arc furnaces.

The only type of equipment which seems to hold out some promise in this application is the combination of sand-bed filters with cyclones as described in Section 3.1.5. These units are known to be insensitive to the temperature of the incoming gas, but their behaviour under actual routine service conditions in calcium carbide plants still remains to be investigated.

6.4 Foundries

The dusts generated at foundries mostly contain a lot of free silica. Consequently, any separators intended for foundry service must be highly efficient at trapping fines in the 1 to 5 microns range. Furthermore, foundries are seldom provided with tall chimneys; usually, their stacks terminate little above roof level, so that much of what they emit returns through the windows into the foundry hall. Unless these fumes are cleaned efficiently, therefore, the dust concentration in this hall will soon exceed any acceptable level. Moreover, this is a dust which is known to cause silicosis, and has therefore attracted some particularly stringent pollution control legislation.

Most of the dusts which constitute the worst health hazard in foundries arise from the moulding sands. Before examining the specific particle separating problems of foundries, therefore, we must first briefly review the properties of those sands and the processes by which dust is raised from them. In general, moulding sands consist of the following chief components:

a) Silica sand grains about 0.1 to 0.3 mm across, which are too large to enter the lungs, and consequently represent no great health hazard.

b) Coarse dust, with particle sizes of 100 to 20 microns, which when inhaled tends to remain in the bronchial system and is either exhaled again or expelled e.g. by coughing. This dust must be suppressed, as far as possible, for technological rather than for hygienic reasons, because it impairs the permeability of the sand mixes.

c) A small proportion of minus 20 micron fines, which play a key role in securing the strength and bonding properties of the sands. Every time the sand is handled, some of this fine dust is dispersed into the air – and its minus 5 microns fraction is small enough to reach the lungs and settle there, causing silicosis.

TABLE 44

Grain size distribution of foundry sands

Sand type	Particle size *a* (microns)								Free SiO_2 in minus 5 microns fraction (%)
	plus 1000	1000 to 500	500 to 300	300 to 200	200 to 100	100 to 60	minus 60	minus 5	
	Retained proportion *Z* (%)								
Fresh 1	—	0.2	0.3	1.7	65.7	25.3	6.8	0.8	11
Fresh 2	0.5	0.4	2.0	5.8	61.6	24.0	5.7	1.6	6
Fresh 3	0.2	0.1	1.1	7.6	76.4	12.4	1.5		16
Re-used 4	0.2	0.5	6.5	19.0	59.1	9.8	4.9	0.6	4
Re-used 5	3.5	36.0	24.9	15.9	17.0	1.3	1.3	1.1	2.1
Re-used 6	0.5	0.2	0.2	0.2	7.9	45.0	26.1	8.2	6

Table 44 lists the typical grain size distributions of some moulding sands, and their percentual contents of SiO_2 particles smaller than 5 microns. Naturally, the amount of dust actually raised is strongly dependent on the humidity of the sand: a moisture content as low as 4 per cent will markedly reduce the dust emission when the sand is handled. The most perilous materials are the dried re-used sand mixes, and the old sand remnants that tend to build up on foundry floors and equipment surfaces. That is why public health authorities go out of their way to ensure that foundries are kept clean and orderly: the secondary dust swirled up from the floor generally represents a much more serious health hazard than the primary dust generated in the course of the process itself.

Since the primary objective of dust trapping equipment in foundries is to capture as much as possible of the minus 5 micron fines, the choice is at present essentially limited to cloth filters and wet scrubbers. The latter predominate in most of the recent installations, although in the United States some foundries now pipe all the exhausted air and fumes into one or two central gas cleaning facilities fitted with electrostatic precipitators. Cyclones, though still in use at some of the older plants, are nowadays falling out of favour. For one thing, the minus 5 micron fines can be trapped at all only in small-diameter units, which are highly selective and generally fail to attain satisfactory overall collecting efficiencies. For another, the dust is often highly abrasive, and the flow velocities are of necessity apt to be high, so that the cyclones are subject to fairly rapid erosive wear.

The cloth filters employed in foundries are mostly grouped into a central gas cleaning facility housed in an annex. This arrangement provides the space needed for the usually rather bulky filters, and also facilitates the removal of the dry dust from them by concentrating all the dust handling operations at one point. The main drawback of this scheme is that the dust-laden air and fumes have to be piped to the filters through long and often intricately branched ducting or piping runs. Especially when the exhausted air is humid, which is not at all rare in foundries, these ducts or pipes are prone to clog up with deposits.

In theory, cloth filters can be highly efficient even when the dust is extremely fine, but in practice far too much depends on the state and maintenance of the filters. Foundry dusts contain large proportions of sharp-edged particles, of coarse and consequently heavy fractions, and of grains which reach the filters while still hot. All of them are apt to damage the filter cloths, adding to the maintenance costs and detracting from the efficiency, reliability and annual utilization time of the filters. Consequently, in modern foundry installations cloth filters increasingly tend to be restricted to some special applications, such as the trapping of bentonite dusts.

The use of wet scrubbing equipment in foundries has undergone a long process of evolution. In Czechoslovakia, the earliest such equipment were Elex-Schneibel scrubbers. They were served by central sludge handling facilities made up either of hydrocyclones or of large settling tanks. These units were large, their sludge handling ancillaries were complicated, and their efficiency in the trapping of the finest fractions left much to be desired. They have therefore been largely replaced by more efficient and compact types of equipment, the first of which were MVA washers. The MVA units were generally set up at or near the points of dust generation; they were linked either to a central sludge handling facility with one or two large settling tanks, or, more frequently, to a smaller facility comprising one or two mechanically raked UNB tanks and the relevant pumps, which served only one group of adjacent MVA washers.

At present, the predominant type of scrubbing equipment in foundries are the Roto Clone N and Tilghman units. In Czechoslovakia, the preferred types are the MHA, MHB and MHG scrubbers, which are fitted with mechanical sludge raking systems and thus form fully self-contained units. That probably accounts for the fact that they are more popular than their sister types of the MHC and MHF ranges, which have to be supplemented with external sludge handling facilities – and in foundries, that means central sludge handling systems, with all their inherent drawbacks. The MHA, MHB or MHG scrubbers are commonly installed in clusters of several units each, to reduce the number of points from which sludge has to be evacuated. The sludge is either raked out directly into trucks or bins, or else, when the scrubbers are set up side by side, onto a conveyor which delivers it to the disposal point.

The bulk of the dust generated in foundries arises at one or another of the following points:

a) The moulding machines,where dust is raised chiefly during the turning of the cope, the vibratory compacting and final compressing of the mould. One obvious precaution is to restrict as much as possible the amount of compressed air escaping from the vibratory equipment, and to deflect this air stream well away from any areas where it could swirl up dust. The vibrating components tend to crush or comminute the sand grains, forming large amounts of highly dangerous fines. In the vicinity of flask turning machines, these fines can contain as much as 30 per cent of SiO_2. It is therefore advisable to provide especially the larger machines with automatically controlled brushes and suction nozzles, which will clear all dust off the horizontal surfaces in the course of every moulding cycle. The present widespread practice of blowing the dust clear with compressed air aggravates the problem rather than solving it.

Many of these machines are still not fitted with any exhausting or particle trapping equipment. Even where dust exhaustors are installed, their suction hoods must generally be designed so as to be out of the way rather than for maximum scavenging efficiency. Hence, the dust concentrations at and around these machines are still usually far too high to be acceptable by present-day standards. That must be taken into account in the design of the foundry ventilation system: those machines should never be anywhere near the point where fresh air is brought into the hall, but should always be as close as feasible to the spent air outlet. That will prevent the dust they emit from contaminating the rest of the foundry interior.

The dust-laden air is usually drawn off from the vicinity of these machines by exhaustors fitted with cloth filters. As a rule, each exhaustor serves a cluster of machines, and is set up in their vicinity. A more recent trend is to clean the air in scrubbers, and especially bubble washers. As these are available only for throughflows from about 4000 m^3 per hour upwards, each of them must necessarily serve several machines, and great care must be taken to ensure that a roughly equal volume of air will be induced at every machine.

The production of cores generally raises very little dust, and is therefore best located in separate premises, so as to protect the personnel of this department from the dust generated elsewhere in the foundry.

b) Cupolas, another major source of dust emissions, are dealt with in Section 6.4.1.

c) Electric arc furnaces, and the gas cleaning problems associated with them, have already been examined in Section 6.2.6.

d) The casting process itself, i.e. the pouring of molten metal into the moulds, creates gas rather than dust pollution problems. The moulding sands contain pulverous carbonaceous materials, which on contact with the hot metal release toxic carbon monoxide. The traditional approach has been to build foundry halls large enough for this gas to be sufficiently diluted by the natural air circulation within them. In modern foundries, however, the casting areas are provided with fume and gas exhausting systems, and continuously supplied with fresh outside air.

e) Shaking out, on grates where the sand is loosened and shaken out of the moulds, creates so much dust that some form of exhausting equipment is essential. The layout of this equipment will depend mainly on the size of the grate and the size of the moulds themselves, i.e. of the castings turned out in them. As long as the flasks are no higher than about 20 cm, the dust can be drawn off from the bottom of the grate only. In that case the minimum air intake per square metre of grate area is some 1300 to 1800 m^3 per hour when the moulds are shaken still hot, and at least 900 m^3 an hour when they are shaken after cooling down. In the production of medium-sized or large castings, however, bottom exhausting alone would be inadequate, as the convection of hot air and of steam generated in the shaking process tends to entrain the dust and raise it above the grate. In these cases, the grate must be flanked by suction hoods on one or on both sides, or completely enclosed. In general, the more completely the grate is enclosed, the less air has to be induced to draw off the dust. As a rule of thumb, we must count on exhausting not less than 3000 to 3700 m^3 of air an hour per square metre of grate area through the lateral hoods running the length of the grate. Startling as it may seem, this estimate may actually be on the conservative side: current regulations in the United States stipulate the extraction of 7000 to 9000 m^3 an hour per square metre of grate area when the hoods are fitted at one side only, and 5000 m^3 an hour through mutually opposed hoods. A certain proportion of the total should always be exhausted from the space beneath the grate, so as to prevent the heavier fractions from settling there. Again as a rough and ready guideline, this bottom intake should not be less than some 700 m^3 per hour for every square metre of grate area. Given the size of the shake-out grates nowadays used in the production of medium-sized and large castings, these figures mean that we shall have to handle huge volumes of dust-laden air. At present, this dust is trapped in bubble washers to the virtual exclusion of any other equipment.

f) The cleaning of the castings is a multi-stage process, with every stage creating its own specific air pollution problems. For a start, the gating and risers must be removed, usually by flame cutting. This is generally done on benches or platforms, where the exhaustor intakes can be set beneath the level of the working area. Only when the castings are tall is it necessary to install suction hoods at their sides too. The dense smoke that arises during flame cutting operations consists chiefly of iron oxides, which constitute no health hazard in themselves. What does make the process dangerous are the remnants of sand, which tend to shoot off the casting surface and burst like small shrapnels. Safety considerations thus dictate that the castings should first be shot blasted to remove the sand. The fumes exhausted from the flame cutting stations cannot be cleaned in cloth filters, because of the high incidence of hot particles which would burn holes in the cloths. These fumes are commonly processed in high-efficiency scrubbers, bubble or bath-type washers such as the MHA, MHB or MHG units, or fluidic devices like the MSA models.

The tumbling or shot blasting operations can generate large volumes of dust as the remnants of moulding sand are broken up and scattered. A deplorable practice which should be avoided is sandblasting with silica sand: the impacts tend to break up the sand grains, forming a fine dust which can contain 80 to 95 per cent of SiO_2. The only really suitable blasting material is metallic grit. The dust-laden air exhausted from the drums or blasting chambers is usually cleaned in wet scrubbing equipment, especially in bubble washers, but cloth filters have been adopted for this duty too, and have proved quite successful in it. It has often proved advisable to install a coarse pre-cleaning stage fitted with dry cyclones, but in view of the highly abrasive nature of the particles, these cyclones must be fabricated in heavy-gauge metal.

g) The preparation of new and recovery of used moulding sands involve a lot of handling operations where, unavoidably, a good deal of dust will be raised. All the conveyor belts, chutes, hoppers, bucket-chain elevatots, vibratory or pneumatic conveyors, telescopic filling tubes, etc., should be provided with suction hoods. And as all these hoods will generally be linked to a single fan and particle separator, the system needs careful engineering if an adequate amount of air is to be induced at each of the points it serves. Much the same applies to the sand drying and cooling facilities, which are usually drums blown with hot combustion products and with cold air respectively. The gas or air induction rate must be high enough to prevent any substantial escape of dust from these devices into their environment, but must not be high enough for significant amounts of sand to be entrained and dumped in the particle separators. The latter are again more often than not bubble or bath-type washers.

In the recovery of spent moulding sands, a special problem is created by their high contents of bentonite particles. In mechanically raked bubble washers as in the settling tanks of central sludge handling systems, the bentonite forms a gel-like suspension which gradually blocks up the flow channels. After anything from a few dozen running hours to a few weeks (depending on the dust concentration in the exhausted air and on the bentonite content of the dust), the deposits will bring the equipment to a total standstill. This effect can to some extent be countered by frequent changes of the water in the system, but even that is a palliative rather than a remedy. The obvious alternative are cloth filters, and special measures to keep their cloths dry, but the velocity at which the air passes through the filter cloths must in these cases be kept far lower than is otherwise normal.

h) Grinding, the final stage in the cleaning of casting surfaces, always produces large amounts of dust. Even when the castings have been shot blasted prior to their grinding, this dust is still apt to contain as much as 8 per cent of SiO_2. When the castings are ground while some remnants of sand still adhere to their surfaces, the SiO_2 content of the dust is often higher than at any other point in the foundry. The SiO_2 of course represents an intolerable health hazard; but even the silica-free fractions of this dust, though not actually harmful to human health, cannot very well be left at large in a plant full of machinery with bearings and sliding surfaces to score and seize up. Every grinding station must be provided with dust exhausting facilities, and the air thus induced must be thoroughly cleaned before it is released into the atmosphere.

i) Paint spraying, enamelling, plastics coating and similar processes should always be carried out in booths or cabins fitted with exhausting and particle separating equipment. Apart from the physiological effects of many of these substances, there are the fire and explosion risks to consider. Usually, the rear wall of the spraying booth, on which the particles impinge after they have missed the casting, is continuously flushed with a film of flowing water, and the casting itself is suspended above a water-filled tank. The air exhausted from such booths is cleaned in a variety of devices, but none of the equipment presently available really meets all the specific requirements of this application. The usual practices are either to pass this air through a curtain of finely atomized water in a spray cabin, or else to clean it in coarse filters provided with fibrous or other expendable cartridges. As these cartridges have to be replaced fairly frequently, they can be made only of the cheapest of materials. Some plants pack their air filters with sawdust and/or wood shavings, which are burned after use.

j) Secondary dust, i.e. dust which has already settled, but is swirled up again by the movements of men or machinery or by air streams, can in foundries be as much of a problem as primary dust. The important word is can, because it need not be a problem at all. Dust will settle mainly on horizontal or only gently sloping surfaces, and will cling where the surface is rough. Consequently, all machinery and plant equipment should ideally be shaped, or provided with shaped covers, in a way calculated to eliminate such surfaces. A lot can be done with paints that form a smooth surface finish. The dust that does accumulate should be removed at regular and short intervals, and always with vacuum equipment. The widespread practice of occasionally blowing the dust away with compressed air does not actually dispose of the dust, but merely displaces it – often into the places where it is least welcome, such as antifriction bearings. The principal areas that have to be kept clean are gangways and transport routes, because that is where the incessant movements of men and vehicles would stir up the dust most intensively. There is a lot to be said for hard-surfaced floors in such areas. Cement floors are not all that expensive, and any damage done to them by spilled hot metal is usually slight and easily repaired. Many foundries have found it worth while to cast their own iron floorplates. Given a regular vacuum cleaning, a hard floor will reduce the secondary dust level in a foundry more effectively than any other single measure.

6.4.1 Cupolas

The cupolas in which foundries melt their iron are worth closer examination, because some of the gas cleaning problems they create are unique to this particular kind of shaft furnaces. Cupolas are charged from above with alternate layers of metal-bearing materials and of coke with fluxing additions, usually limestone, sometimes fluorspar. The metal-bearing burden consists of varying proportions of foundry pig iron, discard generated in the foundry (such as gating or risers), iron and steel scrap, and ferroalloys. Tuyères at the bottom of the shaft supply the air needed for the combustion process. Depending on the temperature at which this air is blown in, we distinguish between cold-blast and hot-blast cupolas.

In a cold-blast cupola, the air is blown in essentially at its ambient temperature. That means the gases evolved in the course of the combustion process are not utilized, and present a disposal problem. Usually, they are led off to a stack with an excess of draught, which induces a certain amount of outside air through the cupola charging port. At the top of the stack, the gases are generally ignited, so as to turn their carbon monoxide content into relatively harmless carbon dioxide. In hot-blast cupolas, the air is preheated in recuperators which as a rule exploit the latent heat of the carbon monoxide either in the top gases (some of which are in this case extracted from the furnace just beneath its charging hole), or in the combustion products (drawn off directly beneath the melting zone). In either case, the gases are mixed with outside air and burned immediately ahead of the recuperator intake, so as to impart a maximum of heat to the blast air.

If we are out to assess the dust concentration in the gases, and the total amount of solids emitted by a cupola, the first step should obviously be to ascertain the volume of gases actually obtained from it. Unfortunately, that is easier said than done. The gases are gained at extremely high temperatures, and in a turbulent stream, which hampers any attempt at accurate measurement, and makes the figures we obtain open to doubt. The gases are mixed with a variable amount of air, and usually burned, before their temperature and turbulence decay sufficiently to permit dependable measurements. Therefore, the usual practice is to establish the rate at which gases are emitted by calculations based on the chemical analysis of the gases and the combustion ratio η. This emission rate, Q_{sp}, works out as

$$Q_{sp} = 5.4 + \left(\frac{3.5\eta}{100}\right)\frac{21}{21 - O_2} \qquad (m_n^3\ kg^{-1}\ C^{-1}) \qquad (104)$$

where $\eta = \dfrac{CO_2}{CO_2 + CO} \cdot 100$, or the combustion efficiency ratio ascertained by the analysis of the combustion gases (in %),

O_2 – the oxygen content of the combustion gases (in %), and

C – the weight of carbon in the charge (in kg).

The formula for the emission rate per unit of time is

$$Q_{sp} = K\frac{k}{100} + \left(\frac{3.5\eta}{100}\right)\frac{21}{21 - O_2} \cdot 10S \qquad (m_n^3\ hour^{-1}) \qquad (105)$$

where K is the percentual coke rate (in kg per 100 kg of iron),

k – the carbon content of the coke (in %), and

S – the output rate of molten iron (in tons per hour).

TABLE 45

Chemical composition of cupola top gases and combustion fumes (% by volume)

	Composition ranges found in literature		Cold-blast cupolas Test analyses		Hot-blast cupolas		
					Top gas		Fumes
	Top gas	Fumes	Top gas	Fumes	Test 1	Test 2	
CO_2	8 to 17	2 to 16	13	10.4	8.96	13	12.6
CO	5 to 21		15.4		12.74	15 to 17	
O_2	0 to 4	3 to 16	0.4	8.8	1.28		3.4
H_2			1.02				
SO_2	0.04 to 0.1	up to 0.08	0.087				0.0016
SO_3			0.0041				
N_2	rest	rest	rest			rest	68
H_2O							16

This calculation procedure rests on the assumption that enough will be known about the analyses of the top gases and combustion gases. However, the compositions quoted in the literature differ so widely as to be little use in practice. For a rough orientation, Table 45 lists the composition ranges encountered in the literature, along with some figures obtained in experimental measurements.

Apart from the solids content, the main pollutant constituents of these gases are carbon monoxide and sulphur dioxide. Fortunately, both of them are fairly easy to control. The sulphur dioxide content can be reduced quite substantially by wet scrubbing, especially in scrubbers fed with alkaline water. The carbon monoxide in the top gases is generally burned, in a large excess of outside air induced through the cupola charging hole, to a residual CO content mostly no higher than 1 per cent. However, ignition is as a rule only spontaneous, so that the stack gases do not burn continuously throughout the cupola operating time. Consequently, the amount of carbon monoxide escaping into the atmosphere is apt to fluctuate quite considerably.

The solids contents of the top gases vary both in quantity and in their composition. The main components, and the factors that govern their amounts and nature, are as follows:

a) Coke dust is formed by abrasion of the as yet uncombusted coke during movements of the cupola charge, in amounts which are inversely proportional to the abrasion resistance of the coke.

b) Fly ashes, arising by the combustion of the coke, are entrained by the gas flow in the area above the melting zone.

c) Limestone or fluorspar dust is again generated by abrasion during the motions of the cupola burden.

d) Silica sand occurs both on the surfaces of some of the charge materials (risers, etc.) and in grains eroded from the refractory linings as the charge sinks lower in the shaft. The fractions of interest to us are those fine enough to be entrained by the gas stream, as the coarser grains sink down to the hearth and are dissolved in the slag.

e) Oxides are present in the charge in the form of scale and rust, and are dislodged from the metal surfaces both during the charging process and as the charge sinks down the shaft. The larger particles again dissolve in the slag, but the fines are picked up by the gas stream.

f) Remnants of other materials coincidentally present in the charge, such as paints, corrosionproofing agents, etc., and/or their combustion products.

The chemical composition of the dust in these gases will naturally differ a great deal from case to case, even in one and the same cupola. A rough picture can be gained from Table 46, which lists the usual and the extreme ranges of values found in the literature, as well as some figures established by experimental analyses. It is clear that the dust emitted by hot-blast cupolas differs very little from that

TABLE 46

Chemical composition of dust in cupola top gases (% by weight)

	Composition ranges found in literature		Cold-blast cupola	Hot-blast cupola		Specific gravity ($g\,cm^{-3}$)
	Normal	Extreme	Test result	Test 1	Test 2	
SiO_2	20—40	10—45	44.6		57.07	2.6
CaO	3—6	2—18	4.2	4.6	4.65	3.4
Al_2O_3	2—4	0.5—25	2.2	1.7	3.26	4.0
FeO, Fe_2O_3, Fe	12—16	5—26	9.4	18.5	16.23	5.3
MgO	1—3	0.5—5			2.33	
MnO	1—2	0.5—9			8.70	4.0
CO_2 (heating loss)	20—50	10—65	27.9		3.29	2.3

given off by cold-blast units, the main component in both cases being SiO_2. What is listed as heating losses is essentially the coke dust escaping from the charge. Occasionally, the dust may prove to contain oxides like SnO_2, ZnO, MgO or PbO, which simply means that the charge materials had not been adequately cleared of extraneous substances.

As for the grain size distribution of this dust, there is ample evidence for assuming that it is much the same in hot- and in cold-blast cupolas. True, what figures are available mostly indicate that the dust from hot-blast cupolas is much finer,

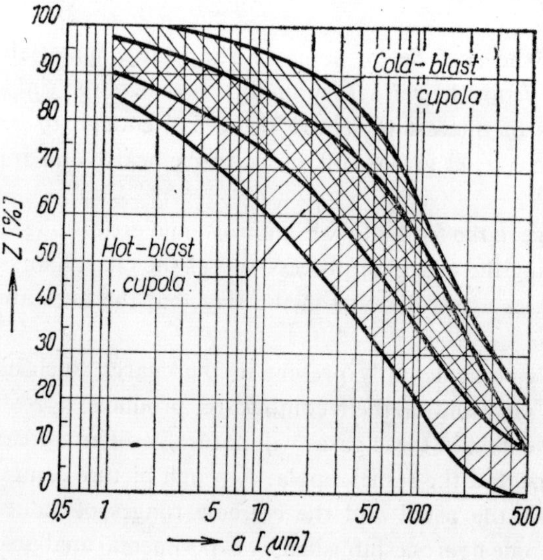

Fig. 167 Grain size distributions of the dusts generated by foundry cupolas.

but that is most likely due to the fact that this dust is of necessity sampled downstream of the blast preheaters, which tend to retain much of the coarser fractions. Fig. 167 shows that the usual ranges of grain size distributions of hot-blast and cold-blast cupola dusts largely overlap each other.

The figure which will actually govern our choice of dust trapping equipment is of course the solids concentration in these gases, but data on this concentration level are surprisingly scarce, and what figures have appeared in the literature often contradict each other. To confuse the issue still further, few of the authors have bothered to specify the technical data and operating conditions of the cupolas on which they ascertained their figures. Besides, the results recorded downstream of blast preheaters are not likely to be truly representative, owing to sedimentation in these units; and measurements at the cupola outlet itself are complicated, by high temperatures and turbulence, to the point where the findings are at best rather dubious. Most authors have quoted dust concentrations ranging from 3 to 20 g m_n^{-3}, i.e. some 2 to 20 kg of dust per ton of molten iron, with the usual dust rate apparently varying between 5 and 12 kg per ton. Table 47 presents both the normal and the extreme dust concentrations, per m_n^3 of top gas and per ton of molten iron, for both hot- and cold-blast cupolas.

TABLE 47

Dust concentrations in cupola top gases

Cold-blast cupolas		Hot-blast cupolas	
Normal range	Extreme range	Normal range	Extreme range
6 to 11 g* 5 to 10 kg**	2 to 15 g* 2 to 12 kg**	6 to 14 g* 8 to 12 kg**	3 to 25 g* 2 to 20 kg**

*per m_n^3 of gas
**per ton of iron

So much for the usual preliminaries to the selection of particle trapping equipment—the business of determining the concentration level and grain size distribution of the dust to be trapped. Apart from all the uncertainties set out in the preceding paragraphs, we must also make due allowance for the fact that both these quantities are likely to fluctuate in the short term and alter appreciably in the long term, since they are dependent both on the composition of the charge and on the way the cupola is operated. A survey of the equipment currently in use abroad shows that it falls into three broad categories. The first consists of coarse pre-cleaning facilities, like settling chambers, large-capacity cyclones, or gas outlet closures fitted with discharge ports for the dust retained in them; these items need not interest us here. The gas cleaning equipment proper can be classified as follows:

a) Medium-efficiency equipment, mainly scrubbers with spray nozzles, which have the advantages of being cheap and requiring relatively little maintenance. Moreover, they can be mounted on the cupola stack or blast preheater, and in that case take up next to no extra space. As long as their nozzles are kept in mint condition, these units can attain efficiencies of the order of 65 per cent; but as this is rather on the low side, there has recently been a distinct tendency to fit them only to the smaller cupolas.

b) High-efficiency units, capable of efficiencies around 99 per cent, where the choice confronting us includes Venturi scrubbers, wet and dry electrostatic precipitators, and, for the final cleaning stage, cloth filters cleared for high service temperatures. These types of equipment differ widely from each other in their first and operating costs, attendance and maintenance needs, and space requirements, so the choice is likely to be decided by a combination of these factors rather than by performance or efficiency considerations.

Of the equipment produced in Czechoslovakia, the spray scrubbers with atomizing nozzles made at the Klatovy plant of the Škoda Works are suitable only for small cupolas with a low annual utilization rate. Larger cold-blast cupolas, and most hot-blast cupolas, can be served with the requisite efficiency level of around 99 per cent by the MSA fluidic washers. In view of the high sulphur dioxide content of the gases, these washers have to be fabricated in stainless steel; but the extra cost is partly offset by the fact that these units can handle the gas after no more than a rudimentary cooling in a long delivery piping or a simple water-spray cooler.

Another type which could afford satisfactory results are the electrostatic precipitators made by the ČSVZ plant at Milevsko. The dry precipitators need a lot of ancillary equipment upstream of them: the gas must first be precooled to about 500 °C, e.g. in a long piping run, and then cooled down to some 150 °C and humidified in a stabilizer, since dry precipitators require the incoming gas to be moist if they are to perform efficiently. Wet precipitators, on the other hand, need no stabilizer ahead of them, and can process the gas after no more than a preliminary cooling by water sprays. Also, the gas flow velocity can be somewhat higher in a wet than in a dry precipitator, so the wet units tend to work out rather more compact. These advantages, however, are not always worth the extra costs and complications of the sludge handling systems associated with wet precipitators.

Cloth filters for this application must always be provided with fibreglass filter cloths, and must be capable of sustained operation at 250 to 300 °C. The gas can usually be brought down to this temperature interval by means as simple as a long delivery piping.

It must be borne in mind that the cupola itself is not merely a furnace, but a whole complex of various items of equipment, with the numbers, layout and mutual interlinking of the individual items differing widely from one plant to the next. Now add the wide choice of separator types and their possible combinations, as outlined above, and you will begin to appreciate the enormous diversity of various installations in use at present. This variability is even greater in hot- than in cold-blast cupola plants, but in either case it is sufficient to make the task of sifting the available data, and deriving recommendations from them, a next to hopeless one.

The designers of particle separating equipment for cupolas should always work hand in hand with foundry production engineers, and should always, no matter what other pressures are brought to bear on them, adopt a comprehensive rather than a fragmented or partial approach to the problem. When this latter rule is neglected, the outcome is often wildly irrational. For instance, at some existing hot-blast cupolas a lot of money and effort is devoted to cleaning the gas that passes through the recuperators with a close to 100 per cent efficiency, while the almost equal volume of gas that by-passes the recuperators is blithely released into the atmosphere after cleaning with an efficiency around 60 per cent. If the gas cleaning specialist fails to bear in mind that those two volumes of gas cause one and the same atmospheric pollution, then as a rule no one else will do so for him.

6.5 Production of building materials

6.5.1 Cement factories

Cement factories are notoriously among the worst sources of air pollution. Which is no wonder, because both their raw materials and their product are finely pulverous substances, and their process equipment is one long succession of dust generators. To make one ton of cement, the average plant has to crush and grind to a fine powder about 2.8 tons of starting material. In the course of its further processing, this material is repeatedly exposed to cold air or hot gas streams powerful enough to entrain it; a single particle may be trapped and recycled several times in succession during its passage through the plant.

The studies so far published suggest that the dust generated at cement factories does not attack the human lung, but it certainly is harmful to the bronchial system and, probably because of its alkaline nature, also damages the skin. Being so fine and abundant, it permeates the whole vicinity as well as the plant itself, and can literally smother plant life in the area. On account of its pale grey colour, layers of this dust are easily visible, and therefore tend to rouse conservationists more than many a more harmful but less conspicuous deposit ever does.

The dust emitted by the rotary furnaces of cement factories consists almost entirely of the raw materials which are being processed, i.e. calcium silicates and aluminates, slates, marls, and some fly ashes. However, it can also occasionally contain some hexavalent chromium compounds from the furnace linings, and can in that case cause serious mechanical damage within the nostrils. Moreover, both the fuel and the raw materials are liable to contain some sulphur, which will appear in the airborne effluents in the form of sulphur oxides. Fortunately, some of these oxides are neutralized by the calcium carbonate used in the process, and some more are rendered harmless in the combustion products, especially in the wet cement-making process. Sulphur input/output balances for various types of cement-making furnaces indicate that some 79 to 98 per cent of all the sulphur entering the process ends up chemically bound in the clinker and dust in the combustion products, mostly in the form of sulphates. So one of the few kind things that can be said about the atmospheric pollution created by cement factories is that sulphur oxides are not one of the major problems involved.

The sources of dust in a cement factory fall into two categories. The first comprises the various items of process equipment, especially the furnaces; these generate the bulk of the dust, and practically all the fractions which spread into the wider area around the plant. The pollution in the immediate vicinity of the plant is largely caused by the other equipment in this category, which includes the raw material crushers and mills, the clinker coolers, cement mills, and slag drying facilities. The other category is made up of dust sources with a predominantly local effect, which pollute the plant rather than its environment. Examples are the storage and handling facilities for the various dry pulverous materials, such as clinker and slag; the raw material and cement silos; the chutes, hoppers, conveyors and grab cranes; the pneumatic delivery and bag filling devices which handle the cement; etc. The amount of dust generated at and escaping from these latter sources depends not only on the design and workmanship of the equipment, but equally on its maintenance, the effectiveness of its covers and hoods, the condition of its gaskets and seals, etc.

With that much said, let us now examine the major sources of dust and the separating equipment suitable for each of them. As the overall efficiency of this equipment will always be heavily dependent on the specific service conditions, which vary a great deal from case to case, we shall also have to review the various factors which affect the dust concentration in the exhausted gases. We shall further have to survey the physical and chemical properties of the dust and of the carrier gas, insofar as they influence the efficiency and serviceability of the separating equipment. The data in the Tables that accompany this section apply to the more recent types of plant equipment. The volumes of gas exhausted from the various dust sources are all referred to 1 kg of clinker output.

A. *Rotary furnaces for the wet process*

These are long furnaces, fed through pipes with raw material in the form of a sludge. The fuel, generally pulverous coal or oil, is injected at the opposite or exit end of the furnace. Hence, the combustion gases stream through the furnace in the direction opposed to that of the material flow. The preheating zone of the furnace is fitted with chains, or arrays of bars, which serve to disperse the incoming sludge so as to enhance the heat transfer to it. These fixtures are the more effective, the more closely spaced they are. Their locations and lengths are usually chosen so as to ensure that the heated material will still have a residual moisture content around 10 per cent as it passes the last chain of the zone; this residual humidity safeguards the furnace fixtures against overheating. Sometimes this zone is preceded by preheating chambers, where grids or spoke-like arrays of bars rotate within the furnace shell. The chamber inlets and outlets are arranged in a way which causes the bars to lift the incoming sludge, mixing it intimately with the opposed flow of combustion gases. After this preliminary heating, the raw material is divested of its carbon dioxide content, and then, at the downstream end of the furnace, sintered into clinker.

Apart from its key role in the cement-making process, the preheating zone also acts as a dust trap, arresting a proportion of the solids in the combustion gases. However, it will do so at all effectively only when the furnace is operating in a steady mode and at or close to its design parameters. Under unfavourable circumstances, it can actually contribute to the dust emission rather than suppressing it. For instance, variations in the calorific value of the fuel, or a sudden rise in the furnace throughput rate, can result in the gases becoming too hot. The sludge will then be deprived of all its moisture while still in the preheating zone; the lumps or cakes that form there will burst; and the last chains of the zone will crush the granules to a dust, which will then be largely entrained by the combustion gases leaving the furnace. This example underlines the need for stable furnace operation, under closely controlled process conditions, if dust emission is to be kept within acceptable limits. Needless to say, this optimum mode of furnace operation is as much a matter of correctly designed and properly maintained equipment as of a fully competent furnace crew.

Given correct furnace operating conditions, the chains or bars can capture the dust so effectively that many of the older, smaller rotary furnaces are still run without any gas cleaning ancillaries. However, the modern high-output furnaces generate far too much dust to be left without some such provisions. Usually, the gases emerging from them are cleaned in dry electrostatic precipitators; when properly maintained and operated, these can restrict the final dust

TABLE 48

Technical data of long rotary furnaces for the production of cement

Output in 24 hours	300 to 1500 tons
Specific loading over 24 hours	0.4 to 0.8 tons m^{-3}
Heat input	5020 to 5860 kJ kg^{-1}
	1200 to 1400 kcal kg^{-1}
Charge material	Sludge
Moisture content of charge	32 to 40 %
Output of combustion products*	3.2 to 4.5 m_n^3 kg^{-1}
CO_2 content of gases	18 to 25 %
O_2 content of gases	4 to 8
CO content of gases	0 %
Temperature of combustion products*	120 to 200 °C
Dew point of combustion products	65 to 75 °C
Dust concentration in gases*	3 to 25 g m_n^{-3}
Proportion of minus 10 micron fines	40 to 60 % (Fig. 168)
Chemical composition of the dust	cf. Table 49
Separator type	Electrostatic precipitator
Particle trapping efficiency	96 to 99 %

*At furnace exit

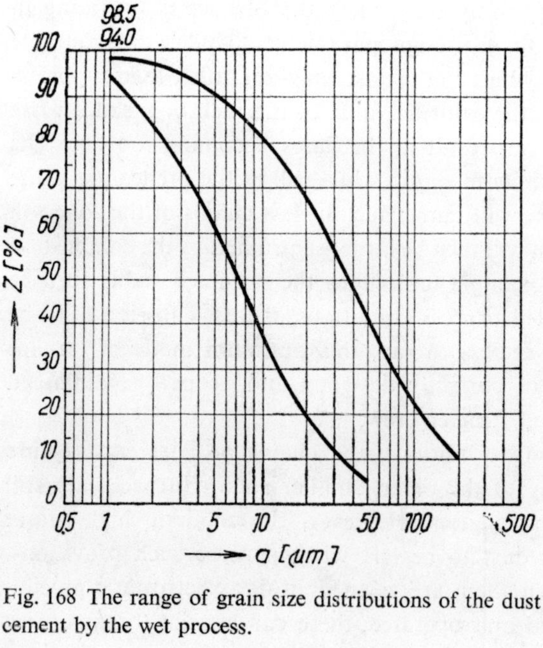

Fig. 168 The range of grain size distributions of the dust emitted by rotary furnaces producing cement by the wet process.

emission quite sufficiently even without any pre-cleaning stage upstream of them. The cyclones sometimes used in these pre-cleaning stages have also been tried out on their own, as the sole dust trapping equipment, but have mostly proved a disappointment on account of their low collecting efficiencies.

Table 48 lists the principal technical data of typical long rotary furnaces for the wet production of cement. The size distribution of the solids in the combustion gases that leave these furnaces is plotted in Fig. 168, and the chemical composition range of this dust is presented in Table 49.

TABLE 49

Chemical composition of dusts emitted by cement furnaces (% by weight)

SiO_2	10 to 25	SO_3	0.3 to 10
Al_2O_3	3 to 12	S	0.03 to 0.4
Fe_2O_3	1 to 4	K_2O	0.5 to 10
Mn_2O_3	0.07 to 0.3	Na_2O	0.1 to 2.5
CaO	30 to 52	CO_2 (heating loss)	5 to 35
MgO	0.7 to 2.5		

B. *Rotary furnaces for the dry process*

In the dry cement-making process, the comminuted raw material is preheated in a system of cyclones and pipings. The latter are blown by a fan with a countercurrent of combustion gases drawn from the furnace. Upstream of each cyclone stage, the material is picked up by the gas stream and heated; it is then separated from the gas in the cyclones, and passes down to the next stage, where the process is repeated. The last stage delivers the material, at temperatures which may run as high as 800 °C, down a chute into the rotary furnace, where it is sintered into clinker. The various preheater types differ in the design, number and layout of their successive stages, but mostly release the combustion gases at some 300 to 350 °C.

In view of the screening effect of the sequence of cyclones, the gases passing up the system pick up a gradually increasing proportion of the finest raw material particles. When they emerge from the first upstream stage, usually more than 90 per cent of the particles they carry are in the minus 10 microns range. The dust concentration in these gases, though always high, depends on the cyclone type, and responds to every change in the state of the cyclones. For instance, a blocked-up dust discharge port at any one of the cyclones instantly produces a marked increase in the final dust concentration.

Table 50 summarizes the principal technical data of these furnaces, and Fig. 169

shows the grain size distribution interval of the dust in their combustion gases.

This dust is nowadays trapped almost exclusively in dry electrostatic precipitators; in fact, this is the application which has triggered off much of the development work performed on these precipitators in recent years. True, the gases emerging from the preheaters are too hot and dry for the precipitators to handle, but then they are usually, in the interests of economy, exploited in the raw material drying drums, where they cool down and pick up enough moisture to fit them for electrostatic treatment. When the gases are not utilized in drying equipment they must be cooled and humidified ahead of the precipitators. This can be done either by injecting water into the penultimate preheating stage, or in a separate gas conditioning unit (also known as a stabilizer). There, water is sprayed in by automatically controlled nozzles and fully evaporated, so as to bring the gas down to a practically constant temperature around 150 °C.

TABLE 50

Technical data of rotary furnaces with preheaters for the dry production of cement

Output in 24 hours	300 to 2500 tons
Specific loading over 24 hours	1.4 to 1.8 tons m^{-3}
Heat input	3140 to 3768 kJ kg^{-1} 750 to 900 kcal kg^{-1}
Charge materials	Dry, pulverous
Moisture content of charge	under 1 % by weight

Combustion products	Unit	Material preheated by combustion products	Not preheated by combustion products	
			With water injection	Without injection
Amount	$m_n^3\ kg^{-1}$	2.2 to 2.5	1.7 to 2.0	1.5 to 1.8
CO_2 content	vol. %	14 to 22	20 to 30	20 to 30
O_2 content	vol. %	8 to 13	3 to 9	3 to 9
CO content	vol. %	0	0	0
Temperature	°C	90 to 150	150 to 200	260 to 320
Dew point	°C	45 to 55	50 to 60	35 to 45
Dust concentration*	$g\ m_n^{-3}$	30 to 800	15 to 40	15 to 50
	$g\ m^{-3}$	40 to 1050	20 to 60	30 to 80
Minus 10 micron fines	%	90 to 99.5	Fig. 169	Fig. 169
Dust composition		Table 49		
Separator type		Electrostatic precipitator		
Trapping efficiency	%	99 to 99.9	98 to 99.9	98 to 99.9

*At furnace exit

If the temperature and humidity of the gas are such that its dust content displays only a low specific resistance, or if the precipitators are specially designed to handle hot and dry gases, the combustion products leaving the preheaters can be led off to the gas cleaning plant without any previous cooling or humidifying.

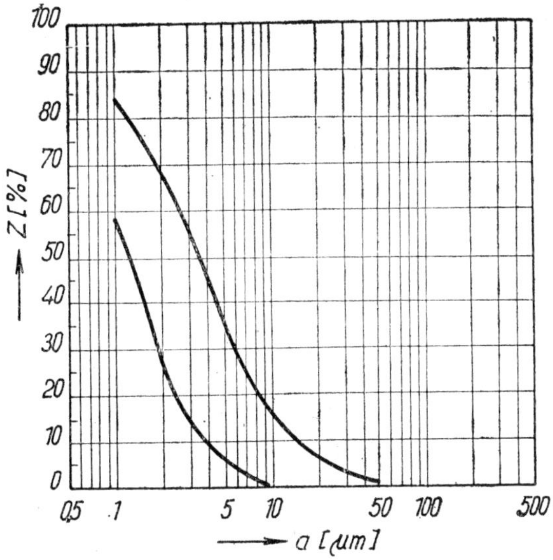

Fig. 169 The range of grain size distributions of dust emitted by rotary furnaces with preheaters, in the production of cement by the dry process.

This, however, is a rare exception. A more usual case is for only part of the total volume of combustion gases to be utilized in the raw material drying facilities. In that case, it is vital for the two volumes of gas – the part which has been cooled and moistened in the drying drums, and the part which has by-passed the drums – to be intermixed as thoroughly as is feasible ahead of the precipitator entry. Even then, it is often impossible to prevent the gas inlet temperature from building up to some 180 to 250 °C, which is too much for the precipitators to function efficiently. When that happens, an increased proportion of the dust will evade trapping and escape into the atmosphere.

C. *Shaft furnaces*

Shaft furnaces are charged with either compacted or granulated lumps of a mixture of raw material, coke or coal, and a certain amount of water. An automatic rotary dosing device spreads this charge evenly over the entire cross section of the shaft. The lumps are first dried and preheated in the upper part of the furnace, and then sink to a lower zone where the temperature is high enough to ignite the fuel they contain. Combustion in the sintering zone next turns the lumps or granules into clinker. The material further drops to successively cooler zones, where it gradually cools down before leaving the furnace through a three-stage closure. The combustion air is blown in at the bottom of the shaft, at pressures ranging from about 8000 to 18,000 Pa, to be preheated in the cooling zones. The hot gases generated by the combustion process in the sintering zone then preheat the moist charge materials at the top of the shaft. Typical technical data of these furnaces are presented in Table 51.

TABLE 51

Technical data of shaft furnaces

Output in 24 hours	120 to 300 tons	
Specific loading over 24 hours	25 to 40 tons m^{-3}	
Heat input	3140 to 3978 kJ kg^{-1}	
	750 to 950 kcal kg^{-1}	
Charge materials	Lumps, granules	
Moisture content of charge	8 to 14 % by weight	
Output of combustion products	2.0 to 2.8 m_n^3 kg^{-1}	
CO_2 content of gases	10 to 26 vol. %	
O_2 content of gases	5 to 10 vol. %	
CO content of gases	2 to 6 vol. %	
Temperature of gases	45 to 125 °C	
Dew point of gases	40 to 55 °C	
Dust concentration in gases	2.0 to 7.0 g m_n^{-3}	
Proportion of minus 10 micron fines	15 to 40 % (Fig. 170)	
Chemical composition of dust	Table 49	
Separator type	Dry precipitator	Wet precipitator
Particle trapping efficiency	96 to 99 %	96 to 99 %

The dust concentration in the gases emerging from the top of the shaft depends on the filtering effect of the humid lumps or pellets in the uppermost zones. Under stable operating conditions, relatively little dust penetrates these layers to escape from the furnace top. The trouble is that in these furnaces combustion hardly ever proceeds uniformly. And as the air is blown in at a considerable pressure, random variations in the combustion process may cause it to puncture the uppermost charge layers, at one or more points, carrying large amounts of coarse dust with it. The flames that then appear at the shaft top signal an imminent abrupt rise in the dust emission rate. These disturbances can to some extent be forestalled by precise dosing of the raw materials and fuel, but an excessive furnace loading can substantially increase the dust emission even when the uppermost charge layers remain intact.

The grain size distribution of the dust emitted by shaft furnaces is plotted in Fig. 170. The coarse fractions of this dust can be trapped in dust arresters or large-capacity cyclones, but the fines can be captured only in electrostatic precipitators. In this application, dry precipitators will function dependably only if the temperature is kept reasonably constant at slightly more than 100 °C, and if the unit is protected against the ingress of both hot gas and carbon monoxide streams.

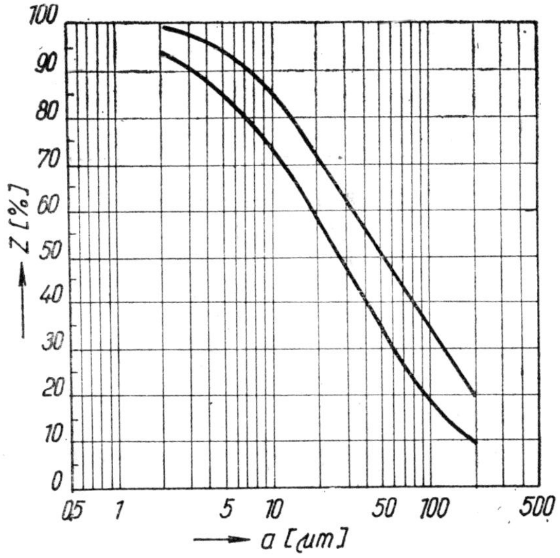

Fig. 170 The grain size distribution interval of dust emitted by shaft furnaces.

To this latter end, we can increase the fuel ratio in the charge, or post-heat the combustion gases with oil burners, or else mix them with hot gases drawn off from the rotary furnaces. An advantage of this latter scheme is that the mixing chamber, where gases from the various furnaces are mixed together, can also do double duty as a coarse pre-cleaning stage where at least part of the largest dust fractions is retained. In all these installations, the dust discharge chutes, seals and closures must be thermally insulated, or sometimes even heated, to prevent the dust from picking up moisture and forming adhesive deposits. All these precautions, including the post-heating or mixing of the combustion gases, can of course be avoided by the use of wet precipitators, but only at the cost of the inevitable sludge handling and water recirculation systems.

D. *Drying drums*

The raw materials and re-usable slag are generally dried in drums fired with coal, gas, oil, or heated by the combustion products from the rotary furnaces. The typical technical data of such an installation are listed in Table 52. These drums are as a rule inclined, and rotate slowly, so that the processed material gradually flows down their lengths. Various internal fittings are provided to intensify the heat transfer. The amount of dust in the gases leaving those drums depends both on the physical properties of the material that is being dried, and on the internal fittings, which obviously tend to raise the dust. Even the best-designed of fittings cannot avoid the formation of large quantities of dust when soft and crumbly materials are handled. The range of grain size distributions of this dust is charted in Fig. 171.

TABLE 52

Technical data of drum-type drying facilities

Throughput rate		8 to 60 tons h^{-1}	
Output of combustion products		0.8 to 2.0 $m_n^3\ kg^{-1}$	
Temperature of gases		70 to 150 °C	
Dew point of gases		40 to 70 °C	
Dust concentration at drum outlet		30 to 90 g m_n^{-3}	
Proportion of minus 10 micron fines		40 to 70 % (Fig. 171)	
Separator type	Cloth filter	Electrostatic precipitator	Sand-bed filter
Trapping efficiency	over 99 %	over 99 %	over 99 %

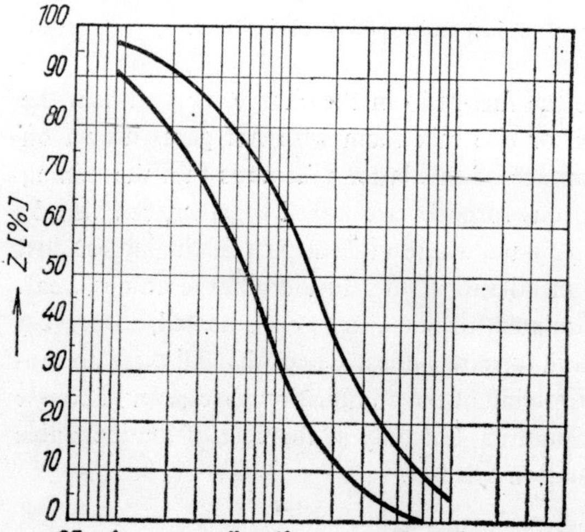

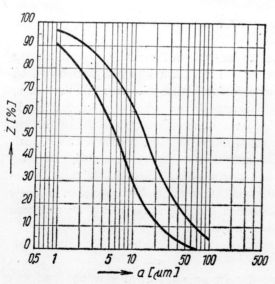

Fig. 171 The grain size distribution interval of dust trapped at the drying drums of cement factories.

As cement factories are often equipped with electrostatic precipitators anyway, the dust emitted by the drying drums is sometimes led off to these precipitators. Even when separate gas cleaning facilities are provided for the drying section precipitators are sometimes chosen simply to avoid the complications of operating two different kinds of cleaning equipment side by side. However, this gas can just as well be cleaned in sand-bed filters or, if its temperature is roughly in the 70 to 140 °C interval, in cloth filters. As the gas is apt to be humid, both the cloth filters and the delivery pipings that feed them must be thermally insulated to prevent the condensation of moisture. Often it is also advisable to heat the reverse blowing air with which the filter cloths are cleaned of deposits. When the dust concentration in the gases is exceptionally high, it is always preferable to install a pre-cleaning stage, fitted with large-capacity cyclones, rather than risk the accelerated wear and frequent clogging of the principal gas cleaning equipment.

E. *Crushers*

The first step in the comminution of the raw materials is their crushing. The amount of dust generated in this process depends on the crusher type as much as on the type, initial and final lump sizes, and moisture content of the material itself. The exhaustors which draw off the dust-laden air commonly have to induce some 50 to 200 m^3 per ton of crushed raw material. This air is usually cleaned either in high-efficiency cyclones or in cloth filters. As the dust always contains a high proportion of fines, and sometimes is difficult to handle on account of its physical properties, this is an area where separator clogging problems can be acute and intractable.

F. *Ball mills*

In cement factories, ball mills are used to grind down both the cement and the previously crushed raw materials and coal. Some typical technical data of such

TABLE 53

Technical data of ball mills for cement, cement-making raw materials, and coal

Throughput rate	5 to 150 tons h^{-1}	
Air outflow to separator	0.2 to 0.8 $m_n^3\ kg^{-1}$	
Air temperature	70 to 120 °C	
Dew point of air	20 to 60 °C	
Dust concentration in air	30 to 200 g m^{-3}	
Proportion of minus 10 micron fines	40 to 80 %	
Separator type	Cloth filter	Electrostatic precipitator
Trapping efficiency	over 99 %	over 99 %

mills are presented in Table 53. Air has to be drawn through these mills to cool, both the unit itself and the processed materials, so it can obviously be exploited to carry off the dust that would otherwise escape into the outside atmosphere. The air is then cleaned either in cloth filters or in electrostatic precipitators; the latter are particularly suitable in cases when the air can be humidified by water injection upstream of them. High dust concentrations call for a pre-cleaning stage equipped with cyclones.

G. *Clinker coolers*

The clinker produced in rotary or shaft furnaces is usually cooled down in drums, or on grates, blown with cooling air. Part of this air is then generally driven into the furnaces as secondary combustion air, the rest is exploited in the drying equipment, and subsequently cleaned in the separators serving that equipment. As a rule, the air leaves the coolers at something like 200 to 250 °C, with a highly variable dust content that depends on the grain size, temperature, and mineralogical composition of the clinker as well as on the flow rate of the air. The dust concentration at the cooler outlet is highly sensitive to any irregularity in the operation of the furnaces. For example, the penetration of unreacted raw materials past the sintering zone of a rotary furnace will cause both an abnormally high dust concentration and an abnormally high proportion of fines in the hot air emerging from the coolers. That need not matter very much when all the cooling air is exhausted into the furnaces or drying drums; but it does call for efficient cleaning equipment if part or all of that air is to be released straight into the atmosphere.

Since this dust is usually fairly coarse, it has up to now mostly been trapped in small cyclones of some 150 to 250 mm diameter. Not that they are in any way ideal for this duty: they cannot always afford an adequate collecting efficiency, are apt to suffer rapid abrasive wear, and are prone to clogging when large inlet dust concentrations are unevenly distributed among the individual cyclones of a battery. Until quite recently, however, there has simply been no other type of equipment to replace them. Cloth filters, which may one day prove ideal for this purpose, have so far been ruled out by the lack of any filter cloth that could withstand the constant impacts of hot and coarse particles. The heavy dust grains would be liable to puncture the cloths even if the gas were cooled before its admission to the filters. Besides, the gas temperature at the cooler outlet is apt to fluctuate so pronouncedly that if cooling were adopted, it would be difficult to avoid the risk of moisture condensation and the resultant clogging of the cloths. Lately, however, a variety of sand-bed filters have been tried out in this application, and what reports have been published so far suggest that these units may soon become the preferred type of equipment for cleaning the air from clinker coolers.

H. *Materials handling equipment*

Practically all the devices used in cement factories to convey pulverous or dust-bearing materials are liable to pollute their immediate vicinity with fairly high dust concentrations. Consequently, the bucket-chain elevators, worm and flight conveyors, pneumatic conveying troughs, and the hoppers and transfer chutes of belt conveyor systems must all be provided with dust exhaustors, and a lot depends on the design and location of the suction intakes. The air intake rates at these points commonly range from 5 to 20 m^3 per ton of conveyed material. Many of these dust generating points have to be fully hooded or enclosed, in which case it is usually difficult to reconcile the conflicting requirements of permanent tight sealing and of adequate access for operating, inspection and maintenance purposes. Moreover, both the enclosures and the dust extractors that serve them must usually be a compromise: while it is vital to restrict the workplace pollution level, it is equally important to do so economically. Since the volumes of air to be cleaned at the individual installations are relatively small the prevalent equipment for these duties are cloth filters or, lately, sand-bed filters.

I. *Silos and other storage facilities*

The silos which hold powdered or dust-laden materials are often kept at a slight internal pressure, in which case they must be hermetically sealed, and provided with at least simple filter bags on their venting outlets, to prevent the escape of dust. Some of these silos are fitted with pneumatic filling, discharging or mixing systems, which inevitably release considerable volumes of air. That air must then be cleaned, which is usually done in cloth or sand-bed filters. Sometimes the concentration level or physical properties of the dust necessitate the installation of cyclones as a pre-cleaning stage.

The clinker, slag, raw materials and coal storage facilities can be a major source of pollution, but their dust emission rates can quite often be kept down by relatively simple measures. Probably the most troublesome of these materials is the clinker produced in shaft furnaces, which is easily swirled up by the slightest wind or draught. Open-air stockyards can be kept much cleaner by the simple expedient of roofing them over. In enclosed storage premises, it is often worth while to step up the air extraction rate whenever the material is actually being handled, so as to create and maintain a controlled air flow pattern. The air thus exhausted is generally cleaned in cloth filters.

J. *Filling and loading devices*

The finished cement is mostly either loaded in bulk into special-purpose containers or vehicles, or filled into bags, usually by automated machinery. The dosing, filling and bag closing machinery, as well as the vibratory screens or screening drums that serve it, and the filling pipes or hoses used for bulk loading must all be provided with dust exhaustors. The air drawn off from these points is as a rule cleaned in cloth filters, which have proved entirely satisfactory for this application.

6.5.2 Quarries

The dust problems created by the actual quarrying operations are usually neither as acute nor as permanent as those that arise in the further processing of the stone, especially at the crushing, granulating and screening equipment, and in the manual or machine cutting, grinding and polishing processes. If the dust level at a quarry is to be kept within reasonable limits, the first rule is therefore to enclose the machinery whenever this is feasible. The amounts and grain size distributions of the dusts exhausted from this equipment, and the volumes of air which have to be induced to exhaust the dust, will largely depend on the efficacy of these enclosures. Effective sealing in itself is not difficult to achieve; the trouble is that the requirements it imposes are often incompatible with ease of access for the operation, inspection and maintenance of the machinery. That explains why the conventional sheet-metal covers are increasingly giving way to plastic foils, which can be transparent to allow visual inspection of whatever they enclose, and are simply stripped away to gain access to the equipment. Their service life is naturally limited, especially under the arduous conditions in a quarry, but then they are usually cheap and easy to replace.

The dust generated at quarries generally has a high content of free silicon, which can be in excess of 75 per cent, and contains a large proportion of very fine particles. Microscopic analyses of samples taken from the dust exhaustors at several quarries have revealed that 74 per cent of the particles were smaller than 2 microns, and 43 per cent were in the minus 1 micron bracket. Since the free silicon constitutes a serious health hazard, this means that the particle separators used at quarries must achieve a high degree of efficiency in the trapping even of these fine fractions. That rules out cyclones, which can seldom trap more than half of the particles in the 2 microns range. Cyclones can be used to good effect to prevent the spread of coarser dust fractions into the surroundings of the quarry, but cannot keep the concentration of fines at the quarry workplaces down to anything like the level required to protect the personnel from silicosis.

This local pollution level is best controlled by means of wet scrubbers or cloth filters. In the former category, the units best suited for these duties are the MHB washers. They can dependably afford the requisite collecting efficiency; incorporate suction fans which are adequately rated for this application; need no separate sludge handling facilities; and get along on so little make-up water that they can be topped up only intermittently, say two or three times during each shift of operation. This is a combination of features not found in other scrubbers, such as the MVB types or the various bubble washers. Those types mostly need external sludge handling and water recirculation systems, and are less efficient than the MHB units in the trapping of one-micron fines. The main reason why the MHB types have not been more widely adopted in quarries is that, like any other water-filled equipment, they are sensitive to cold weather. As long as the ambient temperature remains around zero, these units can be kept going by heating the water in their casings, by infra-red heating of their sludge raking or scraping ancillaries, and/or by housing the whole unit e.g. in a wooden shed. Once the outdoor temperature drops below some −3 °C, however, the efficacy of all these measures becomes doubtful. There is some consolation in the fact that frosty weather is likely to limit the emission of dust from the usually moist material even when the separators are shut down, but that could be cold comfort if the equipment did freeze up solid.

Cloth filters are much less susceptible to sub-zero temperatures, and, given proper upkeep and operation, can consistently attain efficiencies around 99 per cent, which is fully sufficient for the purpose in view. Unfortunately, few quarries can afford such expensive equipment. In fact, the first costs and running costs of the more efficient separator types have so far discouraged most quarries from installing anything but the more rudimentary cyclone units.

6.5.3 Expanded pearlite production

Expanded pearlite, a material which the building and construction industries are nowadays using in steadily increasing quantities, is made from natural pearlite. The latter is a non-crystalline glassy substance of volcanic origin, with the chemical composition set out in Table 54, and with a moisture content ranging from 1 to 6 per cent. It is gained in opencast (strip) mines; crushed; dried; screened into the minus 0.5 mm, 0.5 to 1.0, and 1.0 to 1.5 mm fractions; and then usually filled into sacks and despatched for further processing.

TABLE 54

Composition of typical raw pearlite (% by weight)

SiO_2	70	$MgO + CaO$	3
Al_2O_3	15	$Na_2O + K_2O$	6
Fe_2O_3	2	CO_2 (heating loss)	4

Natural pearlite is expanded by exposure to a temperature of 1150 °C over an interval which grows with the initial grain size. One to ten seconds of this exposure causes the volume of the grains to increase seven- to fifteenfold. The expanded material, with a grain size of up to 4 mm, displays microscopic pores, is chemically almost inert, non-inflammable, non-hygroscopic, and not subject to any further volume changes. Its bulk weight varies between 60 and 200 kg m^{-3} in dependence on the grain size and on the selectivity with which the grains have been screened.

Large amounts of dust are generated during the crushing, drying, screening and handling of the raw natural pearlite. All of this is mostly done out in the open rather than indoors, although sometimes the machinery is roofed over. The choice of the particle separating equipment is governed by the fact that both the combustion gases emerging from the drying equipment and the air exhausted from the other machinery contain substantial proportions of very fine particles. That and the predominantly outdoors location of the process equipment narrow the choice down to wet scrubbers, cloth filters, or cyclones.

Scrubbers and washers like the MHB units could be the ideal equipment for this duty in warm climates, but are far less attractive where the units and/or their make-up water supplies are liable to freeze up in winter. This risk might be tolerated in the processing of materials which tend to freeze up and release next to no dust in winter; but natural pearlite, in view of both its inherent properties and the techniques by which it is processed, generates almost as much dust in winter as in summer.

Cloth filters cannot deal with the combustion gases from the drying equipment, both on account of their high temperatures and because the high moisture content of those gases involves a risk of condensation. This risk is especially acute when the equipment is operated intermittently. Even the cold air extracted from the other dust-generating points in these plants is difficult to clean in cloth filters. The air must be induced regardless of the momentary atmospheric humidity, and rain or early morning fogs can moisten it to the point where the filter cloths would promptly clog up with slime. True, the filters can be thermally insulated, and even heated during the critical humidity periods, but that entails more in the way of costs and complication than most pearlite quarries are prepared to face.

That leaves cyclones as the only generally practicable alternative. In view of the efficiency requirements in this particular application, however, it often takes a two-stage or even a three-stage cyclone installation to provide satisfactory results. To prevent the formation of adhesive dust deposits in the last downstream cyclones, the latter should not be mounted in the inclined position, and should preferably not be less than some 300 mm in diameter. Given well-selected equipment in a suitable configuration and layout, such installations have proved fully capable of keeping the dust concentrations at these plants within the acceptable limits.

The pearlite is subsequently expanded in special furnaces, which are usually heated by oil burners. The combustion gases passing down the furnace pick up large amounts of the expanded pearlite, so that the separators which trap those particles are really part of the process equipment. In the most usual arrangement, the gases are first pre-cleaned and cooled down in a settling chamber which induces some outside air, and are then divested of the finer fractions in cyclones. The final cleaning is generally left to cyclones too. Scrubbers could of course perform it much more effectively, but are normally ruled out by the high sulphur dioxide content of the gases. Even if the scrubbers were fabricated in stainless steel to overcome the corrosion problems, it would still be difficult to devise sludge handling systems for them, since most of the solids in these sludges are lighter than water. Cloth filters are better avoided in this application: the gas temperatures fluctuate so widely that, especially during the running-up periods, it would be practically impossible to preclude moisture condensation in the filters. That is why the prevalent final gas cleaning equipment at pearlite expanding plants are cyclones of 300 to 600 mm diameter. In view of the physical properties of the expanded pearlite, and particularly its low specific gravity, the size of these cyclones must be picked so as to keep the flow velocity within them down to about 7 m s^{-1}, a good deal less than is usual in other applications. That means using high-efficiency cyclones with relatively low circumferential flow velocities and consequently small pressure drops across the units.

6.6 Chemical industries

6.6.1 Carbon black production

Carbon black is nowadays a key raw material for the rubber and allied industries, for the production of paints, fillers, printing inks, etc. Apart from the perpetual rise in the overall output rate of this material, there is also a distinct change in progress in the pattern of its output. An ever increasing proportion is formed by the highly disperse active and semi-active grades, where particle sizes are commonly no more than a few hundredths of a micron. Carbon black has a relatively simple chain structure, a high pH value, a low content of volatiles, and a high electrical conductivity. It arises by the incomplete combustion or thermal decomposition of hydrocarbons, and is produced industrially in special reactors. There, carbonaceous materials gained in the processing of oil, coal or natural gas are burned in the presence of less air than would be needed for their complete combustion. The dust trapping facilities form an integral part of the process equipment, as the recovery of the escaping fines is as important for the economic survival of the plant as the control of its pollutant emission.

If carbon black is exceedingly fine, tends to form chains, and is an excellent electrical conductor, then all that applies particularly to its active and semi-active grades—and none of it makes these particles any easier to trap. Moreover, the concentrations emitted into the atmosphere must usually be kept down to far less than 100 mg m_n^{-3}. The only approach so far evolved that has proved at all satisfactory is the one illustrated schematically in Fig. 172: the gases are first pre-cleaned in high-capacity cyclones, and then filtered down to the requisite low solids content in cloth filters. This scheme has been almost universally adopted in the production of the active grades.

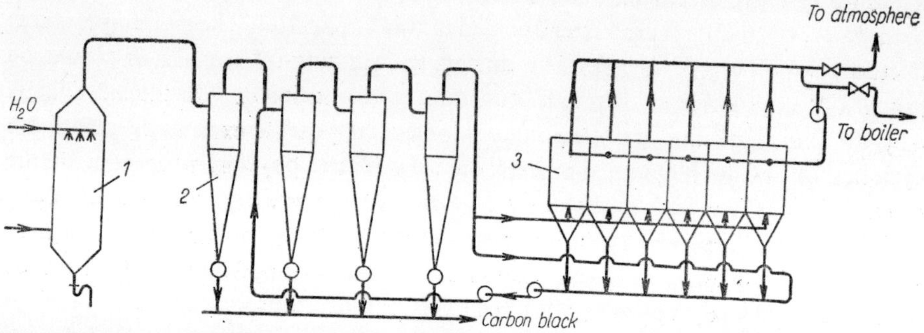

Fig. 172 Scheme of a typical equipment configuration for trapping carbon black particles: *1* — spray tower; *2* — cyclones; *3* — cloth filters.

The production process is fairly simple. The raw material, mostly oil, is first preheated and filtered to remove any dirt or solids. It is then fed to the spray nozzles of a battery of parallel reactors, which are also supplied with fuel (e.g) natural gas) and preheated air. In the automatically controlled reaction within these vessels, carbon black particles are formed at roughly 1400 °C, and are sprayed with clean water immediately beyond the combustion zone to prevent the fine primary particles from clustering into larger grains. The dry gases led off to the particle separators generally contain some 440 to 500 g of water vapour and 55 to 60 g of carbon black per m^3; their usual chemical composition range is listed in Table 55. The carbon black particles have a specific surface area of

TABLE 55

Composition of dry gas from reactors producing carbon black (% by volume)

CO_2	3.7 to 4.4	N_2	63.8 to 65.7
CO	14.9 to 17.0	H_2S	at most 0.3
H_2	13.5 to 15.0	O_2	0.1 to 0.6
CH_4	0.2 to 1.1		

about 85 to 70 $m^2 g^{-1}$, which corresponds to a primary particle size of 0.02 to 0.035 microns.

The particle-laden gases leaving the reactor are first preheated to temperatures as high as 600 °C, but before entering the separators are cooled in spray towers to roughly 270 or 300 °C. Cooling supports the natural tendency of these particles to agglomerate, and thereby improves the efficiency of the subsequent separating processes. The top of the spray tower is commonly fitted with a remote-controlled relief valve, which releases the gases into the atmosphere while the separator is running up after a shutdown. That helps to avoid the condensation of moisture in the cyclones, and especially in the cloth filters. The oils which form the raw materials often contain quite a lot of sulphur, and in that case the gases leaving the reactors are highly corrosive. Their dew point is approximately 130 to 150 °C, which means that the gas temperature in the separators should never be allowed to drop below 180 or 200 °C. That in turn implies that the cloth filters in the final cleaning stage will have to be fitted with fibreglass filter cloths. It also means that the gas temperature at the spray tower outlet will have to be closely controlled, by automatic regulation of the spray water supply, and that the gases emerging from the tower must not be exposed to the risk of substantial and probably uneven cooling in long piping runs upstream of the separators.

These gases, which leave the tower at pressures often as high as 6000 Pa, are first admitted to a primary cleaning stage. In the existing installations, this stage mostly comprises either four high-capacity cyclones or two cyclone banks in tandem. The cyclones, run at some 220 to 250 °C, trap about 90 per cent of all the carbon black formed in the reactors. The gas leaves the cyclones at typically 2500 to 3000 Pa, which means that the pressure drop across one cyclone stage, including the piping runs that link the stages, is roughly 700 Pa. The pressure drop and efficiency figures might seem incongruous with each other, but then the efficiency of those cyclones is greatly enhanced by the tendency of carbon black particles to link up into long chains, large enough to be trapped even in high-capacity cyclones. In fact, in this application the large units can actually do better than the small-diameter high-efficiency cyclones, because in the latter the high circumferential flow velocities in the centrifugal chambers would tend to break up the particle chains and cause a lot of re-entrainment. The high-capacity types with their relatively low circumferential flow velocities avoid this risk, as they largely avoid the condensation and particle adhesion problems which might prove insuperable in the small-diameter units.

The gases emerge from the cyclones with a residual carbon black content of 5 to 6 $g\,m^{-3}$, which is brought down to a mere 20 to 40 $mg\,m^{-3}$ in cloth bag filters. On the face of it, this concentration looks slight enough to cause no complaints. However, even the slightest speck of carbon black can form a large and often almost indelible smudge, so releasing this residual content into the atmosphere is not likely to improve the public relations of the plant which does so. The best way of disposing of these gases is to exploit their heat content, and burn the remaining carbon black in them, in waste heat boilers.

In the filters, the bags of fibreglass fabric call for reverse blowing combined with vibratory rapping, for some 15 to 40 seconds, roughly every 15 minutes. Naturally, the individual filter sections are blown and rapped consecutively, in a fixed rota. That and the frequency of the operation make it worth while installing automatic controls of the section shut-off flaps and reverse blowing valves. The reverse blowing must always be done with clean gas, not with air, as that would involve explosion risks. Since carbon black particles are strongly inclined to cling, the reverse blowing fan must deliver at an adequate pressure margin to dislodge them. In view of its mechanical properties, the fibreglass fabric must not be exposed to bending, and must not be allowed to rub even on well-rounded edges or corners. These constraints must be borne in mind in the design of the rapping provisions.

The carbon black is discharged from the filters, through rotary air locks and collector pipings, usually into a pneumatic conveying system. The latter is generally run on gas extracted downstream of the last cyclone stage. A widely recommended arrangement is to recycle this material into the piping from the first to the second cyclone stage; this is said to make the most of its potential for assisting the particle agglomeration process, and thus enhance the collecting efficiency of the cyclones downstream of the re-infusion point.

The carbon black retained in the cyclones is discharged, again through rotary air locks and collector pipings, to a pneumatic or worm conveyor system for delivery to the further processing stage. As a rule, this next stage is the granulation treatment. The air locks, and the general tightness of all the separating equipment, are crucial: since the ingress of air could cause an explosion, the entire gas cleaning system must be operated at a sufficient internal pressure to prevent air from leaking in.

Another inherent restriction of this scheme is the service life of the filter bags, which averages about 12 months. This figure is naturally affected by a number of variables, like the flow velocity, the grade of carbon black being handled, the reverse blowing intervals and intensity, the state of the incoming gas, the manner in which the bags are mounted and secured in place, and the design and workmanship of the bags themselves. Triple stitched bags have been found to last longer than other types. Both process considerations and the need to reduce the wear rate of the bags dictate that the specific gas loading of the filter cloths should be kept very low. A typical recommended figure is around 0.36 $m^3\ m^{-2}$ per minute, which yields a flow velocity of about 0.6 $cm\ s^{-1}$ through the interstices of the cloths.

An innovation in this field, too recent to be fully evaluated as yet, are large-diameter filter bags some 280 to 300 mm across and 7.5 to 9.2 m long. Filters fitted with these bags are reportedly expected to prove capable of trapping all the particles on their own, without any cyclones upstream of them, but that claim so far remains to be substantiated in practice.

6.7 Food industry

6.7.1 Flour mills

Flour mills present some quite specific dust trapping problems. The dusts generated at various points in the mill differ considerably in their size distributions and physical properties, and therefore call for different types of separating equipment. Moreover, most of this equipment has to be specially adapted to meet the requirements of these applications. Finally, now that most mills employ pneumatic or vacuum systems for the in-process conveying of the grain and products, the separators at the ends of the individual conveying lines form an integral part of the process equipment, and their performance and reliability vitally affect those of the whole mill.

The first source of dust in a flour mill are the conveyors, elevators, transfer chutes, etc., which deliver the grain to the storage silos and remove it from them for further processing. The dust generated at these points is made up chiefly of soil particles and grain and chaff fragments. As it is neither particularly abrasive nor as a rule adhesive, it is mostly captured in cyclones. Since this is purely a question of local dust level control, not of recycling the trapped material for further treatment, cyclones offer all the efficiency needed for this particular purpose.

Once withdrawn from the silos, the grain is first subjected to coarse cleaning in drums or pneumatic devices. The large amounts of dust released by these units are similar in composition to the dust exhausted from the conveyors, but also contain a proportion of coarser particles like chaff and inferior corn grains, and lighter particles such as the seeds of various weeds. Heterogeneous as these solids are, they can be trapped even in larger-diameter cyclones, say 600 mm across.

Things become far more complicated in the further processing stages, starting with the fine cleaning facilities. The mechanical conveying equipment that used to predominate in these installations has now largely given way to pneumatic or vacuum systems. These move the grain through a sequence of processing units which vary in type, number, and system configuration in dependence on the kind of grain they are to treat. Wheat has to be divested of soil remnants, weeds, and inferior grade grains. Once that is done, it has to be brushed, moistened, peeled, etc. Each of these treatments generates its own kind of solid emissions – dirt, or light-weight particles – which have to be trapped in separators that form an integral part of the process equipment. These "universal" separators capture roughly 60 per cent of the extraneous solids. The secondary air cleaning stages downstream of them are usually fitted with medium-diameter cyclones, mostly some 400 to 650 mm across. Such cyclones, however, cannot retain the fine chaff, grain dust, fragments of the outer fibrous husks, and other such fines, so there would be no point in equipping a third cleaning stage with similar units. In fact, the most promising equipment for the final cleaning of the air about to be released into the atmosphere would seem to be cloth filters. Unfortunately, relatively little is so far known about the correct choice of filter types for this duty, and about such technical details as the optimum rapping arrangement. This dearth of directly applicable experience has so far discouraged most mills from embarking on what they no doubt feel would be a risky and costly experiment.

The universal separators are of course not the only source of air that calls for further cleaning. Air laden with undesirable solids has to be exhausted from practically all the other process equipment too. All these dusts resemble that gained from the universal separators, and are again trapped in cyclones, which should preferably be supplemented with cloth filters in the final cleaning stage. In view of their relatively low flow rates and dust concentrations, these air flows can sometimes be left to ordinary exhaustors that have filter cloths built into their outlets.

Rye differs from wheat in that the cleaning provisions tend to be simpler. Apart from the mechanical or pneumatic conveying system, the main points from which air has to be extracted are the universal separators and the peelers. This air is again generally pre-cleaned in cyclones of 400 to 650 mm diameter, and then passed on to a final cleaning stage that is usually fitted with smaller cyclones – although, as explained previously, cloth filters would be much more suitable for this duty. The various minor sources of dust-laden air, like the winnowers, graders, magnetic filters, sifters, etc., are mostly served either by cyclones or by cloth filters.

Once cleaned, the grain reaches the mills themselves. Let us first examine the situation in a rye mill with a pneumatic conveying system. Typically, the grain is processed in about five coarse and two fine passes, each of which delivers to one or more cyclones. The products retained in each cyclone are screened and sifted for recycling or further processing in the individual passes; this again yields proportions of oversize and of undersize fractions that have to be retained in cyclones. Similar sifting then sorts the final products out into the various grades. We are thus faced with an intricate network of pneumatic conveyor pipings terminating in cyclones. The air leaving those cyclones contains high concentrations of fine and usually strongly adhesive flour; it is usually induced, through a central collector main, by a high-pressure fan which delivers it into cloth filters. The filters commonly operate under a partial vacuum, maintained by further fans at their outlets. Sometimes cyclones are installed as a pre-cleaning stage ahead of the filters, but the fine clinging flour is apt to clog them with adhesive deposits so quickly that these cyclones are often more trouble than they are worth. Omitting them, however, aggravates the clogging of the filters, which none of the presently available filter types can fully avoid. So far, the only satisfactory way of operating cloth filters in this application is to restrict the flow velocity through them to something like 1.5 cm s^{-1}, and blowing the cloths with intense reverse air blasts at frequent intervals.

Wheat mills are much more complex than rye mills, since the wheat grains are processed in three successive stages comprising several passes each. Moreover, some of this equipment performs two different duties: for instance, the rolls that break down the grains may also process the husk fractions screened out at the sifters. The conveying equipment is again generally pneumatic. Both this equipment and the separators in which its various branches terminate naturally reflect the increased complexity of the mill, although that does not mean that they must necessarily differ in principle and general layout from the arrangements described in the preceding paragraph. The products leaving the various stages are again scavenged pneumatically and retained in cyclones, which discharge them into sifters. The air emerging from the cyclones is led off, through a collector main, into a high-pressure fan and on into cloth filters, in which a partial vacuum is maintained by secondary fans at their outlets. The choice of filter types and filter cloths is dictated by much the same considerations as in rye mills, and is no easier either. One problem specific to wheat mills is the dust generated at the succession of semolina screens. The latter must of necessity be blown with air, which is then extracted by a fan with a filter cloth at its outlet. The trouble is that this dust, while predominantly a good deal coarser than the fractions exhausted from the pneumatic conveying system, also contains a fair proportion of exceedingly fine particles which arise by the comminution of the husks. Only few types of separators can cope with such a mix of coarse and extremely fine particles with any degree of efficiency.

Altogether, then, flour mills create large amounts of dusts which are both too fine and too adhesive to be trapped efficiently in anything but cloth filters. Unfortunately, no filter design and no filter cloth at present in existence can fully meet all the requirements of this application. This is an area where much work remains to be done, and considerable rewards await those who get it done quickly and well.

7. SEPARATOR AND SOLIDS EMISSION MEASUREMENTS

Ing. J. Kurfürst, CSc.

Particle separators, like most other items of plant equipment, are apt to reserve one kind of behaviour for the laboratory and an entirely different kind for everyday practice. If we are out for meaningful figures, the efficiency and general performance of these devices has to be checked by measurements under actual plant conditions. These are not to be confused with measurements of solids emission rates. In the latter, we are out to ascertain the amount of solids released into the atmosphere per unit of time, or of plant output, or per unit volume of the gas that carries the solids. Sometimes, the measurements are also intended to establish the size distribution, chemical composition, or other specific properties of the solids. The distinction between separator performance and solids emission measurements is often overlooked, because all these investigations are based on one and the same principal technique – that of determining the solids concentration in a stream of gas or air. That involves measuring several different parameters of the gas stream, and several properties of the dust, throughout the cross section of the stack or piping, under actual operating conditions. Obviously, such measurements will be complicated, technically exacting, and costly. Just as obviously, their inherent errors will always be greater than those associated with the measurements of most other physical quantities.

7.1 Measurements on separators in service

Measurements performed on separators in the course of their operation can follow any of three often overlapping objectives:

a) To ascertain the performance of the units in actual operation, in order to compare them with the specified or design parameters.

b) To check the functioning and general state of the equipment after it has completed a certain period of operation.

c) To determine the amount of dust escaping into the atmosphere (cf. Section 7.3).

Irrespective of their purpose, these measurements and the calculations which complement them will generally be intended to establish some or all of the following quantities:

a) *The performance parameters of the separator*: its overall, partial and/or fractional collecting efficiencies; its pressure drop; its power consumption; and its consumption rates of water, compressed air, steam, etc.

b) *The properties of the gas* at the point of measurement: its flow velocity, dynamic and static pressures; its volumetric flow rates in the main stream and in any extracted or by-passed branch streams; its temperature, humidity, specific gravity, and its chemical composition.

c) *The properties of the dust*: its concentration levels in the main or branched-off gas streams ahead of and beyond the separator; the amount trapped in the separator; its specific gravity, grain size distribution, chemical composition, and moisture content; its content of combustible constituents; its wetting properties; its specific electrical resistance; and its abrasive and adhesive properties.

d) *Environmental factors*: the ambient temperature, relative humidity, barometric pressure and specific gravity of the air around the equipment.

7.1.1 Overall collecting efficiency

To determine the overall collecting efficiency of a separator, i.e. the proportion by weight of the incoming solids which it traps, we must ascertain two out of the following three weights of these solids (or the relevant dust concentration levels): the weight or concentration of the dust entering the unit (G_p or k_p); the weight or concentration of the dust emerging from the unit untrapped (G_v or k_v); and the weight or proportion of dust retained in the unit (G_0 or k_0). However, the total weights of dust are not a really suitable criterion for assessing the functioning of a separator. A much better picture can be gained if we base our examinations on the weights of dust per unit volume of gas, i.e. on the g m^{-3} or mg m^{-3} concentrations rather than on the mere weight figures. In that case, the overall efficiency is calculated by relations similar to those used when we work with the absolute weight figures:

$$O_c = \frac{k_0}{k_p} 100 = \frac{k_p - k_v}{k_p} 100 = \frac{k_0}{k_0 + k_v} 100 \tag{106}$$

The overall efficiency of a two-stage separating installation is then defined as

$$O_c = O_{c1} + O_{c2}\left(\frac{100 - O_{c1}}{100}\right) \tag{107}$$

A common case, in these efficiency investigations, is for the G_0 value to be ascertainable directly, by the simple expedient of weighing the dust trapped in the unit, while the magnitudes of G_p and G_v can be derived only indirectly. We can either determine the amount of dust in a certain part of the gas and then extrapolate the dust content of the entire gas flow, or else establish the dust concentration per unit volume of gas and thus obtain the k_p and k_v rather than the G_p and G_v values.

Things are much easier when the gas temperature and pressure, as well as the total volume of the gas, remain unaltered during its passage through the separator. The last of these stipulations implies that there must be no leakage from or induction of outside air into the unit, nor any expansion in the volume of gas e.g. by the evaporation of water. In that case, we can work with a known magnitude of Q, and, given a G_0 figure, can calculate the value of k_0. The k_p and k_v concentrations are determined by the well-established technique of extracting or bleeding off part of the gas stream, and trapping its dust content in an absolute filter.

Usually, however, the state of the gas alters within the separator, for example by cooling, which means that the volumetric flow rate at the separator outlet will differ from that at the inlet. In those cases, the dust concentrations are generally referred to the normal state of the gas, at 0 °C and 101,323 N m^{-3}, in other words to m_n^3 of gas. That is fully equivalent to referring the concentration levels of the dispersed admixtures to the units by weight of the gas. Since calculations neglecting the change in the state of the cooled gas would yield apparent efficiency figures substantially lower than the real efficiency of the unit, it is always safer to convert all the data obtained by measurements so that they apply to the normal state of the gas. Similarly, leakage into or out of the equipment must be taken into account if the results are to be valid. For instance, when some outside air is drawn into equipment run at less than atmospheric pressure, we must relate the m_n^3 of gas leaving the unit to the m_n^3 of gas entering it, else we shall end up with an efficiency figure which will be unrealistically high. This is best demonstrated on the extreme case of a separator which traps no particles at all, but induces some outside air into the gas flow: the particle concentration at its outlet will be lower than at its inlet, even though not a single particle is retained in the unit.

So much for the errors that can result from a wrongly selected evaluation procedure. Further errors will be caused by the inaccuracies that inevitably creep into the results of each of our measurements. The next step, therefore, is to clarify how markedly the resultant efficiency figure will be affected by inaccuracies in our findings on each of the individual quantities. A theoretical basis for these considerations is provided by the error propagation formula. In his detailed analysis of this problem, PRAŽÁK resorts to what he calls accuracy coefficients, as follows:

When the quantities we measure are k_p and k_0,

$$X_1 = O_c$$

When what we measure is k_p and k_v,

$$X_2 = 100 - O_c$$

And when we measure k_0 and k_v

$$X_3 = O_c\left(1 - \frac{O_c}{100}\right)$$

Pražák further examined the formulae for calculating the mean absolute error in the overall efficiency figure, $\overline{\Delta O}_c$, and demonstrated that no matter on which two of the three quantities k_p, k_0 and k_v we base our calculations, the expressions for $\overline{\Delta O}_c$ will differ only by the error coefficient that enters them. In other words: given the same degree of accuracy in each of our measurements, the resultant $\overline{\Delta O}_c$ value will depend on which two of the k_p, k_0 and k_v figures we start out from, as that decides which of the X_1, X_2 or X_3 coefficients will apply. Fig. 173 shows how the magnitudes of these three coefficients vary with the overall efficiency level.

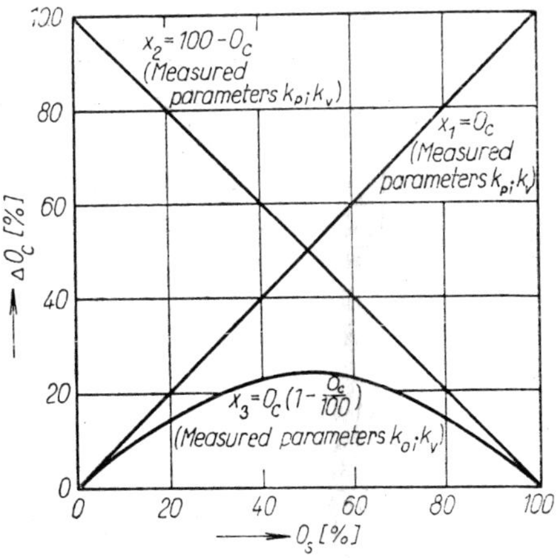

Fig. 173 How the mean absolute errors in calculations of overall collecting efficiency depend on the parameters on which the calculation is based, and on the level of that efficiency.

It is evident from this chart that the preferred combination of starting values for our calculations will always be k_0 and k_v, as this is the combination that entails the lowest mean absolute error in the resultant overall efficiency figure. When we work with k_0 and k_v values ascertained by measurements, the mean absolute error in the final efficiency figures will be greatest around $O_c = 50\,\%$. At high overall efficiency levels, the mean absolute error $\overline{\Delta O}_c$ will be much the same no matter whether we proceed from k_0 and k_v or from k_p and k_v values. This latter combination of starting data should be avoided especially when the overall efficiency under investigation is low. At $O_c = 50\,\%$, the $\overline{\Delta O}_c$ inherent in this combination is twice as high as in calculations based on k_0 plus k_v data. On the other hand, at very low overall efficiency levels it makes little difference whether we start out from k_0 and k_v or from k_p and k_0 data, as the $\overline{\Delta O}_c$ will in both cases be roughly equal.

At efficiencies higher than about 75 per cent, we can choose between measuring either k_0 and k_v, or both k_p and k_v. The decision will obviously depend on which of the quantities k_0 and k_p is easier to establish by measurements. At overall efficiencies lower than 75 per cent, however, we must always measure both k_0 and k_v. Otherwise, the mean relative errors of roughly 10 per cent, inherent in the data from which the concentration levels are calculated, would involve the risk of an unacceptably high mean absolute error in the calculated efficiency figure. Unfortunately, there are some cases when it is simply not feasible to measure k_0 and k_v, and we have to measure another pair of quantities instead k_p and k_v, or even k_0 and k_p. There is then no option but to reduce the mean relative error in these concentration figures to one per cent or even less, else we could never keep the $\overline{\Delta O}_c$ within tolerable limits. That sort of accuracy is not at all easy to attain, and may call for sophisticated techniques and costly instrumentation. Since measurements have to be paid for, all these factors should be carefully considered in advance. At the outset of our investigation, we should be perfectly clear on the magnitudes and sources of the errors likely to affect the individual measurements; on the degree of accuracy required in each of these measurements if we are to obtain a valid overall efficiency figure; and on the costs, complications, and resultant accuracy of each of the alternative procedures open to us.

7.1.2 Pressure drop across the separator

The pressure drop across a separator is the difference between the total pressures ahead of and beyond it. The total pressure means the sum of the static and dynamic pressures at the investigated cross section of the piping or duct. Pressures are now quoted in Pa (1 Pa $=$ 1 N m^{-2}), but in most of the existing literature are expressed either in kp m^{-2} or in mm of water column (1 kp m^{-2} equals 1 mm of water column or 9.80665 Pa). The term "water column" of course refers to a water-filled vertical U-shaped tube, in which the water level is displaced by the difference between the pressures that act on the two open ends of the tube.

Industrial gas cleaning plant often has to handle gases which differ in specific gravity from air at normal temperature and pressure – because of their composition, or their elevated temperature, or both. It was therefore at one time customary in this field to quote the pressure drops in metres of the gas column of whatever gas was being handled. The advantage of this unit was that the value remained unaffected by any changes in the specific gravity of the gas. Now that we are tied to the SI system of units, the corresponding quantity would be the specific energy loss across the separator,

$$H_z = 9.80665 \frac{\Delta p}{\varrho} \tag{108}$$

where H_z is the specific energy loss (in Pa m^3 kg^{-1}),
Δp – the pressure drop (in Pa), and
ϱ – the specific gravity of the gas (in kg m^{-3}).

The relationship between H_z and the formerly so common H_p (in metres of gas column) is 9.80665 $H_z = H_p$.

Generally speaking, the pressure drop across a separator is governed by the following factors: the pressure energy losses caused by friction and turbulence within the unit; the amounts of mechanical energy supplied to or dissipated by the unit; and the difference between the inlet and outlet levels of the separator. The mechanical energy input applies e.g. in the case of rotary separators. When there is neither any input nor any output of mechanical energy, which is by far the most common case, the pressure drop depends only on the friction and turbulence losses and the vertical distance between the inlet and outlet of the unit. The importance of each of those factors varies with the design of the separator. In some units the inlet and outlet lie more or less in the same horizontal plane, so the pressure drop is practically identical to the internal losses of pressure energy. On the other hand, in some vertical electrostatic precipitators these internal energy losses account for only a slight proportion of the total pressure drop, which is caused mainly by the elevation of the outlet above the inlet ports.

The internal pressure losses (and, in separators where the influence of the other factors is negligible, the total pressure drop too) follow a distinct pattern. Within the range of flow velocities and flow rates common in present-day separators, they are roughly proportional to the square of that velocity or flow rate. That permits us to calculate the way the pressure drop across a given separator will respond to a change in the flow velocity or flow rate, as follows:

$$\Delta p_2 = \Delta p_1 \frac{v_2^2}{v_1^2} = \Delta p_1 \frac{Q_2^2}{Q_1^2} \quad \text{(in Pa)} \tag{109}$$

where the indices "1" and "2" denote the state before and after the change respectively.

7.2 Dust concentration measurements

A dust concentration can be expressed either by the number of particles per unit volume of gas or by their aggregate weight, so we must distinguish between numerical and weight-based concentration figures. We must furthermore discern between concentration measurements in a moving gas stream, and those performed in what for this purpose can be considered as a stationary environment, such as the outside atmosphere.

The weight concentrations of dust in a moving gas stream are nowadays usually ascertained by the gravimetric method. First, however, a fixed volume of gas must be extracted from the point of measurement, in a way which ensures that the solids content of the sample will be representative of that in the whole gas stream at that point and instant. Which means that not only the concentration of solids in the sample, but also their chemical compositions and grain size distributions must be fully identical to those in the gas stream. A tall order indeed: we shall never even come near to meeting this requirement unless the sample is drawn off sokinetically, i.e. at a velocity and in a direction which coincide with those of the gas stream itself.

Moreover, the sampling point must be carefully selected, especially when the gas contains some coarser dust fractions (as it usually will do upstream of the separators). Ideally, the dust samples should be taken from a straight part of a vertical pipe or duct, at a point where the flow is as smooth and even as possible, and well away from any features which affect the flow pattern. That means well away from the intakes or outlets of fans or separators, from pipe bends or arches, from any constrictions or expansions of the flow channel, from valves or fittings or guide vanes and the like. The length of straight piping upstream of the sampling point should be longer than that downstream of it; some authorities recommend that the ratio between these two lengths should be 4 : 1. Furthermore, the gas velocity at the sampling point should not be less than about 3 m s^{-1}. And if all this seems to narrow the choice of sampling points down to nil, then some of the more recent regulations on the subject are even more explicit and restrictive.

Most of these new specifications are based on D_e, which is the internal diameter of a round pipe or, in rectangular-section ducts, usually the hydraulic diameter

$$D_e = D_h = 2\frac{A \,.\, B}{A + B} \tag{110}$$

where A and B are the sides of the rectangular section. The Swedish "Guidelines for Air Pollution Control at Stationary Sources" of 1973 define the equivalent diameter D_e of a rectangular duct as

$$D_e = \frac{1}{2}(A + B) \tag{111}$$

The D_e thus defined equals D_h when the section is square, and is 13 per cent greater than D_h when the ratio between the sides of the rectangle is 1 : 2. Since in practice most ducts are close to square in section, this D_e generally differs only slightly from D_h, but is much easier to calculate.

The same Swedish guidelines lay down that the lengths of straight piping must be at least $2D_e$ upstream of the sampling point, and not less than $0.5D_e$ downstream of it. Moreover, when these straight lengths are shorter than $5D_e$ ahead of and $1D_e$ beyond this point, the number of sampling points spread over the cross section of the duct or piping must be increased. Similar provisions are incorporated in the American 1971 "Standards of Performance for New Stationary Sources".

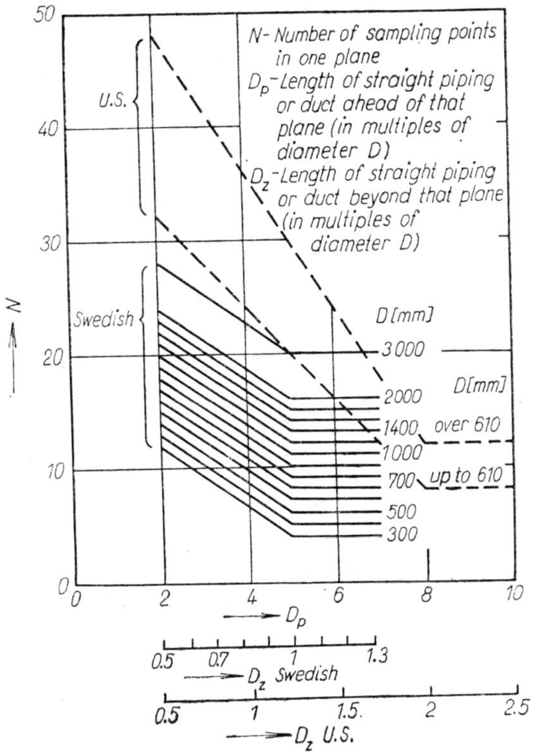

Fig. 174 Numbers of sampling points in each plane of measurement, in dependence on the straight piping or ducting lengths upstream and downstream of that plane, as specified in the recent Swedish and American regulations.

Fig. 174 shows the mandatory numbers of sampling points, in one and the same cross-sectional plane, for various D_e values and various straight lengths of piping or ducting upstream and downstream of that plane, as stipulated in the Swedish and in the American regulations.

Nor is the location of the sampling points in the plane of measurement left to chance. The details of course depend on the size and shape of the cross section, but the latter must always be divided into the specified number of areas of equal size, with a sampling point in the centre of each of those areas. Circular sections are divided into a number of annular areas, and into four quadrants of 90 degrees each, with all the sampling points lying on two mutually perpendicular diameters of the section, as indicated in Fig. 175a. If we denote the section diameter D and the number of annuli n, the distance of these points from the centre of the section is

$$R_i = \frac{D}{2}\sqrt{\frac{2i-1}{2n}} \qquad (112)$$

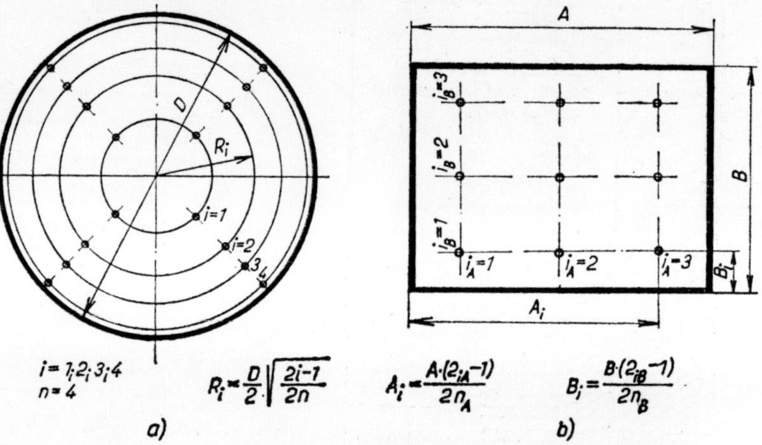

Fig. 175 Specified locations of sampling points in the plane of measurement in: *a*) Circular-section pipes or ducts; *b*) Rectangular-section ducts.

Square or rectangular sections are broken down into geometrically similar areas of equal size, as evident from Fig. 175b, again with a sampling point in the middle of each of them. The rectangular coordinates of those points are

$$A_i = \frac{A(2i_A - 1)}{2n_A} \quad \text{and} \quad B_i = \frac{B(2i_B - 1)}{2n_B} \qquad (113)$$

The number of sampling points in any one cross-sectional plane is of course dependent on the size of that cross section and on the lengths of straight piping or ducting ahead of and beyond it. Typical relationships governing the number of sampling points are presented in Fig. 174.

The relevant Czechoslovak Standard, ČSN 12 4010, stipulates that the number of sampling points must be increased when the dust is coarse and/or when there is a velocity gradient across the plane of measurement. The minimum numbers of sampling points for various cross-sectional areas are specified as follows:

0.25 to 2.5 m^2	9 points (8 in circular sections)
2.50 to 4.0 m^2	16 points
4.00 to 9.0 m^2	25 points (24 in circular sections)
over 9.0 m^2	36 points.

Areas smaller than 0.25 m^2 can be covered by a lesser number of sampling points, or even by one point only, provided that the solids concentration in the flow is lower than 5 g m^{-3} and absolutely uniform over the entire cross section, and that none of the particles have a free falling velocity in excess of 5 cm s^{-1}. The second of those three stipulations obviously tends to relegate the whole of this clause into the realm of pure theory.

If the straight lengths of piping or ducting are too short, or if there are any features in the vicinity which deflect or create eddies in the gas stream, there will be a substantial velocity gradient over the cross section. In that case, the number of sampling points has to be increased roughly by half. If the piping or ducting is horizontal instead of vertical, or if the solids are coarse enough for most of the particles to have free falling velocities in excess of 5 cm s^{-1}, the number of vertically superimposed sampling points must be raised by half even if the flow velocity throughout the cross section is fully uniform.

This Standard thus has two features which are not likely to endear it to practising engineers. Firstly, the number of sampling points needed to meet its requirements may in extreme cases be as high as 72. Secondly, we shall often have to install more probes than the standard number laid down for a cross section of the given size, but the need to do so will come to light only after a set of measurements has actually been performed in the plane in question. Therefore, the more recent regulations restrict the basic number of sampling points in any one cross section to more reasonable levels like twelve (in the American "Standards of Performance" of 1971) or twenty (in the Swedish "Guidelines", or the West German "Erste

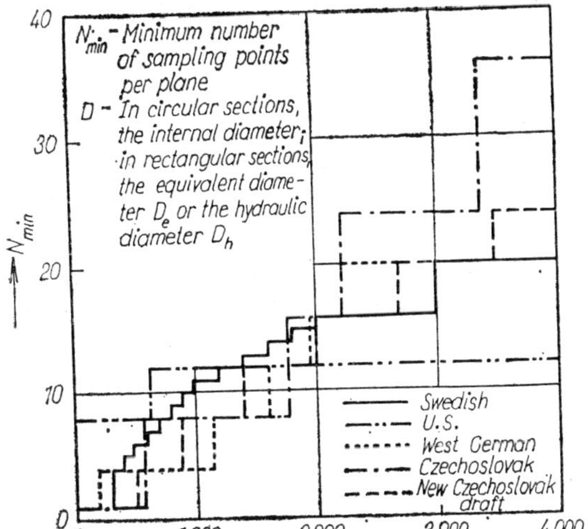

Fig. 176 The minimum numbers of sampling points in one plane of measurement, as specified for various piping diameters in some recent regulations.

allgemeine Verwaltungsvorschrift zum Bundes-Immissionsschutzgesetz" of 1974). Fig. 176 shows the number of sampling points needed, for any given cross-sectional area, to ensure compliance with the Swedish, American, West German and Czechoslovak regulations mentioned above.

With the number of sampling points thus defined, the next question is what sort of probes should be installed at them. Essentially, the choice facing us is as follows:

a) A plain tubular probe will extract a sample of the gas and solid admixtures, but will not do so isokinetically. An isokinetic relationship can be established by calculations relating the flow velocity in the sampling tube to that in the main gas stream, but this is very much a second best solution.

b) An isokinetic (or "zero differential") probe will, while sampling, also measure the static pressures within and around it. The static pressure of the extracted gas sample can be adjusted to equal that in the main gas stream, which will ensure fully isokinetic sampling.

c) A velocity detecting probe can combine sampling with measurements of the static and total pressures of the main gas stream, but an isokinetic extraction ratio can again be ensured only by flow velocity calculations, which are an obvious source of potentially serious extra errors.

d) Composite probes are available which combine the functions of an isokinetic and a velocity detecting probe; like most attempts to combine the best of two worlds, they can often end up doing neither job properly.

e) Probes with built-in dust traps can greatly simplify the measurement technique where the particle concentration is known in advance to be low, e.g. in investigations of the residual amounts of dust released into the atmosphere.

Once a sample has been gained, its solids content must be captured with as near as makes no difference a 100 per cent efficiency, and weighed. The volume of the gas sample must be recalculated to ascertain the volume actually extracted, at the state of the gas which prevailed at the time and point of extraction. That yields the data from which we can compute the solids concentration in the gas stream, on the implicit but not necessarily correct assumption that the mean concentration at the time and point of sampling equalled that in the sample. When these measurements are carried out under actual operating conditions, it is also essential to sample the gas at a temperature which will rule out any distortion of the results by the subsequent condensation of its vapour content. Moreover, the process conditions of whatever process is generating the dust-laden gas must be kept as constant as possible throughout the sampling period. The latter requirement is easy enough to formulate, but often virtually impossible to enforce in the face of hostile operating staffs and production-oriented managements.

To establish the concentration by weight of solid particles in a gas stream, we have to ascertain the mean values of the following quantities:

a) The total weight of solids in the gas sample (in g).

b) The total volume of the gas sample, recalculated to apply to the mean state of the gas in the sampling plane at the time of sampling (in m^3).

c) The volumetric flow rate in the sampling plane (in $m^3\ s^{-1}$), and the flow velocity in that plane (in $m\ s^{-1}$), at the time of sampling.

d) The gas temperature in the sampling plane (in °C).

e) The absolute gas pressure in the sampling plane (in Pa).

f) The chemical analysis of the dry gas (in per cent by volume).

g) The moisture content (or absolute humidity) of the gas (in $g\ m^{-3}$).

Furthermore, we shall usually also need to know the grain size distribution and specific gravity of the solid particles.

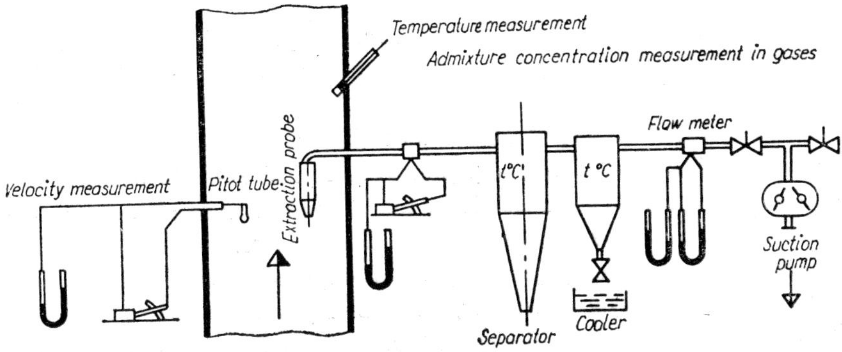

Fig. 177 Scheme of an apparatus for determining the solids concentration in gases.

Fig. 177 is a scheme of a typical apparatus for measuring dust concentrations in gases. One of the key requirements on any such apparatus is that it must effectively prevent any losses or changes in the state of the dust between the sampling point and the filter where the solids are retained. The dust will tend to settle upstream of the filter unless the flow velocity, from the probe tip to the filter inlet, is kept up to some 30 to 50 $m\ s^{-1}$. The sampling times at each of the sampling points must be closely identical, and should never be less than three minutes at a time; normally, these times range from 5 to 20 minutes in dependence on the number of sampling points. The object is to sample all the points in any one cross-sectional plane within one to two hours. The weight of dust obtained from the sample should never be less than one per cent of the weight of the receptacle in which that dust is weighed. Usually, however, we need a good deal more dust than is required for the concentration measurements, because we shall also be investigating the size distribution and chemical composition of the particles. A final constraint is that the sample should preferably equal at least 0.005 per cent of the total throughflow through the sampling plane during the sampling time, and must never represent less than 0.001 per cent of that throughflow.

There would be no point in any such measurements if we failed to trap to all intents and purposes the entire amount of dust in the gas sample. What filtering equipment we pick will naturally depend on the assumed concentration level and particle sizes, on the flow rate and temperature of the gas stream under investigation, and on the degree of accuracy we are out to attain. In practice, our choice is generally restricted to one or several of the following types:

a) Paper or fine cloth filters.

b) Glass or ceramics filters.

c) Filters packed with mineral fibres or wool – glass, asbestos or slag-based.

d) Cyclones with an ultra-high efficiency in the trapping of even extremely fine particles.

e) Electrostatic precipitators.

f) Thermoprecipitators, where the dust-laden gas passes through the narrow gap between a hot and a cold plate, and thermal effects cause its solids content to settle on the latter plate.

When no single type is adequate, there is nothing for it but to resort to various combinations of particle trapping equipment. A typical example would be say a small-diameter cyclone to knock down the concentration level, a cloth filter downstream of it, and a paper filter at the end of the line.

The volume of the extracted gas sample is mostly ascertained by means of an aperture-type or nozzle-type flowmeter, but ordinary gas meters and rotameter-type instruments are used for this purpose too. In the latter, the gas lifts and spins a free-floating plug within a slightly tapering tube; the height at which this plug stabilizes within the tube is indicative of the flow rate.

7.3 Investigations of solids emission rates

The amount of solids released into the atmosphere, from any particular source, is commonly investigated for either of two reasons: to ensure compliance with public health legislation, or to gather more detailed data of a technical nature. In the former case, the object is simply to check whether the actual emission rate is within the legal limit. In the latter case, we may be out to assess the requisite height of a newly planned chimney stack; to check the efficiency of an existing gas cleaning plant; to evaluate the performance of a combustion or other dust-generating process; or to assemble data on dust dispersal patterns, for which we need a total emission figure to correlate with the results of spot pollution level measurements in the vicinity. No matter which of these purposes we have in view, we can generally choose any of four alternative procedures:

a) Material input-output balances of the dust-generating process.

b) Mean emission coefficients, determined for the specific case in hand.

c) Spot checks on the emission rate.
d) Continuous measurement and recording of the emission rate.

7.3.1 Material input-output balances

As a tool for establishing solids emission rates, these balances can be drawn up for only some of the many industrial processes which pollute the atmosphere with solids. The main reason why they are compiled at all is that they yield reasonably reproducible emission rate data for some of the most common and worst sources of air pollution – the combustion equipment at boilerhouses, steam power stations, district heating plants, etc. However, the resultant figures for plants that fire solid fuels are subject to relative errors as high as ± 60 per cent, which is nothing like the accuracy attainable in even the most routine of direct measurements. Consequently, this indirect approach is not really dependable even for a first rough orientation. All it really does is to take some of the guesswork out of preliminary estimates, but it can do that so quickly and conveniently that there is still some room for it in the rough and ready sifting of a large number of alternatives. By way of an illustration of this technique, Fig. 178 presents a nomogram for the estimation of fly ash concentrations in flue gases. Obviously, the ease and speed with which such nomograms can be applied make up for much of their other shortcomings.

7.3.2 Mean emission coefficients (factors)

This is another indirect and very approximate method for establishing the total solids emission from a number of discrete sources. It is used e.g. by municipal or regional planners who have to put up with say a large number of small boilerhouses in their area, and by government authorities or corporations out to establish how much dirt a particular industry or division contributes to the overall desecration of the environment. The procedure is to average out the emission rates of a large number of sources of one and the same type and general size. That yields an average emission figure per unit of output or capacity, per ton of products or raw materials, etc. For instance, the emission coefficients for boilerhouses are expressed in kg GJ^{-1}, those for cement factories in kg per ton of clinker, etc.

Some of the more recent regulations, such as the Swedish "Guidelines for Air Pollution Control at Stationary Sources" of 1973, or the American "Standards of Performance for New Stationary Sources" of 1971, specify the maximum permissible emission limits in terms of similar coefficients. Instead of laying down one rigid limit common to all pollution sources, they relate the acceptable pollution level to the output rate or some other characteristic parameter of the source. When the actual emission of that source is then checked by direct measurements, the findings are related to the magnitude of that characteristic parameter at the time of measurement. That yields a relative emission rate which is required to lie within the specified limit.

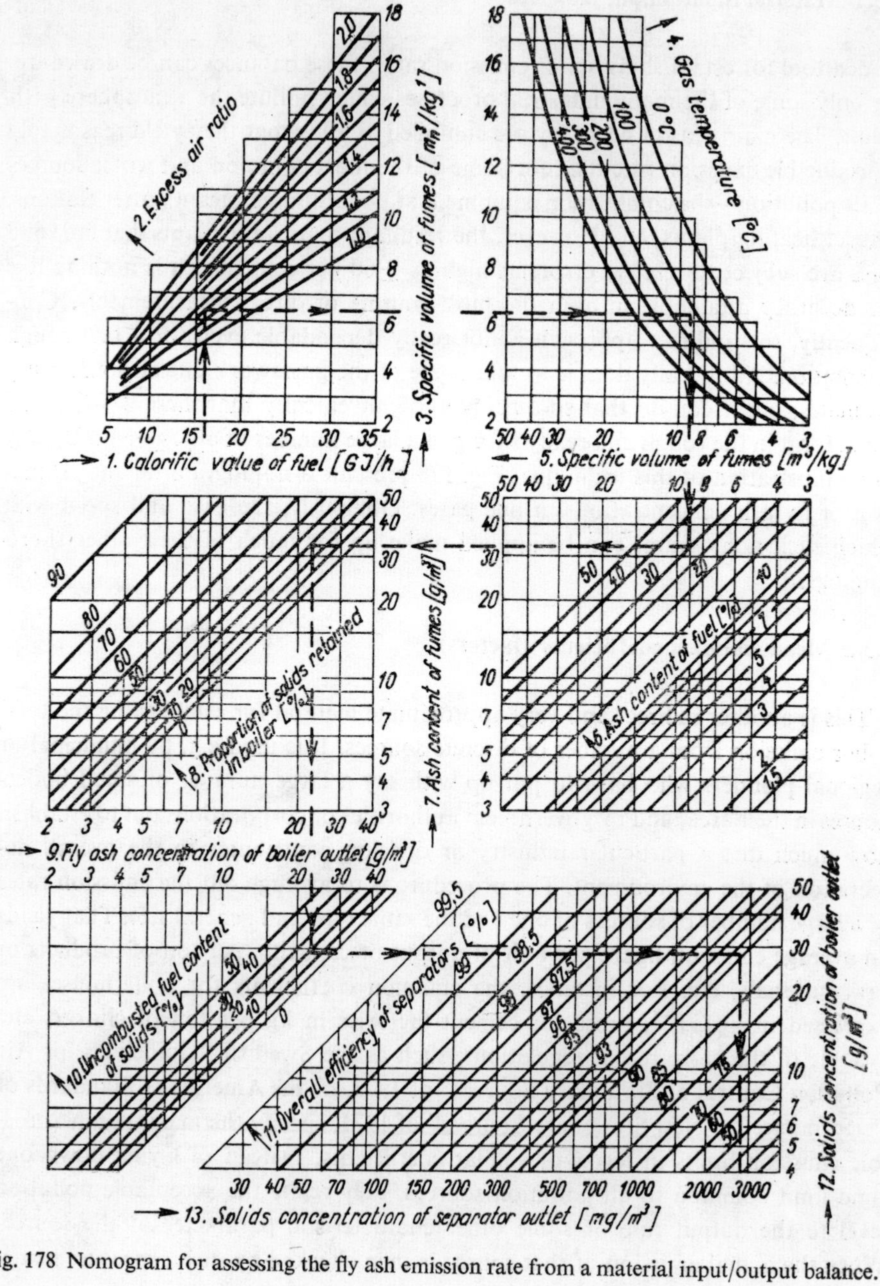

Fig. 178 Nomogram for assessing the fly ash emission rate from a material input/output balance.

7.3.3 Spot checks on emission rates

Spot checks can be performed to determine either the amount of solids emitted per unit of time, or the concentration of solids per unit volume of the gases or fumes, or else a relative emission rate as described in the preceding paragraph. No matter what their object, however, the actual measurements essentially establish the dust concentration in a gas stream, in the way explained in Section 7.2.

7.3.4 Continuous recording of emission rates

The term continuous means only that the relative or absolute emission rate is followed and recorded throughout a substantial period of operation of the pollution-generating equipment. It does not refer to the measuring technique itself, which may be either continuous or continual; in other words, the sampling and evaluating apparatus may function either permanently or intermittently, at intervals and for periods which render a quasi-continuous record.

There are signs that continuous recording provisions will soon become a compulsory feature of major industrial pollution-generating equipment. For instance, the West German "Erste allgemeine Verwaltungsvorschrift zum Bundes-Immissionsschutzgesetz" of 1974 stipulates that all steam generators with outputs exceeding 5.6 tons of steam per hour must be equipped for continuous measurement of the smoke density. Those with outputs in excess of 38 tons per hour, and all equipment that releases more than 15 kg of dust an hour into the atmosphere, must be instrumented for continuous measurement and recording of the amount of solids actually released.

These instruments will detect and record any changes in the solids emission rate, but can generally offer no clue to the reasons for those changes. To help establish the causes and decide on countermeasures, they are usually supplemented with further instruments which monitor other parameters of the emission or of the process that generates it. A typical set of instrumentation for this purpose comprises devices for the following functions:

a) Continuous measurement of the solids emission rate.

b) Continuous measurement of the CO_2, CO and/or O_2 contents.

c) Continuous measurement of the gas temperature.

d) Continuous measurement of the SO_2 or NO_x emission rate.

e) A multi-channel pen recorder for synchronous plotting of the data from all the above instruments, as well as of other relevant data—e.g. the momentary boiler output in GW or GJ per hour or in tons of steam per hour; the current fed to the electrostatic precipitator which cleans the flue gases; etc.

The owners of boilers must install and operate all this instrumentation under the supervision of special field measurement teams and of the national supervisory authorities, which also calibrate the devices and check the data they record.

This brings us to the essential difference between spot checks and continuous measurements. Spot checks usually render fairly accurate results, but errors creep in as soon as we attempt to extrapolate these findings so as to cover longer time spans. Continuous measurements do show how the emission rate varies in the course of time, but are subject to much larger inherent errors. The latter can to some extent be kept down by duplicating the instrumentation, e.g. fitting each duct with two instruments set up perpendicular to each other. Moreover, these instruments must be recalibrated, by a precise spot check, at least once a year, to counter their usual tendency to drift.

The instruments themselves fall into several categories. The simplest of them are gravimetric devices, which determine the concentration by weight of the dust. For continuous measurements, they must be fully automated; perhaps the best known such apparatus are the Gast dust scales.

Radiometric methods have one outstanding advantage over gravimetric ones,

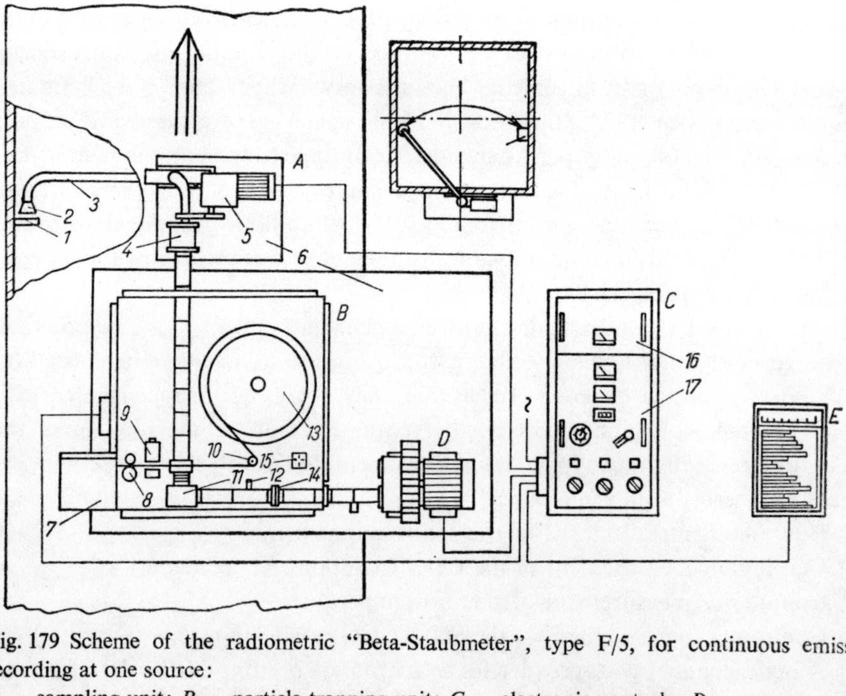

Fig. 179 Scheme of the radiometric "Beta-Staubmeter", type F/5, for continuous emission recording at one source:
A — sampling unit; *B* — particle trapping unit; *C* — electronic controls; *D* — vacuum pump; *E* — pen recorder; *1* — extraction probe shield; *2* — intake nozzle; *3* — probe piping; *4* — swivelling tube bend; *5* — probe reciprocating drive; *6* — base plates; *7* — filter strip receiver; *8* — strip drive rollers; *9* — radiation source and detector; *10* — strip drive reversal; *11* — flow control unit; *12* — temperature sensor; *13* — reel of filter strip; *14* — flow rate meter; *15* — flow rate adjustment; *16* — programme control unit; *17* — control and power feed cabinet.

in that the results become available while the measurement is still in progress. Since the readings are proportionate to the mass of the solid particles, these instruments also share the advantage of gravimetric devices that they do not require recalibrating for every particular dust type or set of service conditions. Fig. 179 shows the radiometric Beta-Staubmeter made by Krohne (formerly known as Verewa). In this device, the gas stream is sampled by a probe which incessantly swings back and forth over the entire cross section of the duct, so as to ensure truly representative sampling. The dust content of the sample is trapped on a strip of filter paper; the latter unreels from a drum which normally holds about a month's supply. The thickness and specific gravity of the dust deposit are then evaluated from the way this layer attenuates the beta rays from a permanent radiation source. The unit is fitted with electronic programme control and regulating circuits, and has found widespread application in power stations, cement factories, and metallurgical plants.

Optical methods underlie a number of different measurement techniques, and a large assortment of different instruments now on the market. In principle, they all exploit the fact that light rays passing through a gas are to some extent absorbed and dispersed by its solids content. Some of the instruments, like the Swiss Sigrist apparatus, measure the amount of light reflected by diffraction processes. Others measure the attenuation of the light beam; *the Lambert-Beer relation*, correlating the degree of attenuation with the dust concentration, then enables them to yield a relative figure indicative of the actual dust content of the gas. Briefly, on the assumption that all the other variables remain constant, the dust content is directly proportional to the extinction factor, i.e. to the logarithm of the relative attenuation of the transmitted light beam.

The photometric method most widely used in actual practice relies in scanning the whole cross section of the duct with a light beam, so that no sample of the gas has to be extracted. Since Siemens performed the basic calibration of an apparatus of this type as long ago as 1928, and instruments of this description have been in routine use in West Germany for more than two decades now, it is hardly surprising that these units have been developed to a state approaching perfection. Most of the development work was intended to evolve ways of compensating out the inherent errors of the methods and all the inaccuracies introduced by changes in the process variables. Especially, most of the instruments of this kind nowadays incorporate provisions for cancelling out the following effects:

a) Changes in the intensity of the primary light beam, caused by ageing of the light source and/or mains voltage fluctuations;

b) Changes in the response characteristics of the photocell, caused by temperature fluctuations and/or ageing of the cell;

c) Soiling and/or mechanical damage of the optical surfaces exposed to the dust in the flue duct;

d) Environmental factors such as dust deposits, sunlight or other incidental light sources, low ambient temperatures in outdoors locations, vibration, deflection of the primary light beam from the axis of measurement, etc.

Most of these devices employ only a single light beam, but there are several dual-beam designs, such as those made by Sick or Durag. Figs. 180 and 181

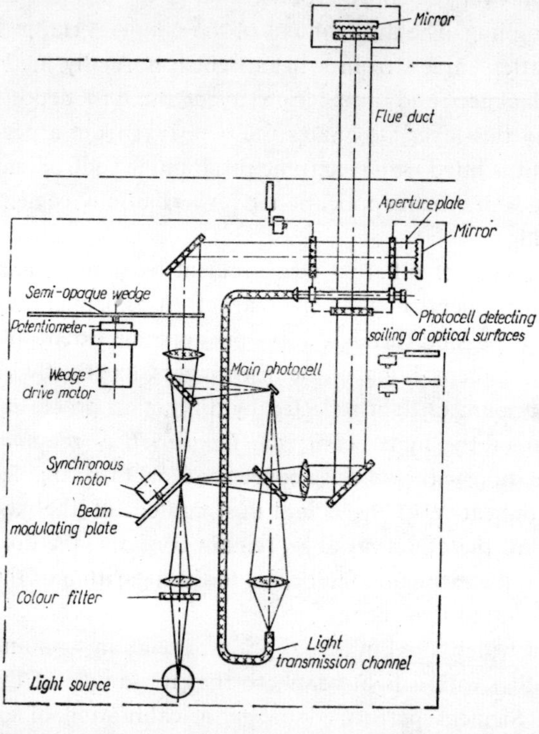

Fig. 180 Scheme of the optical D-R 110 instrument made by Durag Apparatebau GmbH.

show the Durag D-R 110 model, designed to cover the whole cross section of a flue duct. Apart from monitoring the solids emission rates, it is also used for indicating incomplete combustion and checking that optimum process conditions are being maintained, especially in steam generating equipment. The unit is adapted to signal the need for cleaning or repairs of the process equipment, when the light beam is being attenuated beyond a certain limit, and can also raise the alarm when the smoke density or dust concentration overstep a pre-set maximum value. Its makers claim that there are currently more than 7000 of these instruments in operation, and that some of them have been in service for more than two decades now.

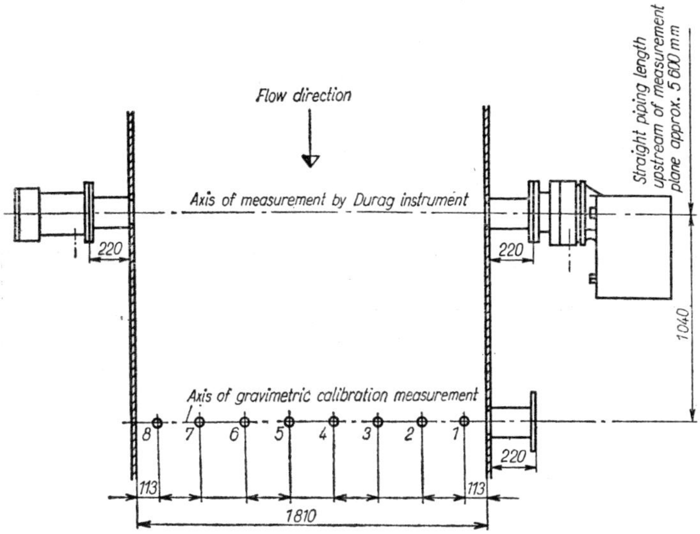

Fig. 181 Instrumentation layout for gravimetric spot checks conducted to calibrate a Durag D-R 110 apparatus after its installation at the pollution source.

7.3.5 Special measurement methods

The overwhelming majority of both spot checks and continuous emission rate measurements are nowadays performed by the three methods outlined above – the gravimetric, radiometric and optical techniques. However, several other methods have been reported or at least put forward in the literature. Some of them are derived from the techniques normally used for measuring the amount of solid pollutants in the atmosphere, and are therefore applicable only where the solids content of the gases is known to be very low.

Only one of these methods is employed on any scale, chiefly for rough and ready spot checks where speed and ease of evaluation are more important than accuracy. This is the method based on *the Ringelmann scale*, which, for all the well-based objections voiced against it, is still in use in a number of countries. The Ringelmann scale is no more than a set of reference patterns for comparison purposes, i.e. an aid for subjective assessment procedures, and cannot serve for any quantitative evaluations. Its underlying idea, that of qualitatively judging smoke plumes by their colour intensity, was proposed as far back as 1898. Nevertheless, it is still utilized e.g. in West Germany, in the United States, Great Britain, and France, mostly for checking the smoke plumes at the tops of boilerhouse chimneys.

The Ringelmann scale, as shown in Fig. 182, is a set of six white fields cross-hatched with progressively increasing numbers of black lines, so that the black areas of the individual fields form certain fixed proportions of their total areas.

In the individual fields, these proportions are 0, 20, 40, 60, 80 and 100 per cent respectively. In most of the countries that still use the device, the smoke is required to be no darker than the 20 per cent field of the scale. In West Germany, it is expected to be lighter than the 40 per cent field, except for the first five minutes after the fire is lit; during those five minutes, it must at least be lighter than the 60 per cent field of the scale.

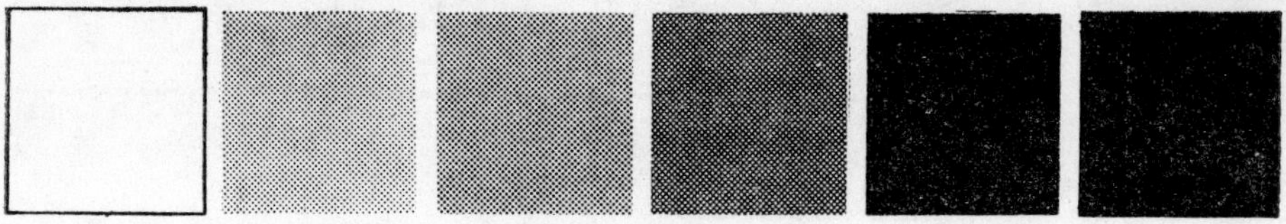

Scale field

0 1 2 3 4 5

Black line coverage [%]

0 20 40 60 80 100

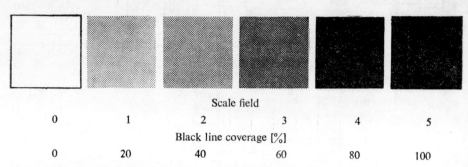

Fig. 182 The Ringelmann scale, showing the percentage of area covered by the black cross-hatching in each of the scale fields.

The Ringelmann scale is normally fixed to a plate that is held up so that the observer sees the scale next to the top of the investigated chimney, with the same patch of light or cloudy sky forming the background to both. The plate must be held far enough from the eye for the individual lines of the cross-hatching to be indistinguishable. To facilitate these comparative assessments, the Dr. Wöhler company produces a prismatic telescope with a 3 by 30 rating, where half the field of vision remains clear for observation of the smoke plume while the other half is taken up by filters representing the Ringelmann scale. The same scale also forms the basis of the Evans Electroselenium unit, a British instrument for continuous solids emission measurements within flue ducts or chimney stacks.

For all the hard things the experts have to say about this method, the fact remains that many advanced industrial countries still consider it good enough for quick checks on boilerhouse smoke emissions, where it can aid rapid interventions to counter incorrect or fluctuating combustion conditions. Czechoslovakia is about to adopt this method as a compulsory procedure for assessing the dark smoke caused by mismanagement of the combustion process.

Literature

Beth: Handbuch Staubtechnik, 2nd edition, Lübeck.

Böhm, J.: Elektrické odlučování a odlučovače (Electrostatic Precipitation and Precipitators) Prague 1958.

Cadle, R. D.: Particle size, New York 1965.

Cooper, H. B. H.—Rossano, A. T.: Source Testing for Air Pollution Control. Environmental Science Services Division, USA 1971.

Fuks, N. A.: Mekhanika aerozoley (Aerosol Mechanics), Moscow 1955.

Geck, W. H.: Zündfähige Industriestäube, Düsseldorf 1954.

Glushkov, L. A.: Obespylivanie oborudovaniya drobilno-razmolnykh otdelenii (Dust Clearing from Crushing and Milling Equipment), Sverdlovsk 1957.

Gordon, G. M.—Peysakhov, I. L.: Kontrol pyleulavlivayushchikh ustanovok (Checking of Dust Trapping Equipment), Moscow 1961.

Gordon, G. M.—Peysakhov, I. L.: Pyleulavlivanie i ochistka gazov (Dust Trapping and Gas Cleaning), Moscow 1958.

Hašek: Čistota ovzduší (Air Pollution), Prague 1968.

Havelka, M.: Prašnost v průmyslu, zejména hutním (Dust in Industry, especially in Metallurgy), Prague 1960.

Hawksley, P. G. W.—Badzioch, S.—Blackett, J. H.: Measurement of Solids in Flue Gases. BCURA, Leatherhead 1961.

Holmes, R. G.: Air Pollution—Source Testing Manual. Los Angeles County Air Pollution District, USA 1965.

Huml, J.: Mechanické odprašování kouřových plynů (Mechanical Trapping of the Dust in Flue Gases), Prague 1953.

Klobouk, B.: Údržba a opravy elektrických odlučovacích zařízení (Maintenance and Repairs of Electrostatic Precipitating Equipment), Prague 1962.

Konašinskij, D. A.: Elektrické filtry (Electrostatic Filters), Prague 1955.

Krebs, R.: Staubgehaltmessungen in strömenden Gasen. LURGI, Frankfurt 1967.

Kubíček, L.: Odstraňování průmyslových škodlivin (Removal of Industrial Contaminants), Prague 1954.

Kucheruk, V. V.: Ochistka ot pyli ventilatsionnykh i promyshlennykh vybrosov (Dust Removal from Ventilation-System and Industrial Exhalations), Moscow 1955.

Kucheruk, V. V.—Krasilov, G. I.: Nové způsoby čištění vzduchu od prachu (New Ways of Removing Dust from Air), Prague 1952.

Lutyński, J.: Elektrostatyczne odpylanie gazov (Electrostatic Gas Cleaning), Warsaw 1965.

Magill, P. L. et al.: Air Pollution Handbook, New York 1956.

Martinec, E.: Umělé tahy a odpopílkování komínů kotelen (Forced Draught and Fly Ash Trapping in Boilerhouse Chimneys), Prague 1962.

Meldau, R.: Handbuch der Staubtechnik, Vol. I, II, Düsseldorf 1956.

Pergler, B.: Odprašovací zařízení v průmyslu kamene (Dust Trapping Equipment in the Stone Industry), Prague 1964.

Pražák, V.: Čištění plynů (Gas Cleaning), SNTL Prague, 1962.

Prechistenskiy, S. A.: Tsentrifugirovanye aerozolej v CRP (Centrifuging of Aerosols), Moscow 1960.

Rameš, J.: Parametry bezpečnosti hořlavých plynů a par (Safety Parameters of Inflammable Gases and Vapours), Prague 1958.

Ryazanov, V. A.: Predelno dopustimye kontsentratsii atmosfernykh zagryaznenii (Maximum Permissible Concentrations of Atmospheric Pollutants).

Ryszka, E.: Pomiary zapylenia gazov w przewodach (Dust Lading Conditions in Gas Pipings), Katowice 1969.

Schmidt, K. G.: Staubbekämpfung in der Giesserei Industrie.

Smolík, J. et al.: Technika prostředí (Environmental Technology), Prague 1957.

Spurný, K.—Jech, Č.—Sedláček, B.—Štorch, O.: Aerosoly (Aerosols), Prague 1961.

Šimeček, J.: Zneškodnění prachu v dolech (Rendering Harmless the Dust in Mines).

Sneerson, B. L.: Elektricheskaya ochistka gazov (Electrical Gas Cleaning), Moscow 1950.

Štorch, O.: Průmyslová odlučovací zařízení (Industrial Dust Trapping Equipment), SNTL Prague 1957.

Urban, J.: Pneumatická doprava (Pneumatic Conveying), SNTL Prague 1964.

Uzhov, V. N.: Borba s pyliu v promyshlennosti (Combatting Dust in Industry), Moscow 1962.

Uzhov, V. N.—Myahkov, B. Y.: Ochistka promyshlennykh gazov filtrami (Cleaning Industrial Gases by Filters), Moscow 1970.

White, H. J.: Entstaubung industrieller Gase mit Elektrofiltern, Leipzig 1969.

Žižka, J.: Pěnové kolony pro odlučování prachu a pro jiné účely (Bubble Columns for Dust Trapping and for Other Purposes), Prague 1961.

Proceedings: Pyleulavlivanie i ochistka gazov v tsvetnoy metalurgii (Dust Trapping and Gas Cleaning in Non-ferrous Metallurgy), Moscow 1963.

Proceedings: Ochrana čistoty ovzduší (Protection of the Clean Air), STIO Prague 1968.

Proceedings: Exhalace z koksoven a kuploven (Exhalations from Coking Plants and Cupolas), ČTIO Prague 1970.

Guidelines for Air Pollution Control at Stationary Sources in Sweden. National Environment Protection Board, Solna, Sweden, 1973.

Standards of Performance for New Stationary Sources. Federal Register, Washington, D.C., Vol. 36, No. 247, of December 23, 1971.

Erste allgemeine Verwaltungsvorschrift zum Bundes-Immissionsschutzgesetz (Technische Anleitung zur Reinhaltung der Luft) of August 28, 1974. Gemeinsames Ministerialblatt, Bonn, No. 24, of September 4, 1974.

Czechoslovak Standard ČSN 12 4010 (Regulations for Measurements on Separators and Separating Equipment in Industry). Vydavatelství úřadu pro normalizaci a měření, Prague 1964.

VDI 2066 — Staubmessungen in strömenden Gasen. VDI — Verlag, Düsseldorf 1974.

Entstaubungsnorm und Messreglement. Inspektorat für Emissionen, Verein Schweizerischer Zement-, Kalk- und Gips-Fabrikanten, Zürich, 1964.

Subject Index